AF726394

Environmental Soil Science: A Sustainable Perspective

Environmental Soil Science: A Sustainable Perspective

Edited by Frankie Casey

SYRAWOOD
PUBLISHING HOUSE
New York

Published by Syrawood Publishing House,
750 Third Avenue, 9th Floor,
New York, NY 10017, USA
www.syrawoodpublishinghouse.com

Environmental Soil Science: A Sustainable Perspective
Edited by Frankie Casey

© 2022 Syrawood Publishing House

International Standard Book Number: 978-1-64740-082-8 (Hardback)

This book contains information obtained from authentic and highly regarded sources. Copyright for all individual chapters remain with the respective authors as indicated. All chapters are published with permission under the Creative Commons Attribution License or equivalent. A wide variety of references are listed. Permission and sources are indicated; for detailed attributions, please refer to the permissions page and list of contributors. Reasonable efforts have been made to publish reliable data and information, but the authors, editors and publisher cannot assume any responsibility for the validity of all materials or the consequences of their use.

Trademark Notice: Registered trademark of products or corporate names are used only for explanation and identification without intent to infringe.

Cataloging-in-Publication Data

Environmental soil science : a sustainable perspective / edited by Frankie Casey.
p. cm.
Includes bibliographical references and index.
ISBN 978-1-64740-082-8
1. Soils--Environmental aspects. 2. Soil science. 3. Sustainable agriculture. I. Casey, Frankie.
S591 .E58 2022
631.4--dc23

TABLE OF CONTENTS

PREFACE

The purpose of the book is to provide a glimpse into the dynamics and to present opinions and studies of some of the scientists engaged in the development of new ideas in the field from very different standpoints. This book will prove useful to students and researchers owing to its high content quality.

The field of science that addresses the interaction of humans with the pedosphere, the lithosphere, hydrosphere, biosphere and the atmosphere is known as environmental soil science. This field encompasses the study of surface water quality, soil degradation, vadose zone functions, movement of viruses and bacteria in water and soils, etc. It is also involved in developing strategies for remediating adverse soil conditions through land treatment of wastewater, erosion control, restoration of wetlands, nutrient management, bioremediation, besides others. Environmental soil science is an upcoming field of science that has undergone rapid development over the past few decades. Most of the topics introduced in this book cover new techniques and applications of environmental soil science studies. Coherent flow of topics, student-friendly language and extensive use of examples make this book an invaluable source of knowledge.

At the end, I would like to appreciate all the efforts made by the authors in completing their chapters professionally. I express my deepest gratitude to all of them for contributing to this book by sharing their valuable works. A special thanks to my family and friends for their constant support in this journey.

Editor

1

THE APPLICABILITY OF DISINTEGRATION TESTS FOR COHESIVE ORGANIC SOILS

Anne-Katrin GROßE[a, b], Stefan CANTRÉ[b], Fokke SAATHOFF[b]

[a]*Department of Maritime Systems, Faculty of Interdisciplinary Research, Universität Rostock, Justus-von Liebig Weg 6, 18059 Rostock, Germany*

[b]*Chair of Geotechnics and Coastal Engineering, Universität Rostock, Justus-von-Liebig Weg 6, 18059 Rostock, Germany*

Abstract. The use of ripened fine-grained organic dredged materials as construction materials, e.g. as top soil on slopes such as landfills or dikes, is an important contribution to environmental engineering science. The materials are legally considered a waste and need to be beneficially re-used. Therefore, not only standard geotechnical parameters have to be determined but also their erosion resistance which is a particularly critical environmental parameter. There is a variety of different tests to determine the flow dependent erosion resistance of soils, such as the erosion function apparatus (Briaud *et al.* 2001). In this study, however, the focus lays on the aggregate stability as an indicator for the erosion resistance under static loading, which can be determined using wet sieving and disintegration tests. The disintegration tests after Weißmann (2003) and Endell (RPW 2006) have a similar setup; however, the specific boundary conditions for the tests as well as the evaluation procedures are different. Weißmann proposed his test to determine the erosion stability of dike cover materials while the Endell test should be used for mineral sealing liners in navigation channels. In this study both tests have been used to evaluate the aggregate stability of fine-grained organic dredged materials that have been installed in large-scale research dike facilities and in the recultivation layers of different landfills. The materials showed good visual performance with respect to rainfall induced erosion so far; however, problems in determining erosion and aggregate stability indices limit the value of the studies: both disintegration tests investigated have major limitations with respect to the organic soils tested. Particularly the evaluation methods are not suitable for the soils but also some boundary conditions are critical and are discussed in this paper. The gained knowledge is a valuable basis for the development of standard characterisation methods for dredged materials in environmental and geotechnical applications.

Keywords: soil erosion, disintegration tests, aggregate stability, waste management technologies, cohesive organic soils, dredged material, marsh clay, marl, environmental sustainability.

Introduction

Soil erosion plays an important role both in agricultural and geotechnical environments. In agriculture the workability and long-term availability of arable soils are very important issues while the erosion stability of slopes, (e.g. at dams and dikes) is of great interest in the geotechnical context. Most importantly, international soil protection legislation states that the top soil should always be protected against relocation by erosion. Coastal dikes are often constructed in environmentally sensitive areas and in spite of their flood protection function and the associated maintenance as a coastal protection structure dikes are large environmental earth constructions that characterise the coastal landscape. They are usually made of a sand core, a cohesive top layer with high erosion resistance, and a grass cover. At the German Baltic Sea coast the cohesive layer is usually made of marl while at the North Sea coast it is mostly made of marsh clay which sometimes has considerable organic contents. During the last few years also dried fine-grained organic dredged materials have been investigated to be used in dike and landfill construction as protective and/or recultivation layers. These materials can be characterised as young cohesive organic soils, which is the focus of this paper.

Dredging means the removal of sediments from water bodies by excavators or hydraulic dredgers. It is a

Corresponding author: Anne-Katrin Große
E-mail: anne-katrin.grosse@uni-rostock.de

necessary operation in waterway and harbour maintenance but also in hydraulic construction works and ecological revitalisation projects. The major amount of the removed sediments is usually relocated within the water body which means the sediments are transported to another location and dumped to underwater reservoirs. However, due to environmental sustainability the relocation of contaminated and cohesive organic sediments is restricted in most areas of the world. In the Baltic Sea there is, for example, a general understanding under the HELCOM convention only to relocate clean sands (regarding precaution values usually provided by national law of the Baltic Sea bordering states).

Cohesive organic dredged sediments are usually taken ashore to be deposited on special land-fills (mainly if contaminated) or to be re-used beneficially according to the respective environmental regulations, e.g. the soil protection law, ground water protection law, and other soil-water related regulations (in Germany: BBodSchG 1998; BBodSchV 1999; WHG 2009). The re-use of fine-grained organic dredged materials in agriculture and landscaping has become an acknowledged method (Sigua *et al.* 2004; Canet *et al.* 2003). Usually the materials are processed before re-use. In most cases the materials are therefore pumped into containment areas where they are dewatered. After some time, the mud is taken out of the dredging polders and put on heaps on adjacent drying fields where the ripening process starts. The so ripened soil-like material is usually an in-homogenous mix of mineral and organic soil particles, depending on the dredging project. However, sometimes the dredged materials are also separated in so-called classification polders where the coarse and medium coarse sandy material settles around the inlet pipes, followed by different mixed soils with decreasing sand and increasing silt, clay and organic fractions down to the polder outlet where the very fine-grained organic sludge is deposited. Then particular batches can be classified and separately processed to receive better qualities for the re-use of the materials. The Hanseatic City of Rostock successfully runs two such containment areas as waste management technology with classification polders to process defined ripened soils for re-use. The mixed soils and fine-grained organic materials have been used in landscaping, agriculture and recultivation layers on landfills so far.

The project DredgDikes deals with the application of ripened dredged material in dike construction. Within the project three different cohesive organic dredged sediments have been installed in a full scale research dike facility (Cantré, Saathoff 2013). All materials are generally cohesive soils with an organic content of five to ten per cent. Visually, they seem to have considerable erosion resistance when installed and compacted with caterpillars. In contrast, laboratory flume tests showed rather low erosion resistance with respect to overflowing. Particularly for coastal and river dikes the time dependent stability of the erosion protection layer with respect to a static water loading is of great interest because the soil aggregates may decompose due to aggregate explosion and swelling phenomena which may lead to dike failure. Therefore, two different disintegration tests have been investigated to study the aggregate stability of the materials as an index for the erosion resistance under static conditions. However, some important limitations regarding both the test setups and the evaluation methods disparage their conclusiveness which will be discussed in this paper. The results will be used to develop a set of characterisation tools for dredged materials in environmental and geotechnical engineering applications.

1. Basics

1.1. Aggregate and erosion stability

Erosion stability is defined as the stability against "the replacement and the transport of soil particles along the surface" (Blume *et al.* 2010) and can be characterized as a combination of the stability of the overall soil texture and the stability of the aggregates within the texture, which are agglomerations of fines of different mineral grains mixed with organic carbon, lime and other soil components (Le Bissonnais 1996).

Surface erosion can be triggered from water, wind, snow melting or gravitation and is influenced by different processes like rainfall or surface run-off. Here, only water induced erosion is investigated. The main processes of soil erosion by water are soil displacement, soil transport and soil deposition (Blume *et al.* 2010). The soil displacement is the release of erodible material and includes sub-processes like the destruction of soil aggregates and the siltation of bounded particles. The soil structure consists of single grains or aggregates which are characterized by their external shape and size or their internal structure (Blume *et al.* 2010). The stability of soil aggregates differs with respect to water and pressure which influence the mechanical capacity, the siltation and the erosion risk. An aggregate can be described as stable if the particle position does not change by voltage variation. Also, high organic contents improve the stability of aggregates against water immersion due to inter-particle cohesion with organic carbon (Chenu *et al.* 2000; Cantón *et al.* 2009; Dal Ferro *et al.* 2012).

The definition of aggregate stability and soil structure stability is not standardised and both definitions are often used synonymously (Ametzketa 1999). The German DIN standard 19683-16 (2009) defines aggregate stability as an important parameter for the siltation of soils and provides a comparison of the stability of different agricultural farmlands. Here, aggregate stability is defined as the resistance of soil particles to their structure (Blume *et al.* 2010). Hartge and Horn (2009) describe aggregate stability as the resistance of aggregates against destructive triggers which can be expressed by a shear parameter. The determination of the aggregate stability against water is not trivial because it depends on the surface binding energy of the soil particle, the water cohesion and the adhesion forces. Adhesion is caused by suspended solids that may deposit at the contact points of the soil particles and thus glue the particles together. Both the building and reversibility of aggregates depend on the degree of drying and the type of rehydration and thus on the aggregate size, aggregate density and the pathways through intra aggregate pores are important parameters (Hartge, Horn 2009). Le Bissonnais (1996) also describes a high dependence of water infiltration and soil erosion by aggregate stability. In his opinion the erosion rates are the results of the aggregate disintegration divided into disintegration of macrostructure and dispersion. The aggregate disintegration through water can be the result of a variety of mechanisms, four of which are particularly important (Le Bissonnais 1996):

I. Slaking occurs with overpressure through enclosed air during moisture infiltration and appears when dry aggregates are put fast into water. The disintegration decreases with increasing clay content.

II. Swelling and slaking during wetting and dehydration lead to aggregate blasting. The disintegration increases with increasing clay content.

III. Soil fragments are dissolved by the splash effect if the kinetic energy of rain drops is high enough to cause disintegration. This phenomenon occurs most often with wet soils because the aggregates are weaker and can be dissolved more easily.

IV. Physico-chemical dispersion means the reduction of attraction between single colloid particles during drainage of soils and is facilitated with slaking and swelling. It is one of the most effective processes of aggregate destruction. Auerswald (1993) on the other hand focuses on the influence of the initial soil moisture and the resulting erosion process to explain aggregate stability. Investigations about the shear strength from soils or aggregates show lower strengths with increasing moisture. Investigations with a sudden wetting, however, show a lower aggregate blasting at increasing moisture content. Hereby, the initial moisture seems to have an influence on the mechanisms of temperature weakening, impact hardness, air blasting and swelling pressure with respect to aggregate disintegration (Auerswald 1993):

I. With penetrating water a heat front proceeds into the aggregates due to an energy release that weakens the aggregate bounding substances. This mechanism is called temperature weakening and is important for soils with moistures below 10% or above 30% (gravimetric).

II. The soil stiffness exponentially increases with decreasing soil moisture, thus the impulse recurrence from a rain drop to the soil increases.

III. The decrease of soil moisture causes an increase of air filled pores. If water penetrates quickly into an aggregate without air escaping, an overpressure occurs inside the aggregate, leading to air blasting. This phenomenon increases with decreasing moisture content.

IV. Quick and irregular wetting of aggregates causes irregular swelling. The resulting drag stress causes aggregates to break-up.

1.2. Methods to determine erosion stability

To determine the soil erosion stability a variety of methods have been proposed (Table 1) which can be divided into tests to determine the erosion resistance of the whole soil texture (e.g. laboratory flumes or erosion function apparatus after Briaud) and tests to determine the aggregate stability (e.g. wet-sieving methods, flow-rate methods, and water-immersion methods). In the following, different tests to determine the aggregate stability will be briefly reviewed.

For the wet-sieving tests air-dried soil aggregates are put into a sieve-diving apparatus with different sieve sizes to be moved vertically through a liquid medium (water or ethanol). Thereby, the instable aggregates are resolved in the liquid and the remaining grains are automatically sieved. The mass left on the sieve is measured and related to the initial mass.

Table 1. Overview of tests to determine aggregate stability

Method	Test
Wet-sieving	German DIN-standard 19683-16
	Hartge/Horn
	Le Bissonnais
Flow-tests	Pinhole test after ASTM D 4647-06
	Modified pinhole test after BAW
Dispersion-tests	Dispersion test after ASTM D 4221-99
	Crumb-test after ASTM D 6572-06
	Modified crumb test after Haghighi
	Disintegration test after Weißmann
	Disintegration test after Endell (BAW)

The pinhole test after ASTM standard D 4647-06 and the modified pinhole test (RPW 2006) represent direct qualitative measurements of the dispersibility of compacted clay. A defined hole is put into a compacted soil sample and a water flow is realised through the hole. After the test the increase in hole diameter is measured and related to a degree of aggregate stability.

The disintegration of aggregates or compacted samples in water immersion is analysed using dispersion tests. They can be performed using the crumb test after ASTM standard D 6572-06, the modified crumb test after Haghighi *et al.* (2012), the dispersion test after ASTM standard D 4221-99 or different disintegration tests. In this paper the disintegration tests after Endell (enhanced by the Federal Waterways Engineering and Research Institute, BAW) and Weißmann have been studied.

2. Investigated soil materials

For the investigations five different soils were used: three different fine-grained, organic dredged material batches from a containment area near Rostock, Germany and two standard dike cover materials (marl and marsh clay). Apart from their composition the dredged materials differ in their ripening time: materials M1 and M3 had been ripened for 5 years, while material M2 had been ripened for two years at the time of the investigation. For materials M1 and M2 three subsamples were taken respectively. The marsh clay batch was sampled from a dismantled dike cover layer in Hamburg Port while the marl was originally extracted from the Baltic Sea off the coast of the Island of Rügen (Fig. 1).

Fig. 1. Locations from where the soil samples were taken

Table 2. Reliable results of geotechnical characterisation

Material	M1	M2	M3	Marl	Marsh clay
Clay [%]	25–28	22–25	15	5	17
Sand [%]	29–34	40–47	54	89	46
W [%]	61–68	55–73	46	13	21
OM [%]	10–11	9–10	6	0.9	1.7
LC [%]	9–10	8	10	12	0.9
c_u [kPa]	53-132	19–34	120	24	83
φ [°]	28–30	28–31	30	42	66
c [%]	35–47	13–19	59	15	3
k_f [m/s]	5E–08	8E–10	5E–09	4E–08	4E–10

Table 3. Preliminary results of geotechnical characterisation

Material	M1	M2	M3	Marl	Marsh clay
LL [%]	80–98	64–88	52–57	17	36
PL [%]	75–81	54–67	49–54	13	22
SL [%]	58	42–47	51	16	22
PI [%]	4–22	11–24	3–4	4	14
CI [-]	2–5	0–1	2–4	1	1
w_{opt} [%]	40–43	32–35	31	9.2	14
OD [g/cm^3]	1.1–1.2	1.3	1.4	2.1	1.8

All materials were investigated with respect to their soil mechanical characteristics such as water content, grain-size distribution, compaction parameters, shear strength, water permeability and shear parameters, principally according to the German DIN-standards for soil mechanical analyses (DIN… 1998). Because of the high organic content of the dredged materials the grain-size distribution was determined according to DIN ISO 11277 for fine-grained, organic soils. For the organic content the TOC (Total Organic Carbon) value was determined after DIN ISO 10694 with an elemental analyser at temperatures above 1000 °C, because the marine dredged sediments possess high lime contents which influence the results from a 550 °C muffle furnace.

The results of the geotechnical characterization are shown in Table 2 and Table 3. Because of different challenges using the standardized DIN characterization for the specific dredged materials the tables are divided into reliable and preliminary results.

3. Disintegration tests

For both disintegration tests in this study cylindrical compacted samples with different water contents are put into a wire mesh basket which is connected to an electric scale. The disintegration is measured by recording the weight loss caused by the dropping of aggregates at water immersion. The dropping of aggregates is described as "crumbling" in the following while disintegration describes the status of the overall sample. Both disintegration tests use the weight loss as a function of time for further analysis. The principle of the test procedure is shown in Figure 2. In spite of the similar setup of the disintegration tests and similar processes studied the tests differ in several important issues which will be discussed in the following.

3.1. Disintegration test after Endell

The disintegration test after Endell (enhanced by the Federal Waterways Engineering and Research Institute, BAW) is used to estimate the erosion susceptibility of clay to be used in waterways construction. Sealing compounds with higher disintegration are more sensitive to erosion which is especially interesting while installing mineral sealing liners while they are still uncovered (RPW 2006).

For the analysis at least five proctor compacted samples (proctor power of 0.6 MNm/m^3) with a diameter of 2 cm and a height of 4 cm are used. The samples are each put into a wire mesh basket which is not specified with respect to its dimensions. It is connected to an electric scale and put into a water basin with distilled water (Fig. 2).

After the RPW (2006) the weight has to be recorded until a constant weight is reached, but not less than 24 hours. The result of this test is the disintegration number $Z(t)$ as a function of time (Eqn (1)):

$$Z(t)=\frac{A1-A(t)}{A1-A2}, \quad (1)$$

where: $Z(t)$ – disintegration number as function of time; $A1$ – uplift sample weight including mesh basket weight [g] at the beginning; $A(t)$ – uplift sample weight including mesh basket weight [g] at the time t of the test; $A2$ – uplift mesh basket weight [g] without sample.

To compare different samples the disintegration number after 8 hours $Z(8)$ is proposed. Therefore, it must be ensured that all dropped particles drop out of the wire mesh basket; otherwise, the test has to be repeated. According to Endell there is one specific moisture content for every soil at which the most significant disintegration occurs. Therefore, the test has to be carried out with at least five different water contents resulting in different undrained shear strengths (e.g. c_u = 10…70 kPa). The lowest disintegration should be expected at a water content near the liquid limit because of a reduced water immersion in saturated samples.

RPW (2006) specifies a limit value for the disintegration number of $Z(8)$ = 0.05. Above this value the soils are defined as susceptible to erosion while soils with a value $Z(8)$ < 0.05 are defined as erosion resistant. However, there is no standardised limit value for the disintegration number yet.

3.2. Disintegration test after Weißmann

The test equipment of Weißmann's disintegration test was developed independently to that of Endell to determine the erosion resistance of marsh clay embankments on sea dikes. The test facility is similar to that of Endell, except little differences (Fig. 3). The wire mesh basket is defined with meshes of 8×8 mm² and an overall size of 10×10×12 cm³. The soil samples are bigger than those of the Endell test (5 cm for both diameter and height). The water temperature shall have 20 °C during the entire test and tap water should be used. Nine cylindrical samples have to be prepared in a Proctor apparatus with optimum water content (standard proctor power of 0.6 MNm/m³) to guarantee good compaction.

From the nine samples three test series with different water contents are prepared (three repetitions):

(1) Dry sample – dried at 50 °C to a constant mass;

(2) Standard sample – optimum water content;

(3) Wet sample – lagged with filter paper and put into a water basin for ten minutes, then kept in a desiccator for 24 hours.

The result of the Weißmann test is the time t_{30} when a sample lost 30% of its initial weight. The maximum testing time is 24 hours, even if the 30% weight loss cannot be reached; then the material is classified as erosion resistant. According to Weißmann the disintegration time increases with increasing water content and can be described approximately with an exponential curve, shown in

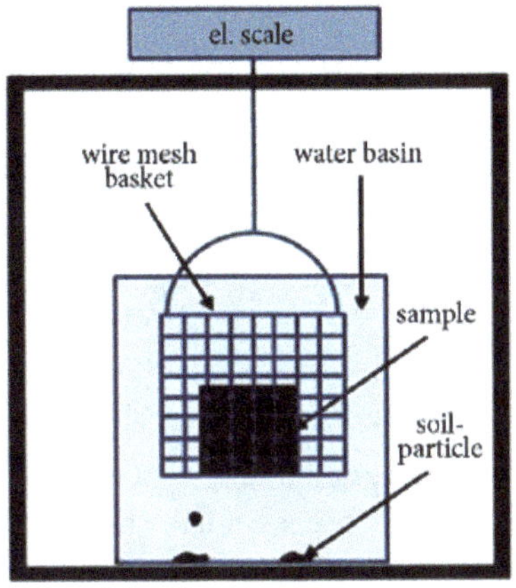

Fig. 2. Schematic of disintegration tests

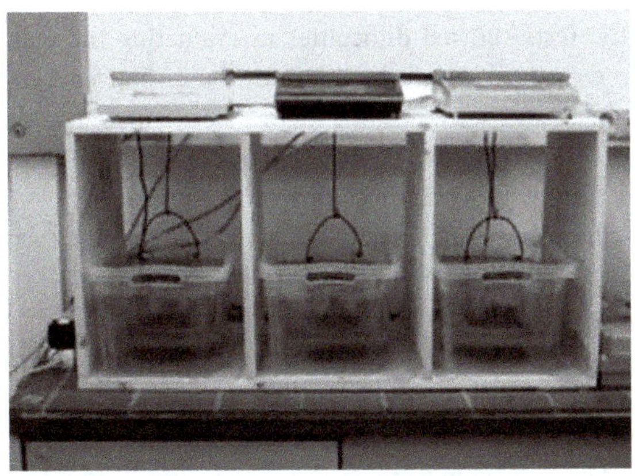

Fig. 3. Disintegration equipment after Weißmann

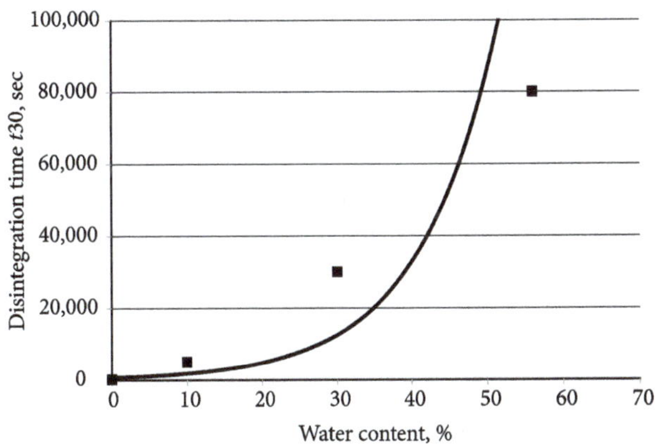

Fig. 4. Diagram of disintegration time depending on water content after Weißmann

Figure 4. After entering the results of the three test series into this diagram the disintegration number $t_{30,V}$ can be calculated with Equations (2)–(5).

For determining the auxiliary parameters A and B (Eqns (3) and (4)) two not defined water contents have to be chosen and the associated disintegration times have to be taken out from the diagram. The reference value $t_{30,V}$ is used at a Consistency Index $Ic = 0.8$ depending on the associated water content w_v, which can be calculated from the liquid and the plastic limit (Eqn (5)). Finally the disintegration time $t_{30,V}$ can be calculated at w_v with Equation (2):

$$t_{30}(w) = t_{30,0} + A \cdot w \cdot e^{B \cdot w}; \quad (2)$$

$$B = \frac{\ln((t_{30(w1)} - t_{30,0}) \cdot w2) - \ln((t_{30(w2)} - t_{30,0}) \cdot w1)}{w_1 - w_2}; \quad (3)$$

$$A = \frac{t_{30(w1)} - t_{30,0}}{w_1 \cdot e^{B \cdot w1}}; \quad (4)$$

$$w_V = 0.2 \cdot w_L + 0.8 \cdot w_P, \quad (5)$$

where: $t_{30,0}$ – disintegration time t_{30} at $w = 0\%$; A, B –auxiliary parameter; w_v – water content at $Ic = 0.8$.

3.3. Modification of the test procedure for the dredged materials

Initial tests showed difficulties in evaluating the results of the dredged materials with both tests. Therefore, some modifications were necessary. For a better understanding of the modifications, the original attributes of the Endell and Weißmann tests are compared in Table 4.

For the Endell test one sample at natural water content w_n and two samples with water contents $w > w_n$ and $w < w_n$ were chosen respectively. Samples with $w < w_n$ were dried step by step until the selected water content was achieved. To produce samples with $w > w_n$ water was added and the material was mixed. Both water addition and sample drying were restricted to values in an interval in which good compaction could be ensured for all samples. The wire mesh basket was made cylindrical having a mesh size of 8 mm and a diameter and height of 3.2 and 5 cm respectively. In this way the small samples would not fall through the mesh and could be installed upright inside the basket.

The experiments after Weißmann were modified after initial tests with the dredged materials. Instead of the optimum water content the natural water content was used for sample preparation. The proposed optimal water content at Proctor's density of the inhomogeneous dredged materials could not be determined reliably. Also, all samples compacted at natural water content showed little spread in the disintegration curve and so ensured comparability of all three repetitions.

For both tests the recording interval with respect to weight loss was chosen to 5 s to ensure a precise measurement of the disintegration during the first few minutes of testing. Every subsample was tested for at least 24 hours.

Table 4. Differences between the tests of Endell and Weißmann

	Endell	Weißmann
Water type	Distilled water	Tap water
Basket	Not defined. Chosen: Size; 3.2×5.0 cm^2, 8 mm mesh	Defined: Size: 10×10×12 cm^3, 8 mm mesh
Sample size	$H = 4$ cm, $D = 2$ cm	$H = D = 5$ cm
Result	Disintegration number $Z(t)$	Disintegration time t_{30}

4. Results

4.1. Disintegration test after Endell

4.1.1. Dredged materials M1, M2 and M3

The results of the disintegration test are the process of disintegration as a function of the time and computed a disintegration number after eight hours. The functions of material M1 and M2 have similar shapes. Figure 5 shows the measured relative sample weight of material M1 relating to the disintegration time.

Already after a short time all samples start to crumble and large aggregates fall off to the bottom of the water basin which can be seen from the recorded scale data. This process generally dominates in a time interval of 10 to 1,000 s. As soon as the samples have lost about 50–70% of their initial weight this process seems to stop; for the rest of the time the recorded weight remains nearly constant. However, there seems to be an exception for samples with low water contents: shortly after the start the dry material absorbs water and the weight increases to about 120% of the initial weight. After some time the increase stops and the recorded weight shows both the largest and fastest decrease among all samples.

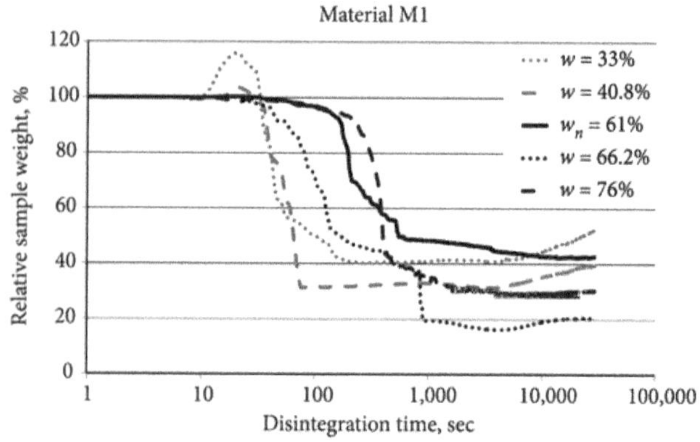

Fig. 5. Disintegration curves (Endell) of material M1

Table 5. Disintegration number $Z(8)$ of material M1 and M2

Material	w [%]	$Z(8)$ after 8 h
M1	33.0	0.7273
	40.8	0.6056
	w_n = 61.0	0.5728
	66.2	0.6954
	76.0	0.6056
M2	30.0	0.7273
	42.0	0.3991
	w_n = 54.0	0.4714
	78.0	0.4784
	79.0	0.6348

Table 6. Disintegration number $Z(8)$ of material M3

Material	w [%]	$Z(8)$ after 8 h
M3	21.9	0.9206
	36.0	0.9958
	36.9	0.7273
	49.1	0.8033
	w_n = 49.7	0.5425
	54.6	0.9524
	57.8	0.9003

Later during the test the initially dry samples seem to start to absorb water again, resulting in a second weight gain. The analysis of the disintegration numbers after eight hours show the lowest disintegration (the best erosion stability) for the natural water content w_n for material M1 and for a water content below w_n for material M2. Table 5 gives an overview about the disintegration numbers $Z(8)$ of the materials M1 and M2.

The more sandy dredged material M3 shows a different set of disintegration curves. Almost all samples show a high disintegration rate. The sample weight reduces to nearly zero within approximately 100 s with the exceptions of the samples with w_n = 49.7% and w = 36.9%. The dry samples also show an initial weight increase to 110% comparable to the materials M1 and M2. Interestingly, there are samples with nearly the same water content which show very different disintegration behaviour: the sample with w = 49.1% shows a significantly higher disintegration number than that with w_n = 49.7%. A similar observation was made for w = 36.0% and w = 36.9% (Fig. 6 and Table 6).

4.1.2. Conventional dike construction material marl and marsh clay

The development of the marl disintegration partly differs from that of the dredged materials, in particular regarding materials M1 and M2. Compared to the dredged materials, the tested marl can only absorb little amounts of water because of its large sand and small clay fraction. The maximum possible water content to produce proctor samples was w = 14.1%. The crumbling of the different marl samples proceeded very fast and in steps (Fig. 7).

The sample with the natural water content shows the lowest disintegration number, a result that was also found for the dredged materials M1 and M3 (Table 7). Some samples even show a full disintegration like material M3.

The investigated marsh clay shows an explicit distinction of the disintegration curves with respect to the water contents of the subsamples (Fig. 8).

The best result was again achieved with the w_n sample. The driest sample crumbled fastest with a total weight loss of 98% and a very high disintegration number $Z(8)$ = 0.9874 (Table 8). The disintegration numbers with $Z(8)$ = 0.2589 and $Z(8)$ = 0.2834 for the samples with w_n and w = 23% are the best results of all materials tested.

Table 7. Disintegration number Z(8) of marl

Material	w [%]	$Z(8)$ after 8 h
Marl	5.6	1.0
	7.8	0.8823
	9.2	0.9821
	w_n = 11.8	0.6838
	14.8	0.807

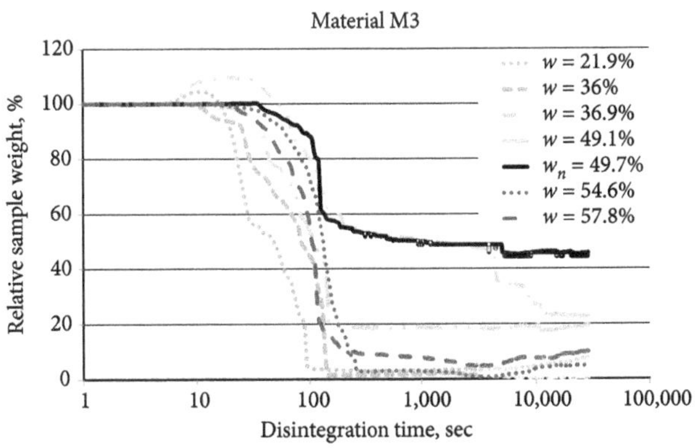

Fig. 6. Disintegration curves (Endell) of material M3

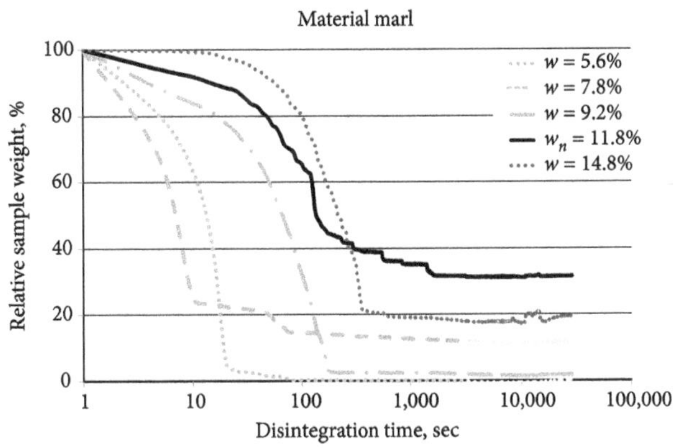

Fig. 7. Disintegration curves (Endell) of marl

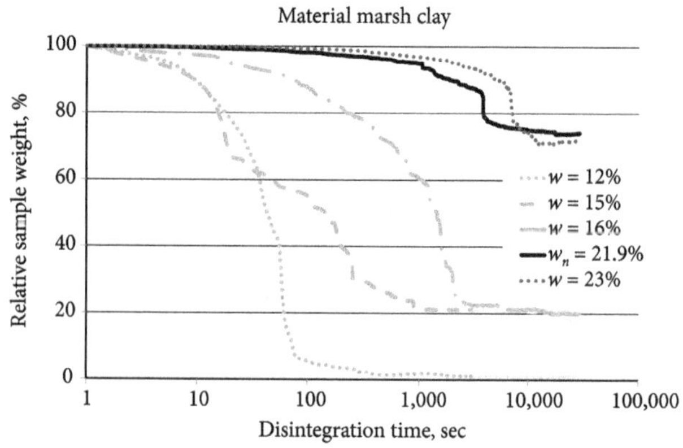

Fig. 8. Disintegration curves (Endell) of marsh clay

Table 8. Disintegration number $Z(8)$ of marsh clay

Material	w [%]	$Z(8)$ after 8 h
Marsh clay	12.0	0.9874
	15.0	0.8216
	16.0	0.8008
	w_n = 21.9	0.2589
	23.0	0.2834

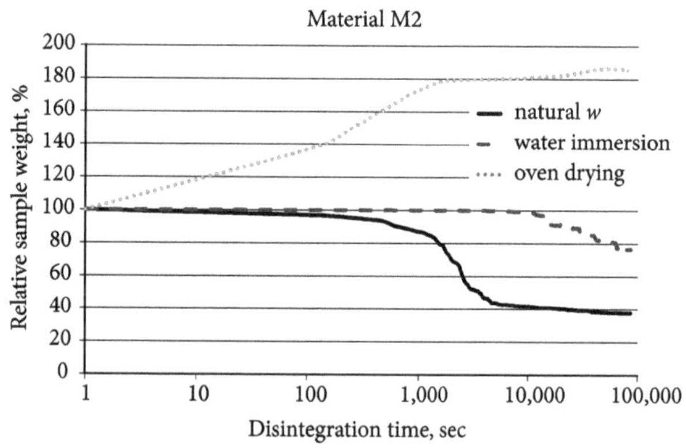

Fig. 9. Disintegration curves (Weißmann) of material M2

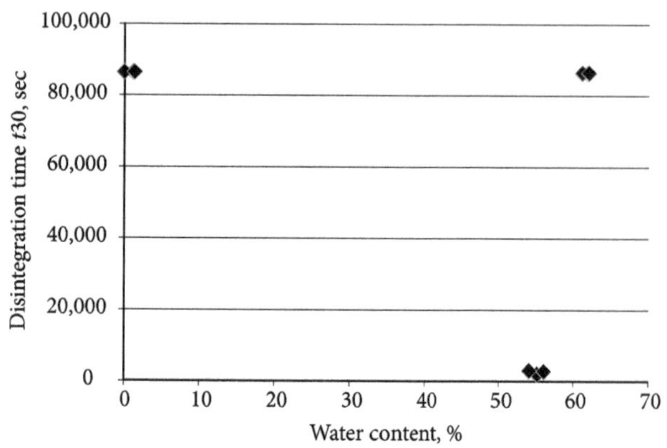

Fig. 10. Disintegration time t_{30} depending on water content for material M2

4.2. Disintegration test after Weißmann

4.2.1. Dredged materials M1, M2 and M3

The results of the disintegration tests after Weißmann are the disintegration curves as functions of time as well as the disintegration time t_{30} for three defined water contents. Depending on the available amount of material two subsamples were chosen for the investigation of materials M1 and M2 and only one for material M3. Since all dredged materials show similar curve shapes for the different sample preparation modes the curves of M2 are presented exemplary (Fig. 9).

The weight of the oven-dried samples started to increase constantly as soon as they were submerged. The weight gain approached up to 260% of the initial weight although aggregates had fallen off the sample. During the test large air bubbles were observed to come out of the soil sample.

The "water immersion" samples which were initially wrapped in filter paper show the highest stability. For a considerable period of time these samples did not lose significant weight before they started to crumble after about 10,000 s. However, there are also intervals in with no weight change was recorded although particles fell of the sample. At the end of the test the water immersion samples started to disintegrate to a relative weight loss of about 20% with no further change.

The w_n samples show the highest disintegration of the three conditions. They usually started to crumble after about 1,000 s and show a final weight loss of about 60%.

Because both the oven-drying and the water-immersion samples show no weight loss of 30% the disintegration time has to be defined to 86,400 s which is why no exponential disintegration curve can be determined (Fig. 10).

The dependency of disintegration and water content cannot be described with this data and consequently no disintegration time $t_{30,V}$ can be calculated. Only the disintegration time after 30% mass loss of the samples with natural water content can be determined and compared (Table 9).

Table 9. Disintegration time t_{30} (w_n) of M1, M2 and M3

Material	Disintegration time t_{30} (w_n) [s]
M1	1.600 and 1.320
M2	2.600 and 4.000
M3	1.120

4.2.2. Conventional dike construction material marl and marsh clay

The marl samples show a different behaviour. After few seconds all samples started to crumble and lost a large amount of mass within a short period of time (Fig. 11). After 2,000 s the disintegration was already finished with a final weight of 20 to 30% of the initial mass.

As distinguished from the investigations with dredged material the optimal water content w_{opt} was used instead of the oven-dried because the oven-dried samples showed only weight increase instead of mass loss. For further results of the disintegration of oven-dried marsh clay samples compare Beyer *et al.* (2012). For the installation of w_{opt} the material was oven dried at 55 °C and rewetted afterwards. The result of the disintegration time at w_v is $t_{30,V}$ = 731s (Table 10).

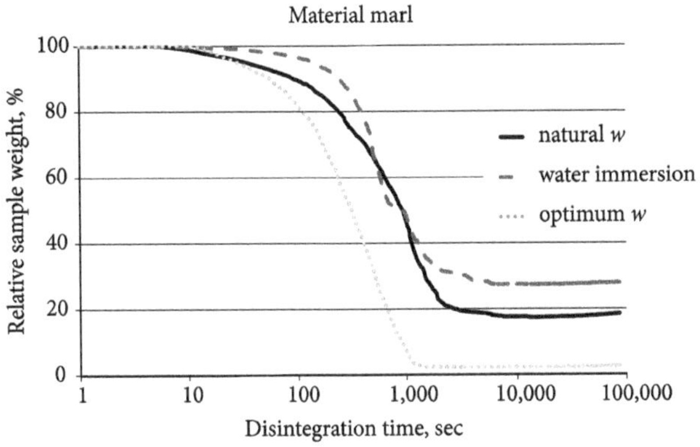

Fig. 11. Disintegration curves (Weißmann) of marl

Table 10. Disintegration time t_{30} of marl

Parameter	w [%]	t_{30} [s]
w_n	11.5	482
w_{im}	12.43	672
w_{opt}	12.0	178
w_v	13.72	731

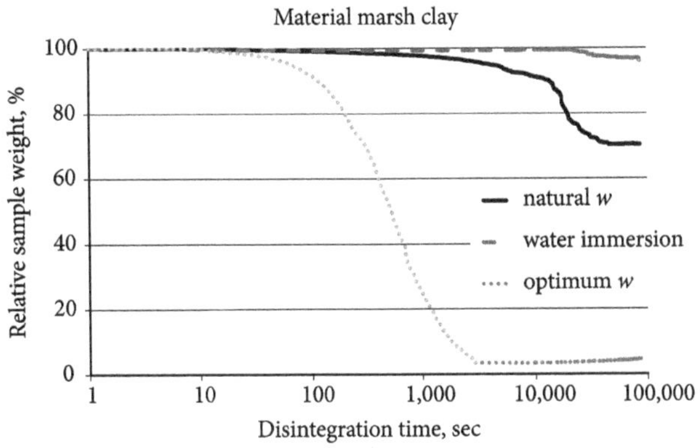

Fig. 12. Disintegration curves (Weißmann) of marsh clay

Table 11. Disintegration time t_{30} of marsh clay

Parameter	w [%]	t_{30} [s]
w_n	24.0	60,098
w_{im}	43.71	86,400
w_{opt}	13.4	373
w_v	25.08	62,438

The disintegration curves of marsh clay show that the crumbling of samples with w_n started late and the disintegration of 30% was reached very late: t_{30} = 60,000 s. The "water immersion" samples show hardly any mass loss during the whole investigation (maximum 2%). The samples with w_{opt} on the contrary showed a result comparable to that of the marl. Here, crumbling started already after 100 s and resulted in an almost complete mass loss (Fig. 12). The disintegration time at w_v is $t_{30,V}$ = 62,438 s (Table 11).

Fig. 13. Total disintegration with different mass loss

5. Discussion

During the investigation both benefits and major limitations of the different disintegration tests could be observed. There is a general problem of both tests with respect to the testing procedure – a sample in a wire basket submerged in water. However, the two experimental setups show test specific limitations also. Finally, the results are discussed in comparison to test results from a dynamic erosion test using a small scale laboratory flume and to data taken from the literature.

5.1. Testing procedure

The testing procedure of both disintegration tests is based on a sample inside a wire mesh basket which is submerged in water. The major limitation of this procedure which was observed in a large variety of test is, that soil aggregates that fall off the sample during disintegration stay in the basket and are not detected as disintegration by weight loss (Fig. 13 left).

This may have different reasons such as the defined small mesh size in case of the Weißmann test together with varying sizes of agglomerates that evolve during disintegration. In Figure 13 both samples are completely disintegrated with the difference that the disintegration is recorded correctly for the right sample and no weight loss is recorded for the left sample. For the Endell test the RPW (2006) proposes to repeat the test until there is one test in which all aggregates fall down. There may be soil samples, however, which always decompose into larger aggregates where this solution will not solve the problem. Weißmann, on the other hand, gives no attention to this problem. The problem may be solved by using baskets with different mesh sizes; this may not affect the Endell method (no definitions with respect to the basket) but an adjustment to the Weißmann method would be needed. To compare different disintegration tests the same mesh size should be used, thus this parameter needs to be recorded for every test.

5.2. Disintegration test after Endell

In RPW (2006) the disintegration number after eight hours $Z(8)$ is used to specify a soil's susceptibility to erosion. The value of $Z(8) = 0.05$ above which a soil is considered erodible has been developed for mineral sealing liners. In this investigation dredged material, marl and marsh clay were tested which may perform differently with respect to the disintegration number and the actual erosion resistance. This may be caused by the materials' lower clay content compared to that of mineral sealing liners.

The lowest disintegration number was determined at w_n for the majority of samples. The most substantive explanation seems to be connected to the sample preparation which was necessary to install the different water contents. Some of the samples needed to be dried, which seems to have a considerable influence on the aggregate stability, particularly for the fine-grained, organic materials. The influence of the drying temperature in soil sample preparation on the aggregation has already been investigated by different researchers (Basma *et al.* 1994; Sunil, Krishnappa 2012). These investigations show that a higher drying temperature leads to an accumulation of fine particles with organic and to larger grains with decreasing specific surface area (Mikutta *et al.* 2005). There is also an irreversible dehydration process of the clay particles causing cementation of fine particles which are then not dispersed (Sunil, Krishnappa 2012). These effects can even be detected in a standard grain-size analysis. The "reduction" of fine particles also reduces their cementing ability and the samples start to crumble faster.

A second explanation may be found in the fast infiltration of the dried materials once they are submerged which leads to a quick displacement of air confined in the soil sample (Rohoskova, Valla 2004). Large air bubbles escape the samples due to over-pressure, usually causing "aggregate blasting" (Auerswald 1993; Le Bissonnais 1996)

For dredged materials the preparation with water addition leads to a higher disintegration. At the beginning higher water contents lead to less crumbling because more micro-pores are pre-filled with water. Then the behaviour changes and more aggregates fall off, possibly because the aggregate cohesion is weakened through the initial water immersion. No dependences of disintegration and water content could be found for the dredged materials (compare Rudat 2012).

According to Dal Ferro *et al.* (2012) and Cantón *et al.* (2009) there is a stabilisation of aggregates by organic carbon. In the investigation no correlation between organic content and disintegration could be validated. Both fine-grained dredged materials M1 and M2 have high organic and clay contents but show lower erosion resistance than marsh clay with less organic content.

To compare the erosion resistance of the different materials the disintegration number $Z(8)\ w_n$ was chosen (Table 12). The conventional dike construction material marsh clay has the lowest disintegration number $Z(8) = 0.2589$, followed by the dredged material M2 with $Z(8) = 0.4714$. The maximum disintegration could be observed for dredged material M1 and marl.

All materials – including the conventional dike construction materials – did not even closely achieve a value of $Z(8) < 0.05$ which is the proposed erosion resistance limit for mineral liners. If the disintegration test after Endell shall be used for dike construction materials the disintegration value needs modification.

Table 12. Disintegration number $Z(8)$, w_n in comparison

Sample	$Z(8)$, w_n
M1	0.5728
M2	0.4714
M3	0.5425
Marl	0.6838
Marsh clay	0.2589

5.3. Disintegration test after Weißmann

In the disintegration tests no dependency of disintegration time t_{30} and water content could be found. All oven-dried samples showed a considerable weight increase of up to 260% of the initial weight instead of a mass loss, although moderate crumbling could be recognized for all samples. There are two reasons for this phenomenon: the high amount of water which is absorbed and the escaping air bubbles which reduce the buoyancy. According to Beyer *et al.* (2012) the influence of buoyancy cannot be eliminated in this test procedure. This may lead to questionable results for unsaturated samples.

Due to the missing dependency of water content and disintegration the disintegration time at $I_p = 0.8$ cannot be determined. Even the modification to use samples with w_{opt} instead of oven-dried samples does not show any improvement. All samples crumbled in a very short time because of the preparation with oven-drying, rewetting and compaction (Lindh, Winter 2003).

To compare all materials the disintegration time at w_n was chosen again (Table 13). For materials M1 and M2 two subsamples were investigated respectively, for material M3, marsh clay and marl only one each.

Table 13. Disintegration time t_{30} at natural water content

Material	$t_{30,wn}$ [sec]
M1	1,320 and 1,600
M2	2,600 and 4,000
M3	1,120
Marl	482
Clay	60,098

Like in the Endell test the conventional dike construction material marsh clay shows the best disintegration behaviour among the tested materials. The best result among the dredged materials was obtained with material M2 with $t_{30,wn}$ = 2,600 s and $t_{30,wn}$ = 4,000 s. The sandy materials M3 and marl show the poorest results.

Based on his disintegration test Weißmann developed an evaluation method to classify the application of marsh clay as dike construction material. Additional evaluation criteria are water permeability (B1), degree of shrinkage (B3), and plasticity index (B4). As a result an evaluation number N is determined (Eqns (6)–(10)):

$$N = \sqrt[4]{B1 \cdot B2 \cdot B3 \cdot B4}; \quad (6)$$

$$B1 = 0.7 - (\log(kf) + 4)/20; \quad (7)$$

$$B2 = 0.2 \cdot \log(t_{30,V}); \quad (8)$$

$$B3 = 1.0 - 1.25 \cdot (V_S - 0.05); \quad (9)$$

$$B4 = 0.3 + 2 \cdot I_P. \quad (10)$$

The evaluation number N can be divided into five suitability classes. Materials with suitability class 1 are very well suited as dike cover layer, while materials with suitability class 5 are not advisable (Table 14). For all materials investigated in this study the evaluation number and suitability class was determined (Table 16). In this case the disintegration time $t_{30,wn}$ was used. The parameters used to compute the evaluation numbers are presented in Table 15, partly determined using the "not reliable" results from Table 3.

All investigated materials can be classified at least as class 4 "limited suitability" and are therefore applicable as dike construction materials. The best suitability was determined for the marsh clay (suitability class 2 "well suitable") closely followed by the dredged material M2 (suitability class 3, nearly "well suitable").

Table 14. Suitability classes with respect to the evaluation numbers N (Weißmann 2003)

Valuation number N	Degree of suitability	Suitability class
$1.00 \geq N \geq 0.85$	Very well suited	1
$0.85 > N \geq 0.75$	Well suited	2
$0.75 > N \geq 0.65$	Suited	3
$0.65 > N \geq 0.50$	Limited suitability	4
$N < 0.50$	Not advisable	5

Table 15. Calculation parameters to determine the evaluation number

Parameter	M1	M2	M3	Marl	Marsh clay
k_f [m/s]	5E-08	8E-10	2E-08	3E-08	5E-10
t_{30} [sec]	1,320	2,600	1,120	482	60,098
V_s [-]	0.33	0.26	0.176	0.026	0.1886
I_p [-]	0.051	0.157	0.028	0.041	0.139

Table 16. Evaluation numbers N and suitability classes (SC)

	M1	M2	M3	Marl	Marsh clay
B1	0.865	0.955	0.885	0.876	0.965
B2	0.624	0.683	0.610	0.537	0.956
B3	0.750	0.750	0.843	1.000	0.827
B4	0.200	0.614	0.200	0.200	0.578
N	0.533	0.740	0.549	0.554	0.815
SC	4	3(+)	4	4	2

5.4. Comparison of disintegration tests with investigations in a laboratory flume

In the project DredgDikes full scale overflowing tests on dike slopes are planned to determine the surface erosion stability of the dredged materials M1–M3. For preliminary tests a small scale laboratory flume was used with dimensions of 2.90 m length and 0.25 m width and a variable slope inclination with a maximum of 3:1. With the associated pumps a flow velocity of 2.3 m/s and an associated shear stress of 210 N/m² can be realised. For the experimental tests 20 samples (ten vegetated and ten non-vegetated samples) with a thickness of 7 cm prepared from the dredged materials M1, M2 and M3 and tested in the flume. The degree of compaction ranged from 0.6–0.85. For reference the conventional dike construction material marsh clay was used. For the non-vegetated samples an erosion rate was determined using laser scan data as the quotient of the eroded soil volume and the volume of water.

The marsh clay showed by far the lowest erosion rate, while material M2 showed the best results among the

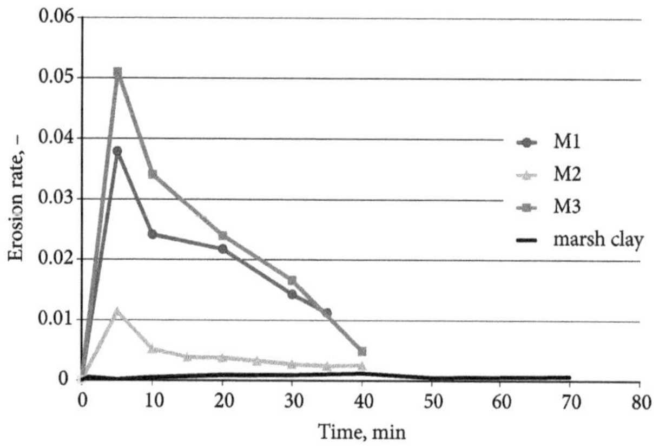

Fig. 14. Erosion rates of dredged material and marsh clay (modified after Lesch 2012)

dredged materials (Fig. 14). The samples of materials M1 and M3 in comparison experienced four to five time the erosion of M2.

When comparing the results of the flume test with those of the disintegration tests the single value of each test is related to the marsh clay values, because the lowest erosion in all investigations have been obtained at these samples. Altogether, similar orders of magnitudes can be recognized especially between the Weißmann test and the laboratory flume (Table 17), whereas the Endell test results show a smaller difference between the disintegration of marsh clay and the dredged materials. The reasons for this may be the small Endell test samples which are quickly wetted, leading to a comparably fast disintegration of all samples.

Table 17. Comparison of orders of magnitudes of disintegration tests and laboratory flume

	M1	M2	M3	Marsh clay
Lab flume	31	9	45	1
Endell	2.2	1.8	2.1	1
Weißmann	38	15	54	1

Conclusions

The use of dredged materials in dike construction is an important contribution to environmental protection engineering since a legal waste is beneficially re-used in environmental constructions. For the characterisation of the materials the erosion stability has to be investigated. During the investigations both benefits and major limitations of disintegration tests could be observed. The following conclusions can be drawn:

1. There is a major problem with the small mesh size in the Weißmann test facility and there are limitations with respect to the varying agglomerate sizes. In case of the Endell test the proposal of RPW to repeat the test until there is one in which all particles fall through the mesh is problematic with respect to the agglomerations that characterise a fine-grained organic dredged material. For the small mesh size proposed there may not be one such result.
2. The sample preparation with drying and rewetting of cohesive organic soils has a considerable influence on the aggregate stability and complete drying should be avoided.
3. The sample preparation methods for disintegration tests need to be modified for cohesive organic soils.
4. In both disintegration test setups no dependency between disintegration time and water content could be found. All oven-dried samples show an increase of the initial weight instead of a mass loss, although moderate crumbling could be recognized. Thus the modification of different test conditions should be considered.
5. Disintegration tests can be used to determine qualitative disintegration differences between materials using samples prepared at natural water content w_n.
6. The disintegration tests after Endell and Weißmann cannot be used to compute the quantitative aggregate stability for cohesive organic soils yet.
7. The results of the study on disintegration tests shows that there is still work needed to establish a laboratory test that allows to determine the erosion stability of fine-grained organic soils to be used in geotechnical constructions in environmentally sensitive areas such as coastal lowlands. The presented determination of benefits and limitations of the existing tests together with an analysis of necessary adaptions and boundary conditions pave the way for environmental engineers to discuss this development.

Acknowledgements

A special thanks to the Faculty of Interdisciplinary Research, Department Maritime Systems (University of Rostock) for a scholarship in the frame of which most of the results were derived.

References

Ametzketa, E. 1999. Soil aggregate stability: a review, *Journal of Sustainable Agriculture* 14(2–3): 83–151. http://dx.doi.org/10.1300/J064v14n02_08

Auerswald, K. 1993. *Relief-Boden-Paläoklima – Bodeneigenschaften und Bodenerosion – Wirkungswege bei unterschiedlichen Betrachtungsmaßstäben* [Relief-Soil-Paleoclimate – soil characteristics and soil erosion], Berlin – Stuttgart: Gebrüder Borntraeger. 208 p. (in German).

Basma, A.; Al-Homoud, A.; Al-Tabari, E. 1994. Effects of methods of drying on the engineering behavior of clays, *Applied Clay Science* 9(3): 151–164. http://dx.doi.org/10.1016/0169-1317(94)90017-5

Beyer, K.; Grabe, J.; Timmers, V. 2012. Zum Verhalten von Deichabdeckungen aus METHA-Material und Klei bei Vernässung [On the behaviour of dike covers made of METHA material and marsh clay with increasing water content], *Siegen Symposium*, 23–24 February 2012, Siegen. 16 p. (in German).

Blume, H.-P.; Brümmer, G.; Horn, R.; Kandaler, E.; Kögel-Knabner, I.; Kretzschmar, R.; Stahr, K.; Wilke, B.-M. 2010. *Scheffer/Schachtschabel – Lehrbuch der Bodenkunde* [Soil science]. Heidelberg: Spektrum Akademischer Verlag. 569 p. (in German). http://dx.doi.org/10.1007/978-3-8274-2251-4

Briaud, J. L.; Ting, F. C. K.; Chen, H. C.; Cao, Y.; Han, S. W.; Kwak, K. W. 2001. Erosion function apparatus for scour rate prediction, *Journal of Geotechnical and Geoenvironmental Engineering American Society of Civil Engineers* 127(2): 105–113. http://dx.doi.org/10.1061/(ASCE)1090-0241(2001)127:2(105)

Canet, R.; Chaves, C.; Pomares, F.; Albiach, R. 2003. Agricultural use of sediments from the Albufera Lake (Eastern Spain), *Agricultural Ecosystems & Environment* 95(1): 29–36. http://dx.doi.org/10.1016/S0167-8809(02)00171-8

Cantón, Y.; Solé-Benet, A.; Asensio, C.; Chamizo, S.; Puigdefábregas, J. 2009. Aggregate stability in range sandy loam soils – relationships with runoff and erosion, *Catena* 77(3): 192–199. http://dx.doi.org/10.1016/j.catena.2008.12.011

Cantré, S.; Saathoff, F. 2013. Investigation of Dredged materials in combination with geosynthetics used in dike construction, *Procedia Engineering* 57: 213–221. http://dx.doi.org/10.1016/j.proeng.2013.04.030

Chenu, C.; Le Bissonnais, Y.; Arrouays, D. 2000. Organic matter influence on clay wettability and soil aggregate stability, *Soil Sciences Society of America* 64(4): 1479–1486. http://dx.doi.org/10.2136/sssaj2000.6441479x

Dal Ferro, A.; Berti, A.; Francioso, O.; Ferrari, E.; Matthews, G. P.; Morari, F. 2012. Investigating the effects of wettability and pore size distribution on aggregate stability: the role of soil organic matter and the humic fraction, *European Journal of Soil Science* 63(2): 152–164. http://dx.doi.org/10.1111/j.1365-2389.2012.01427.x

Haghighi, I.; Martin, T.; Reiffsteck, P.; Chevalier, C. 2012. An "Enhanced Crumb Test" for better characterizing water sensivity of soils, in *6th ICSE Proceedings*, 27–31 August 2012, Paris, France, 1049–1056.

Hartge, K.; Horn, R. 2009. *Die physikalische Untersuchung von Böden* [The physical investigation of soils]. Stuttgart: E. Schweizerbart'sche Verlagsbuchhandlung. 178 p. (in German).

Le Bissonnais, Y. 1996. Aggregate stability and assessment of soil crustability and erosibility: I. Theory and methodology, in *European Journal of Soil Science* 47(4): 425–437. http://dx.doi.org/10.1111/j.1365-2389.1996.tb01843.x

Lesch, S. 2012. *Projekt DredgDikes – Ermittlung der Erosionsstabilität gegen Überströmen von begrüntem Baggergut in einer kleinmaßstäbigen Strömungsrinne* [Project DredgDikes – Determination of the erosion stability against overflowing of greened dredged material in a small-scale flume]: Master's Thesis. Universität Rostock. 100 p. (in German).

Lindh, P.; Winter, M. G. 2003. Sample preparation effects on the compaction properties of Swedish fine-grained tills, *Journal of Engineering, Geology and Hydrogeology* 36(4): 321–330.

Mikutta, R.; Kleber, M.; Kaiser, K.; Jahn, R. 2005. Review: organic matter removal from soils using hydrogen peroxide, sodium hypochlorite, and disodium peroxodisulfate, *Soil Science Society of America* 69(1): 120–135. http://dx.doi.org/10.2136/sssaj2005.0120

Rohoskova, M.; Valla, M. 2004. Comparison of two methods for aggregate stability measurement – a review, *Plant Soil Environment* 50(8): 379–382.

RPW 2006. *Guidelines for the Testing of Mineral Liners in Waterways Engineering*, EU-Notifizierung Nr. 2006/370/D vom 23.10.2006 (in German).

Rudat, M. 2012. *Untersuchungen zur Erosionsstabilität von feinkörnigem Baggergut mittels Zerfallstest* [Investigations on the erosion stability of fine-grained dredged material using disintegration tests]: Master's Thesis. Universität Rostock. 104 p. (in German).

Sigua, G. C.; Holtkamp, M. L.; Coleman, S. W. 2004. Assessing the efficacy of dredged materials from Lake Panasoffkee, Florida, *Implication to Environment and Agriculture* 11(5): 321–326.

Sunil, B.; Krishnappa, H. 2012. Effect of drying on the index properties of lateritic soils, *Geotechnical and Geological Engineering* 30(4): 869–879. http://dx.doi.org/10.1007/s10706-012-9504-7

Weißmann, R. 2003. Die Widerstandsfähigkeit von Seedeichbinnenböschungen gegenüber ablaufendem Wasser [The resistance capacity of landside sea dike embankments against overflowing water], *Mitteilungen aus dem Fachgebiet Grundbau und Bodenmechanik*. University Duisburg-Essen. 147 p. (in German).

Standards and regulations

ASTM D 4221-99 (Reapproved 2005): 2005. Standard Test Method for Dispersive Characteristics of Clay Soil by Double Hydrometer. ASTM International, West Conshohocken, PA, USA.

ASTM D 4647-06: 2006. Standard Test Method for Identification and Classification of Dispersive Clay Soils by the Pinhole Test. ASTM International, West Conshohocken, PA, USA.

ASTM D 6572-06: 2006. Standard Test Methods for Determining Dispersive Characteristics of Clayey Soils by the Crumb Test. ASTM International, West Conshohocken, PA, USA.

BBodSchG: 1998. Gesetz zum Schutz vor schädlichen Bodenveränderungen und zur Sanierung von Altlasten (Bundes-Bodenschutzgesetz-BBodSchG) [Federal Soil Protection Act]. 12 p. (in German).

BBodSchV: 1999. Bundes-Bodenschutz- und Altlastenverordnung (BBodSchV) [Federal Soil Protection and Contamination Ordinance]. 32 p. (in German).

DIN-Taschenbuch 113, 1998. Erkundung und Untersuchung des Baugrundes [Ground investigation and testing], Berlin: Beuth-Verlag. 437 p. (in German).

DIN ISO 10694: 1996. Soil quality – Determination of organic and total carbon after dry combustion (elementary analysis) (in German).

DIN ISO 11277: 2002. Soil quality – Determination of particle size distribution in mineral soil material – Method by sieving and sedimentation (in German).

DIN 19683-16: 2009. Soil quality – Physical laboratory tests – Part 16: Determination of aggregate stability using the method of wet-sieving (in German).

WHG: 2009. Gesetz zur Ordnung des Wasserhaushaltes (Wasserhaushaltsgesetz – WHG) [Federal Water Act]. 56 p. (in German).

Anne-Katrin GROßE. Since 2013, Research Assistant at the Chair of Geotechnics and Coastal Engineering. 2003–2008: University of Rostock, studies in Land Management and Environmental Protection; specialization in Environmental Geotechnics and Landscape Construction. 2009–2013: PhD-scholarship of the Faculty of Interdisciplinary Research, Department Maritime Systems (University of Rostock), topic in dredged materials in dike construction. She is the author of 2 conference presentations with published paper contributions. Research interests include: soil mechanical characterization and erosion stability of ripened dredged materials.

Stefan CANTRÉ. 1995–2001 University of Rostock, education in Land Management and Environmental Protection; specializations in Landscape Planning, Geotechnics, Landscape Construction and Geosynthetics. 1998: University of Natural Resources and Life Sciences, Vienna; specialization in Rural Engineering and Water Management as well as Torrent and Avalanche Control; 1998–1999: Heriot-Watt University in Edingburgh, Scotland, College of Art, School of Planning and Housing and School of Landscape Architecture; Since 2001 Research Assistant at the Chair of Geotechnics and Coastal Engineering. In 2008, Graduation at the University of Rostock, topic: A contribution to the design of geotextile tubes for the dewatering of dredged materials. 2004–2009: University of Hagen (distance learning), Industrial Engineering. Since 2010, Coordinator of the EU-Project "DredgDikes". Author of 13 journal articles and 18 conference presentations with published paper contributions. Research interests: geosynthetics used for the dewatering of dredged materials, geosynthetic tubes and containers, the applicability of dredged materialsin dike construction, erosion control, including geosynthetic erosion control products.

Fokke SAATHOFF. 1976–1984 University of Hannover, studies in Civil Engineering, specializations in Structural and Hydraulic Engineering. 1981–1991 student and Research Assistant at Franzius-Institute for Hydraulic, Waterways and Coastal Engineering. 1991 Graduation to Dr.-Ing., Franzius Institute for Hydraulic, Waterways and Coastal Engineering; topic in geosynthetics in sealing systems: Laboratory investigations on the behaviour of geotextiles and synthetic liners. 1991–1997 Manager Naue Fasertechnik GmbH & Co. KG, Lübecke. 1998–2006 Managing Director at BBG Bauberatung Geokunststoffe GmbH & Co. KG, Lehmförde and Fiestel. Since 2006, Full Professor at the University of Rostock, Chief of the Chair of Geotechnics and Coastal Engineering. Since 2010, Project leader of the EU-Project "DredgDikes". Scientific functions: Vice Dean for Academic Affairs at the Faculty of Agricultural and Environmental Sciences, University of Rostock; Member of the Academic Council of the University of Rostock; Chairman of the Working Group AK 5.4.1: Research Association for Roads and Traffic (FGSV): Erosion Protection; Chairman of the Working Group AK 5.1 of the German Geotechnical Society (DGGT): Synthetics in Geotechnical and Hydraulic Engineering; Chairman of the working group AK 17 "Efficient Revetments at Waterways"; Chairman of the German delegation of the WG 1/CEN TC 189: "specific requirements" and leader of the task group "lining systems". Author of more than 250 publications including journal and conference papers. Uncounted conference participations as well as conference chairman and session leaderships for DGGT, HTG, IGS, etc. Research interest include: Geosynthetics in geotechnics and hydraulic engineering, beneficial reuse of dredged materials, mineral and synthetic filter systems, mineral and synthetic sealing systems, geotechnical and hydraulic aspects of renewable energy systems (e.g. wind power).

2

THE EFFECT OF NITROGEN ON THE MICROBIOLOGICAL AND BIOCHEMICAL PROPERTIES OF ZINC-CONTAMINATED SOIL

Rafał STRACHEL, Jadwiga WYSZKOWSKA, Małgorzata BAĆMAGA

Department of Microbiology, Faculty of Environmental Management & Agriculture, University of Warmia and Mazury in Olsztyn, Plac Lodzki 3, 10-727 Olsztyn, Poland

Abstract. The aim of these studies was to determine the influence of excessive zinc doses on the microbiological and enzymatic properties of soil. Also, an evaluation of the possibility to stimulate remediation processes by nitrogen fertilisation of the soil was attempted. Zinc was applied to loamy sand in the amounts of 0, 250, 500, 750, 1000, 1250 mg Zn^{2+} kg^{-1} DM soil, while nitrogen in the form of urea in doses of 0, 250, 500 mg N kg^{-1} DM soil. Soil samples were incubated at a temperature of 25 °C, maintaining a constant humidity equal to 50% of the maximum water capacity. In the 2nd and 20th week of the experiment, the following factors were determined: activity of dehydrogenases and catalase, and number of organotrophic bacteria, copiotrophic bacteria, oligotrophic bacteria, actinomycetes, and fungi. Zinc inhibited the enzymatic activity of the soil, while causing a slight increase in populations of microorganisms. Only fungi reacted unequivocally positively to contamination of the soil with zinc, therefore demonstrating changes in the biodiversity of microorganisms. Nitrogen fertilisation of the soil resulted in stabilization of the environment contaminated with zinc by stimulation of growth of microorganisms resistant to the influence of this metal.

Keywords: zinc, soil contamination, enzymatic activity, resistance, resilience, microorganisms, biodiversity, biostimulation.

Introduction

Excessive presence of toxic compounds, particularly heavy metals, in ecosystems is one of the biggest problems of contemporary environmental protection and related sciences, particularly agriculture. Environmental pollution is a condition in which an introduction of any substance to the biosphere in an amount exceeding the tolerance of the environment results in a disturbance in its proper functioning and the toxicity of this xenobiotic may lead to the death of organisms (Duruibe *et al.* 2007). The soil contamination depends on, i.a., ground utilisation, the intensity of human activities, the distance from emission sources, and dose of xenobiotics that have been introduced in the past (Luo *et al.* 2012). The adverse influence of a given heavy metal on the environment and human health results from its mobility in the soil and availability for plants, which by an uptake of trace elements, include them into the trophic chain (Butkus *et al.* 2004; Kabata-Pendias 2004). Contamination connected with the presence of heavy metals in the soil affects the size of the harvested crop, possibly leading in the longer term to a loss of food security. Examination of the influence of zinc on biological properties is a complex problem because of the fact that this element in small amounts is necessary for proper growth and functioning of living organisms (Chibuike, Obiora 2014). On the other hand, there are reports in the literature concerning the adverse effects of zinc on the microbiological properties of soil, leading to a loss of homeostasis of the environment, and therefore to a disturbance in the processes of matter circulation in ecosystems (Klimek 2012). The issue connected with high amounts of zinc present in the environment occurs commonly nowadays, particularly in the context of areas with a high anthropogenic pressure (Zheng *et al.* 2010).

Because of their strong relation with cells of living organisms, oxidoreductases are a reliable indicator for the condition of the soil and for changes caused under the pressure of heavy metals (Kızılkaya *et al.* 2004; Oliveira, Pampulha 2006; Wyszkowska *et al.* 2006). Long-term mineral fertilisation may positively affect the biochemical properties of soil resulting from the increased proliferation of microorganisms, or it may be a consequence of stimulation of enzymatic activity by the availability of biogenic elements (Jiao *et al.* 2011).

Corresponding author: Jadwiga Wyszkowska
E-mail: jadwiga.wyszkowska@uwm.edu.pl

Nitrogen is one of the most important biogenic elements that influence soil fertility and plant growth (Abbasi *et al.* 2011). Urea fertilisers may induce changes in the biological and physicochemical properties of soil, which is related with the bioavailability of heavy metals, including zinc. The products of urea hydrolysis may form sparingly soluble complexes with zinc present in soil, thus reducing its bioavailability (Jiang *et al.* 2012). As a result, urea fertilisation may alleviate the adverse effects of zinc contamination on soil biology.

Considering the superior role played by microorganisms in shaping soil fertility, studies on the influence of various doses of zinc on the activity of oxidoreductases, as well as the number and diversity of microorganisms were undertaken, simultaneously taking into account the direction of changes resulting from the retention time of this heavy metal in the soil. Also, suitability of dehydrogenases and catalase activities as an indicator for changes connected with the stress exerted onto the soil microbiome by the presence of toxic substances was examined. Dependencies between the abundance of microorganisms under conditions of a changed environment and changes in the enzymatic activity in time, resulting from their presence in the soil, were considered. In view of the above, our study focused on the biostimulation of zinc-contaminated soil with urea as an additional source of nitrogen. Model experiments were limited to soil fertilization with nitrogen because a more complex fertilization system (e.g. a combination of nitrogen and phosphorus) would make it impossible to reliably evaluate the potential role of nitrogen in reducing the negative effects of the analysed heavy metal. Therefore, the aim of this study was to determine the influence of different nitrogen rates on the microbiological and enzymatic properties of zinc-contaminated soil.

1. Material and methods

1.1. Soil

In the experiment, brown soil with a granulometric composition of loamy sand (sand, 80.5%; clay, 18%; silt, 1.5%) as used, according to WRB (2014) classified among Eutric Cambisols. The soil material was collected from the arable-humic level at the Tomaszkowo Didactic & Experimental Station belonging to the University of Warmia and Mazury in Olsztyn (north-eastern Poland, 53.71610 N, 20.41670 E). The soil was characterized by the following parameters: pH in 1 M KCl – 5.6; $C_{organic}$ – 10.0 g kg^{-1} DM soil; N_{total} – 0.58 g kg^{-1} DM soil; bioavailable cations (mg kg^{-1} DM soil): P – 96.32, K – 179.08, Mg – 50.17; exchangeable cations (mg kg^{-1} DM soil): K – 217.73, Ca – 568.60, Na – 100.34, Mg – 64.52; hydrolytic acidity (HAC) – 18.66 mM(+) kg^{-1} DM soil; sum of exchangeable bases cations (TEB) – 40.00 mM(+) kg^{-1} DM soil; cation exchange capacity (CEC) – 58.66 mM(+) kg^{-1} DM soil; base saturation (BS) – 68.19%; Zn_{total} – 22.68 mg kg^{-1} DM soil; $Zn_{bioavailable}$ – 9.13 mg kg^{-1} DM soil.

1.2. Experimental design

The experiment was carried out under laboratory conditions in triplicate. 100 g of air-dried soil material sieved through a mesh with a diameter of 2 mm was weighed out to each glass beaker with a volume of 150 cm^3. Zinc was applied in the form of an aqueous $ZnCl_2$ solution, in doses corresponding to 0, 250, 500, 750, 1000 and 1250 mg Zn^{2+} kg^{-1} DM soil. Also, urea as a nitrogen source was introduced into the soil samples in the amounts of 0, 250 and 500 mg N kg^{-1} DM soil. During the experiment, the samples were incubated at a temperature of 25 °C, and the humidity was maintained at the level of 50% of the maximum water capacity. In the 2nd and 20th week of the experiment, the microbiological and enzymatic properties of soil were determined.

1.3. Determination of microbial properties and soil enzyme activity

Populations of microorganisms were determined in triplicate by the spread plate method using the following microbiological media: organotrophic bacteria – a medium with an addition of soil extract according to Bunt and Rovira (1955); oligotrophic and copiotrophic bacteria – a medium with a peptone and meat extract according to Onta and Hattori (1983); actinomycetes – Küster and Williams medium with antibiotics (nystatine and actidione) (Parkinson *et al.* 1971); fungi – on glucose-peptone agar with an addition of rose bengal and aureomycine according to Martin (1950).

Enzymatic analyses of soil, carried out in triplicate, pertained to determination of oxidoreductase activity – dehydrogenases (EC 1.1) and catalase (EC 1.11.1.6). Dehydrogenases were determined according to Öhlinger (1996), using 2,3,5-triphenyltetrazolic chloride as a substrate. In the result of metabolic transformations with participation of the tested enzymes, this compound was converted into triphenylformazan. Extinction of the obtained product was measured on a Perkin-Elmer Lambda 25 spectrophotometer using a wavelength of 485 nm. Activity of catalase was determined by a reduction with potassium manganite (VII) of the residue of hydrogen peroxide being a reaction substrate according to Alef and Nannpieri (1998).

1.4. Calculations

Diversity of soil microorganisms subjected to impact of zinc was evaluated by the microorganism colony development index (CD) (Sarathchandra *et al.* 1997) and ecophysiological diversification index (EP) according to the formula proposed by De Leij *et al.* (1993). Abundances of the individual groups of microorganisms was estimated in one-day periods, and based on the changes in the distribution of the developed colonies in time, microbiological diversification of the soil was inferred. A measure for the degree of disturbances caused by the presence of zinc in the soil may consist in determination of resistance index (RS) and index of resilience (RL) proposed by Orwin and Wardle (2004). These indices were calculated based on the oxidoreductase activity.

1.5. Statistical analysis

The obtained results of the studies were statistically processed using the Statistica 10.0 (StatSoft 2012) software package, at a significance level of p = 0.01. According to the requirements of correct statistical inference on the significance of the obtained results, all analyses were carried out in triplicate. Interpretation of effects connected with the presence of zinc in the soil was supported by the following data analyses: percentage of factors of the observed variations η^2, Pearson correlation coefficients (p = 0.01) between the zinc dose and the investigated parameters, cluster analysis (CA) by Ward's method with Euclidean distance in studies of the microorganism populations and principal component analysis (PCA) illustrating the activity of oxidoreductases and community of microorganisms. Homogeneous groups for the colony development index (CD), the ecophysiological diversity index (EP) and the indicators of soil resistance (RS) and soil resilience (RL) were established with the use of Tukey's test and analysis of variance (ANOVA) at a significance level of p = 0.01.

2. Results

2.1. Influence of factors

According to the assumptions made while designing the experiment, the obtained results depended on such factors as the dose of zinc, dose of nitrogen and incubation time of the soil, as well as interaction between the individual factors. Statistical analysis of contribution of the observed variations factors η^2 proved that to the highest degree, the abundance of microorganisms and activity of oxidoreductases depended on contamination of the soil with zinc (Table 1). Only in the case of actinomycetes, the retention time of the metal in the soil turned out to be more important. To a moderate degree, fertilisation with nitrogen affected only growth of organotrophic bacteria, which was conditioned by the interaction between all studied factors to the highest extent. The dependence between the degree of contamination with zinc and the incubation time, also modified the activity of oxidoreductases and population of copiotrophic bacteria.

Table 1. Percentage of the observed variation η^2 in the soil contaminated with zinc

Variable factors	Cat	Deh	B_{org}	B_{cop}	B_{olig}	Act	Fun
Dose of Zn	57.08	47.96	10.99	47.62	37.83	18.35	89.45
Dose of N	2.11	2.80	14.57	5.78	8.32	3.47	0.08
Incubation time	7.79	10.36	5.14	2.22	30.97	43.87	0.49
Dose of Zn x dose of N	1.43	12.10	11.42	4.48	4.90	3.49	1.39
Dose of Zn x incubation time	21.51	22.93	12.01	20.79	0.81	12.19	2.81
Dose of N x incubation time	0.81	0.64	6.43	1.80	1.19	10.29	0.41
Dose of Zn x dose of N x incubation time	8.34	3.06	35.26	7.65	14.28	6.63	4.80
Error	0.94	0.14	4.18	9.65	1.70	1.71	0.57

Explanatory note: Zn – zinc, N – nitrogen; Cat – catalase, Deh – dehydrogenases, B_{org} – organotrophic bacteria, B_{cop} – copiotrophic bacteria, B_{olig} – oligotrophic bacteria, Act – actinomycetes, Fun – fungi.

2.2. Microbial properties and enzymes activity

In soil not fertilised with nitrogen, the colony development indices for organotrophic bacteria and fungi strongly depended on the zinc dose on both dates of the experiment, the CD index value for bacteria being negatively correlated with the zinc dose, while being positively correlated for fungi (Table 2). The decrease in the value of the CD index for organotrophic bacteria in the 20th week indicates an inhibited growth of colonies, caused by the contamination of the soil with this metal. In the 2nd week, zinc caused change in the CD index to a slighter degree, and the index's values suggested that colony development was uniform. On the other hand, while analysing the data pertaining to fungi, it should be stated that zinc improved the development of fast-growing species. After two weeks, the influence of zinc on the growth of actinomycetes was ambiguous, while in the 20th week, zinc generally caused a decrease in the CD value. In the objects not contaminated with zinc, but fertilised with urea, an increase in the CD index occurred for fungi on both dates of the experiment. An inverse dependence was observed for organotrophic bacteria. Additionally, in case of all studied groups of microorganisms, introduction of urea affected the growth rate of colonies positively.

Taking into account the index of ecophysiological diversification of microorganisms, one may ascertain that organotrophic bacteria were developing uniformly during the time of observation (Table 3). It proves a high diversification of the composition of microorganisms. Fertilisation with nitrogen and incubation time of the soil did

Table 2. Colony development (CD) index for microorganisms in the soil contaminated with zinc (the values are means of three replicates)

Dose of zinc, mg Zn^{2+} kg^{-1} DM soil	Microorganisms								
	B_{org}			Act			Fun		
	Dose of nitrogen, mg N kg^{-1} DM soil								
	0	250	500	0	250	500	0	250	500
	Incubation time, 2 weeks								
0	51.444^{a}	46.072^{a}	37.834^{a}	21.146^{a}	22.310^{a}	19.957^{b}	31.759^{b}	39.642^{b}	35.102^{b}
250	41.695^{b}	44.624^{a}	39.224^{a}	25.769^{a}	25.377^{a}	25.356^{a}	45.740^{a}	45.591^{a}	46.941^{a}
500	43.355^{ab}	42.855^{a}	41.849^{a}	21.585^{a}	24.323^{a}	24.511^{a}	45.836^{a}	45.370^{a}	47.189^{a}
750	41.080^{b}	37.796^{a}	38.102^{a}	20.849^{a}	24.369^{a}	21.400^{ab}	47.628^{a}	46.643^{a}	47.301^{a}
1000	35.769^{b}	36.155^{a}	37.315^{a}	23.394^{a}	22.940^{a}	24.718^{a}	45.579^{a}	46.210^{a}	47.301^{a}
1250	40.370^{b}	42.179^{a}	41.031^{a}	22.481^{a}	22.492^{a}	22.249^{ab}	47.441^{a}	48.115^{a}	47.735^{a}
r	−0.707*	−0.544	0.128	−0.022	−0.199	0.142	0.679*	0.782*	0.642*
	Incubation time, 20 weeks								
0	58.409^{a}	22.075^{a}	19.537^{b}	27.617^{a}	21.268^{a}	21.070^{a}	39.884^{b}	43.255^{b}	43.056^{a}
250	37.140^{b}	18.582^{a}	22.648^{ab}	24.043^{a}	19.178^{a}	18.923^{a}	45.503^{a}	20.054^{c}	28.717^{b}
500	24.241^{c}	26.038^{a}	24.286^{ab}	20.042^{a}	16.532^{a}	26.199^{a}	46.949^{a}	47.577^{a}	46.673^{a}
750	26.178^{c}	21.733^{a}	25.971^{a}	18.772^{a}	18.150^{a}	20.370^{a}	46.425^{a}	46.810^{a}	46.606^{a}
1000	21.472^{c}	24.110^{a}	20.132^{b}	18.924^{a}	19.563^{a}	22.320^{a}	46.590^{a}	45.967^{ab}	47.769^{a}
1250	26.579^{c}	26.972^{a}	23.356^{ab}	21.962^{a}	18.943^{a}	21.811^{a}	48.658^{a}	48.239^{a}	47.407^{a}
r	−0.778*	0.489	0.252	−0.539	−0.150	0.105	0.788*	0.499	0.561

Explanatory note: B_{org} – organotrophic bacteria, Act – actinomycetes, Fun – fungi. Letter designations (a–c) describe homogeneous groups formed for each groups of microorganisms, dose of nitrogen and incubation time separately. r – coefficient of Pearson's correlation between zinc dose and the analysed parameters, *p = 0.01, n = 17.

Table 3. Ecophysiological diversity (EP) index of microorganisms in the soil contaminated with zinc (the values are means of three replicates)

Dose of zinc, mg Zn^{2+} kg^{-1} DM soil	Microorganisms								
	B_{org}			Act			Fun		
	Dose of nitrogen, mg N kg^{-1} DM soil								
	0	250	500	0	250	500	0	250	500
	Incubation time, 2 weeks								
0	0.693^{a}	0.794^{a}	0.808^{a}	0.754^{a}	0.765^{a}	0.775^{a}	0.667^{a}	0.509^{a}	0.509^{a}
250	0.718^{a}	0.793^{a}	0.824^{a}	0.816^{a}	0.797^{a}	0.807^{a}	0.282^{b}	0.311^{b}	0.307^{b}
500	0.749^{a}	0.740^{a}	0.606^{a}	0.753^{a}	0.777^{a}	0.781^{a}	0.270^{b}	0.268^{bc}	0.215^{bc}
750	0.607^{a}	0.677^{ab}	0.627^{a}	0.757^{a}	0.803^{a}	0.813^{a}	0.178^{b}	0.227^{bc}	0.196^{bc}
1000	0.702^{a}	0.728^{ab}	0.684^{a}	0.738^{a}	0.733^{a}	0.739^{a}	0.280^{b}	0.269^{bc}	0.209^{bc}
1250	0.682^{a}	0.608^{b}	0.586^{a}	0.739^{a}	0.731^{a}	0.733^{a}	0.199^{b}	0.161^{c}	0.187^{c}
r	−0.186	−0.809*	−0.649*	−0.303	−0.356	−0.497	−0.700*	−0.832*	−0.803*
	Incubation time, 20 weeks								
0	0.721^{c}	0.751^{a}	0.752^{ab}	0.745^{a}	0.553^{a}	0.533^{a}	0.448^{a}	0.389^{b}	0.305^{a}
250	0.872^{a}	0.734^{a}	0.815^{a}	0.774^{a}	0.512^{a}	0.503^{a}	0.311^{b}	0.686^{a}	0.445^{a}
500	0.801^{abc}	0.804^{a}	0.774^{ab}	0.748^{a}	0.565^{a}	0.609^{a}	0.238^{b}	0.195^{c}	0.220^{a}
750	0.815^{ab}	0.745^{a}	0.662^{c}	0.673^{a}	0.590^{a}	0.693^{a}	0.233^{b}	0.227^{c}	0.228^{a}
1000	0.753^{bc}	0.776^{a}	0.735^{ab}	0.622^{a}	0.714^{a}	0.509^{a}	0.231^{bc}	0.248^{bc}	0.180^{a}
1250	0.754^{bc}	0.815^{a}	0.752^{ab}	0.526^{a}	0.686^{a}	0.570^{a}	0.124^{c}	0.152^{c}	0.196^{a}
r	−0.165	0.415	−0.320	−0.726*	0.571	0.111	−0.894*	−0.653*	−0.560

Explanatory note: B_{org} – organotrophic bacteria, Act – actinomycetes, Fun – fungi. Letter designations (a–c) describe homogeneous groups formed for each groups of microorganisms, dose of nitrogen and incubation time separately. r – coefficient of Pearson's correlation between zinc dose and the analysed parameters, *p = 0.01, n = 17.

not affect the changes in EP index of this group of microorganisms significantly. On the first term of the experiment, no significant influence of zinc on the diversity of actinomycetes was found, while a prolonged retention of this metal in the soil not fertilised with nitrogen caused a disturbance in microbiological balance in objects strongly contaminated with zinc. In the 20th week, nitrogen fertilisation of the soil caused a decrease in the diversity of microorganisms unequivocally. In all objects contaminated with zinc, fungi exhibited a small ecophysiological diversity. Low values of the EP index suggest that growth of this group of microorganisms progressed non-uniformly in the studied observation period of colony development.

A decrease in the value of the ecophysiological diversification of fungi, accompanied by an increase in total abundance, suggests growth of species resistant to the impact of zinc. Such changes, occurring as a function of increasing doses of the metal, indicate support for microorganisms with r-type life strategy, being a symptom of a strong disturbance of the given environment.

Based on the created dendrogram by cluster analysis (CA) using the Ward method (Fig. 1), two similarity groups of the response of microorganisms to contamination of the soil with zinc may be distinguished. The first cluster was formed by organotrophic bacteria, copiotrophic bacteria, oligotrophic bacteria and actinomycetes, and the other one by fungi. Within the first main group, subgroups were formed, allowing the ascertainment that on the second term of study, actinomycetes were responding differently from the other groups of microorganisms belonging to the same cluster. It should be emphasised that the similarity

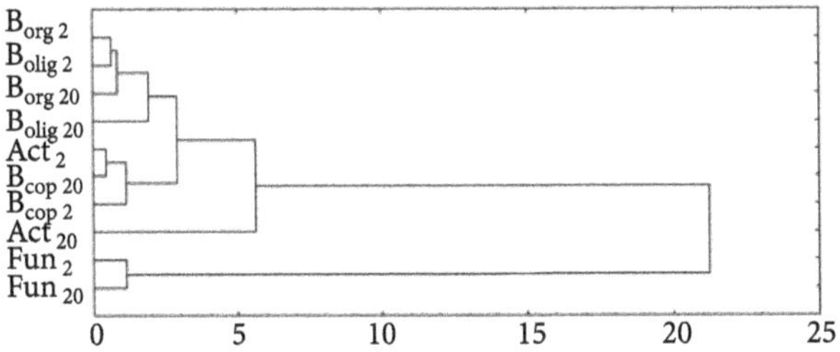

Explanatory note: B_{org} – organotrophic bacteria, B_{cop} – copiotrophic bacteria, B_{olig} – oligotrophic bacteria, Act – actinomycetes, Fun – fungi; incubation time: 2 – 2 weeks, 20 – 20 weeks.

Fig. 1. Cluster analysis (CA) of clusters formed based on the abundance of soil microorganisms subjected to the impact of zinc (irrespectively of nitrogen fertilisation used)

of response to contamination with zinc, between organotrophic bacteria and oligotrophic bacteria was the highest one among the analysed groups of microorganisms. A diversified response of abundances of organotrophic bacteria, oligotrophic bacteria, copiotrophic bacteria and actinomycetes, resulting from the presence of zinc in the soil, was significantly modified by the time the metal exerted its influence. The investigated factors affected the growth of fungi in a different way than in the case of other groups of microorganisms. Various effects of zinc on the individual groups of microorganisms could result from their specific physiological, morphological and genetic properties.

Principal component analysis (PCA) proved that the obtained results of activity of soil enzymes and abundance of microorganisms were modified to more than 46.89% by the first principal component, while the second component explained 20.27% of the data variability (Fig. 2).

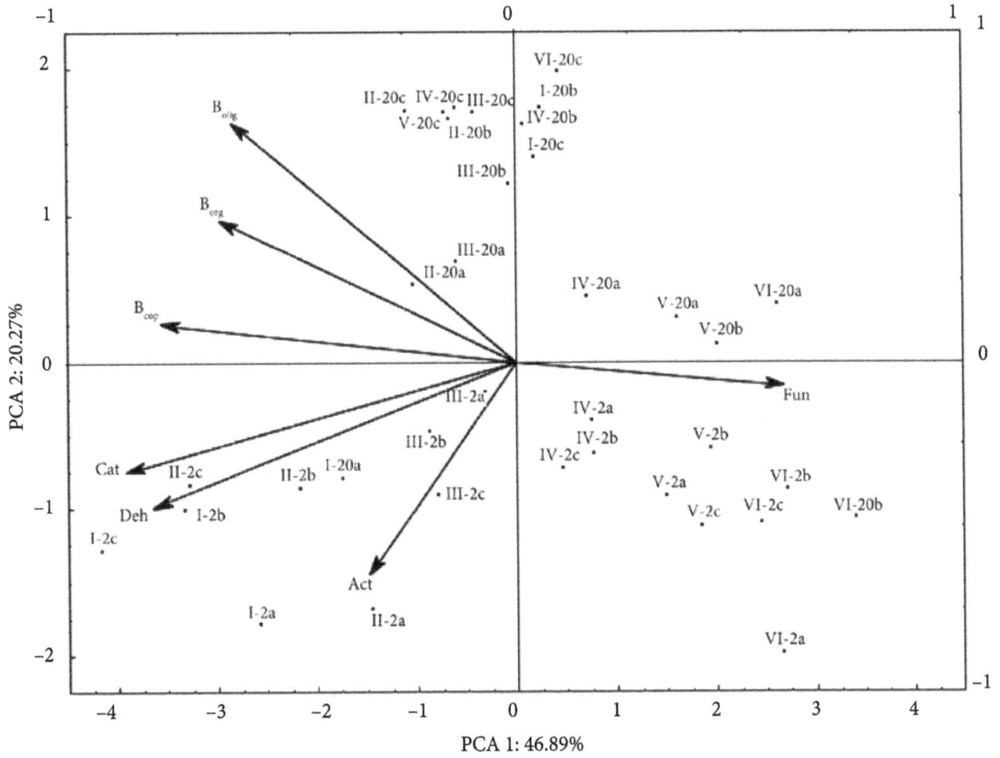

Explanatory note: Cat – catalase, Deh – dehydrogenases, B_{org} – organotrophic bacteria, B_{cop} – copiotrophic bacteria, B_{olig} – oligotrophic bacteria, Act – actinomycetes, Fun – fungi; dose of zinc (in mg kg^{-1} DM soil): I – 0, II – 250, III – 500, IV – 750, V – 1000, VI – 1250; incubation time: 2 – 2 weeks, 20 – 20 weeks; dose of nitrogen (in mg kg^{-1} DM soil): a – 0, b – 250, c – 500.

Fig. 2. Plot of principal component analysis based on soil microbial activity and microbial abundance data

Analysis of microbiological properties of the soil indicated that a short-term influence of zinc lead to an inhibition in the growth of all tested groups of bacteria. Actinomycetes in the soil not fertilised with nitrogen are an exception. In the studied periods, a dose of zinc equal to 1250 mg kg^{-1} DM soil caused a decrease in populations of the analysed bacteria groups. The impact of zinc on the populations of Procaryota is ambiguous, because introduction of a dose in the range of 250 up to 1000 mg Zn^{2+} kg^{-1} led both to a stimulation and an inhibition of bacterial growth. It depended on the group of studied bacteria, the dose being tested, the retention time of zinc in the soil, and the influence of fertilisation with nitrogen. Moreover, fertilisation with urea affected positively the growth of actinomycetes in objects subjected to a two-week incubation. Stimulation of a multiplication of these microorganisms was observed in the soil contaminated with zinc in the amount of 250 to 750 mg kg^{-1} DM soil. However, a prolonged influence of this metal decreased the abundance of this group of microorganisms significantly, and fertilisation with nitrogen did not alleviate adverse effects of zinc presence. Response of fungi to contamination with zinc indicates that they are more resistant to its influence, as their abundance was connected with the used dose of zinc. Abundance of actinomycetes and fungi decreased with an increase in the incubation time of the soil.

The activity of oxidoreductases in the soil indicated a strong influence of the zinc doses applied to the soil. The response of activity of the investigated enzymes was shaped by the studied factors in a similar way, which was indicated by the directions, senses and lengths of vectors describing oxidoreductases. A projection of the cases onto a plane shows that the highest activity of catalase was observed in the 2nd week, while the activity of dehydrogenases was shaped by the whole duration of the experiment. Distribution of the cases in relation to the vectors describing activity of enzymes indicates that the highest activity is exhibited by enzymes in object not contaminated and in those with an addition of zinc in the amount of 250 mg Zn^{2+} kg DM soil. In turn, clusters of cases in the 1st and 4th quadrants suggest that the activity of oxidoreductases was significantly negatively correlated with the zinc dose. Additionally, the distribution of the obtained results was also modified by the duration of the zinc influence and by fertilisation with urea, and the source of nitrogen shaped the biochemical properties of the soil to a lesser extent.

The highest resistance to disturbances in the homeostasis of the soil subjected to the pressure of zinc for 2 weeks, was exhibited by objects contaminated with this metal to the slightest degree (Fig. 3). In case of all combinations, for both tested enzymes, a strong, negatively correlated dependence was occurring between the degree of contamination and the soil resistance. Fertilisation with nitrogen, both in a dose of 250 and 500 mg N kg^{-1} DM soil, increased the resistance of dehydrogenases to zinc significantly. Such dependence was not exhibited by catalase, having a resistance decreased after introduction of urea. Only in the object with the smallest zinc dose (250 mg Zn^{2+} kg^{-1} DM soil), a slight positive effect of fertilisation with nitrogen in the amount of 500 mg kg^{-1} DM soil was observed.

Based on the differences in the activity of oxidoreductases subjected to the pressure of zinc for 20 weeks, it may be ascertained that the metal caused disturbances in the enzymatic balance of the soil. In the objects not fertilised with urea, the zinc dose equal to 250 mg kg^{-1} DM soil exerted a strong adverse effect on catalase, proved by the negative values of the factor RL = −0.162 (Table 4). An average value of the RL index in the objects contaminated with zinc without an addition of nitrogen indicates

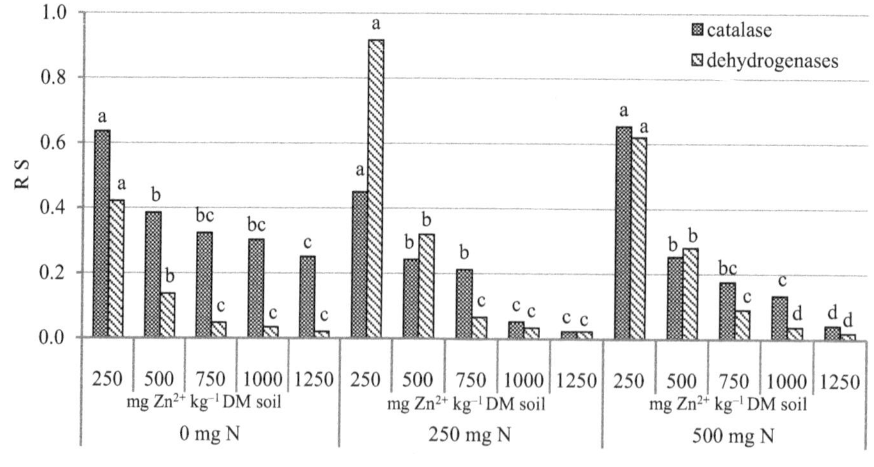

Explanatory note: letter designations (a–d) describe homogeneous groups formed for dehydrogenases, catalase and dose of nitrogen separately.

Fig. 3. Resistance index (RS) of oxidoreductases in the soil contaminated with zinc after two weeks of incubation

a stronger destructive influence of this metal on the catalase than on that of dehydrogenases. Significant disturbances of this enzyme activity are evident in samples contaminated with zinc in the amounts of 1000 and 1250 mg Zn^{2+} kg^{-1} DM soil. In the object subjected to a slighter pressure of zinc, the stabilisation processes occurred more rapidly. Introduction of nitrogen to the soil, both in the amount of 250 and 500 mg N kg^{-1} DM soil, contributed to a faster restoration of balance of the soil. The RL index assumed values in the range of 0.710 to 0.918 for catalase, and 0.185 to 0.959 for dehydrogenases.

Table 4. Index of resilience (RL) of oxidoreductases in the soil contaminated with zinc after two weeks of incubation (the values are means of three replicates)

Dose of zinc, mg Zn^{2+} kg^{-1} DM soil	Catalase			Dehydrogenases		
	dose of nitrogen, mg N kg^{-1} DM soil					
	0	250	500	0	250	500
250	-0.162^{b}	0.810^{ab}	0.814^{a}	0.100^{b}	0.185^{b}	0.788^{a}
500	0.131^{a}	0.885^{a}	0.918^{a}	0.172^{ab}	0.782^{a}	0.854^{a}
750	0.185^{a}	0.855^{a}	0.916^{a}	0.212^{a}	0.741^{a}	0.809^{a}
1000	0.032^{ab}	0.886^{a}	0.889^{a}	0.213^{a}	0.874^{a}	0.590^{b}
1250	0.071^{ab}	0.710^{b}	0.809^{a}	0.201^{a}	0.959^{a}	0.861^{a}
Average	0.051	0.829	0.869	0.180	0.708	0.781
r	0.389	−0.388	−0.090	0.714*	0.801*	−0.168

Explanatory note: letter designations (a–b) describe homogeneous groups formed for dehydrogenases, catalase and dose of nitrogen separately. r – coefficient of Pearson's correlation between zinc dose and the analysed parameters, *$p = 0.01$, $n = 14$.

3. Discussion

The microorganisms are characterised by a high ability to adapt to unfavourable conditions in the environment. Reports on the stimulating effect of moderate doses of zinc on the abundance of microorganisms may be found in the literature (Wyszkowska *et al.* 2013). This is consistent with the obtained results, wherein zinc stimulated the growth of studied groups of microorganisms. It was conditioned by the metal doses used and by fertilisation with nitrogen. Also, in vegetation experiments by Wyszkowska *et al.* (2006, 2007), zinc applied together with other metals, affected the soil microbiota significantly. Changes in the diversity of fungi, observed in this paper, indicate a significant influence of zinc on this group of microorganisms. Differences in the obtained results may be explained by an impact of other factors which may form the bioavailability of the heavy metal. These factors may include: pH of soil, sorption capacity, salinity, organic matter (McBride 2003). Nitrogen is a factor that may affect the shaping of soil fertility (Meng *et al.* 2005) significantly by stimulating the consortium of microorganisms and thus may alleviate adverse effects connected with the presence of toxic substances. As reported in Yu *et al.* (2015), fertilisation with urea affects significantly the change in the structure of microorganisms and its function. The authors found an increase in fungi abundance accompanying the increased availability of nitrogen. However, while considering the phenomenon of biostimulation of the microbiome, one should remember that organic matter seems to be the key factor. Influencing the development of the microbiological consortium of the soil, organic matter affects simultaneously an increase in production of enzymes responsible for the biochemical activity of the soil (Jiao *et al.* 2011). Relating these dependences to the results of microbiological properties obtained in our studies, it should be ascertained that with a low carbon content and a high availability of nitrogen, under conditions of cell stress exerted onto the microorganisms by toxic doses of zinc, proliferation of microorganisms exhibiting a higher resistance to this metal is enhanced. This results from the fact that these microorganisms are able to utilise better the source of available carbon for build-up of their biomass, constituting a labile deposit of this biogenic element in the soil.

The obtained results of the research are consistent with the literature data (Kunito *et al.* 2001; Castaldi *et al.* 2004; Mikanova 2006; Yang *et al.* 2006; Chaperon, Sauvé 2007), which indicate an inhibitive influence of zinc on oxidoreductases. Dehydrogenases are particularly sensitive, and their activity decreases significantly in the soil containing excessive doses of zinc (Chaperon, Sauvé 2007; Kucharski *et al.* 2011; Borowik *et al.* 2014a, 2014b). In an amount of 300 mg Zn^{2+} kg^{-1}, this metal causes a decrease in the activity of these enzymes by nearly a half, and its higher doses lead to intensification of changes, as an introduction of 1300 mg Zn^{2+} kg^{-1} to the soil, inhibits the activity of dehydrogenases by approximately 70% (Moreno *et al.* 2009). In our studies, a significant dependence between the increasing zinc dose and the decrease in the activity of both dehydrogenases, and catalase, was found. Similar conclusions were drawn by Yang *et al.* (2006), who observed a decrease in the activity of catalase in the soil after introduction of zinc in the amount of 100 to 800 mg kg^{-1} DM soil. In model studies, carried out by Borowik *et al.* (2014a), zinc applied in a dose of 300 to 2400 mg kg^{-1} DM soil, decreased the activity of dehydrogenases and catalase, and this dependence was strongly correlated with the dose used. An increase in the soil contamination with zinc was accompanied by an increase in the inhibitive impact of this metal onto the activity of the enzymes, leading to a decrease in the resistance of soil (RS). According to the same authors, EC_{50} amounted to 689 mg Zn^{2+} kg^{-1} for dehydrogenases, and 1255 mg Zn^{2+} kg^{-1} DM soil in case of catalase. Then, in our studies in samples incubated for 20 weeks, ED_{50} for dehydrogenases amounted to 239 mg Zn^{2+} kg^{-1} DM of soil, and for catalase – 719 mg Zn^{2+} kg^{-1} DM soil.

While pondering biochemical and microbiological properties of soil subjected to pressure from toxic elements, it should be remembered that heavy metals may exert both advantageous and destructive impacts on proper functioning of soils. An excessive amount of the metal may lead to disturbance of cellular processes in the result of affinity of the metal for specific biological structures. This leads to deactivation of a given component, and as a consequence, to an imbalance in cellular homeostasis (Clemens 2006). Tolerance for the presence of heavy metals is an individual feature of various microorganisms. Considering this in respect of biological equilibrium of the soil, growth of microorganisms resistant to strong changes in living conditions takes place at the cost of sensitive microorganisms, which may manifest itself in a reduction of diversity of the given environment (Lock, Janssen 2005). As reported by Jiao *et al.* (2011), activity of enzymes actually depends on the chemical properties of the soil. Activity of both dehydrogenases and catalase is positively correlated with contents of C, N and P compounds, and with the biomass of microorganisms. In studies of Romero *et al.* (2010), application of urea did not cause any changes in the activity of dehydrogenases. This lack of stimulation of the microbiome may result from a carbon deficiency, because even with a good availability of nitrogen, growth of microorganisms has been limited by content of carbon source (Bhattacharyya *et al.* 2005).

The values of microbial activity and abundance indicate that nitrogen fertilisation had a minor effect on the proliferation of microorganisms and oxidoreductase activity, but an important role was played by the interactions between nitrogen fertilisation and the other experimental factors. Soil fertilisation with urea contributed to the narrowing of the C:N ratio, which influenced the soil microbiome, thus accelerating the recovery and restoration of soil to a healthy state.

Availability of nitrogen with a comparatively low carbon content in the soil leads to an increased growth of microorganisms, which are more able to adapt to the change in the environment. Shaping of the microbiome's diversity as a function of time may be related to the enzyme activity, because the RL factor in objects fertilised with urea indicates a biostimulation of microorganisms resistant to the effect of zinc, translating into a decrease in disturbances of biochemical activity, caused by the contamination of the soil environment with zinc. It allows for stating that fertilisation with nitrogen affects the intensification of stabilisation of the soil environment subjected to a heavy metal pressure by promoting proliferation of microorganisms adapted to living in such an environment.

Conclusions

The abundance of bacteria and actinomycetes was shaped significantly by the content of zinc in the soil. Small doses of this metal exhibited stimulating effect for growth of these microorganisms, but excessive contamination with this metal affected their abundance unequivocally negatively. Fungi were not exhibiting such a tendency, and their abundance was positively correlated with the zinc dose. The incubation time and nitrogen fertilisation of the soil played a significant role as well. Direction and extent of changes induced by these factors depended on the doses applied. Changes in the diversity of microorganisms were also found, most evident in case of fungi. Their growth was positively correlated with the degree of soil contamination, and the pressure exerted by zinc caused a reduction in ecophysiological diversity, indicating a strong loss of the biological balance of the soil.

In the result of the studies, it was proved that zinc inhibited the enzymatic properties of soil for the whole duration of the experiment. Resistance of enzymes strongly depended on the degree of the soil contamination, which is confirmed by negative correlation coefficients. Fertilisation with nitrogen caused acceleration of the balance-restoration processes in the activity of oxidoreductases.

The biological parameters used in the experiment, describing the condition of soil subjected to a pressure of anthropogenic factors, are a useful tool for monitoring of the environment. Thanks to comprehensive investigations, necessary information could be obtained, useful in restoration of biological balance of soil, disturbed by the effect of toxic substances.

Acknowledgements

The research was supported by the Polish Ministry of Science and Higher Education.

References

Abbasi, M. K.; Hina, M.; Tahir, M. M. 2011. Effect of *Azadirachta indica* (neem), sodium thiosulphate and calcium chloride on changes in nitrogen transformations and inhibition of nitrification in soil incubated under laboratory conditions, *Chemosphere* 82(11): 1629–1635. https://doi.org/10.1016/j.chemosphere.2010.11.044

Alef, K.; Nannipieri, P. 1998. *Methods in applied soil microbiology and biochemistry*. London: Academic Press. Harcourt Brace & Company, Publishers. 576 p.

Bhattacharyya, P.; Chakrabarti, K.; Chakraborty, A. 2005. Microbial biomass and enzyme activities in submerged rice soil amended with municipal solid waste compost and decomposed cow manure, *Chemosphere* 60(3): 310–318. https://doi.org/10.1016/j.chemosphere.2004.11.097

Borowik, A.; Wyszkowska, J.; Kucharski, J.; Baćmaga, M.; Boros-Lajszner, E.; Tomkiel, M. 2014a. Sensitivity of soil enzymes to excessive zinc concentrations, *Journal of Elementology* 19(3): 637–648. https://doi.org/10.5601/jelem.2014.19.2.456

Borowik, A.; Wyszkowska, J.; Kucharski, M.; Kucharski, J. 2014b. Resistance of dehydrogenases, catalase, urease and plants to soil contamination with zinc, *Journal of Elementology* 19(4): 929–946. https://doi.org/10.5601/jelem.2013.18.4.566

Bunt, J. S.; Rovira, A. D. 1955. Microbiological studies of some subantarctic soils, *Journal of Soil Science* 6: 119–128. https://doi.org/10.1111/j.1365-2389.1955.tb00836.x

Butkus, D.; Paliulis, D.; Baltrenaite, E. 2004. Sorption of heavy metals from polluted water and their migration in the system soil-tree, *Journal of Environmental Engineering and Landscape Management* 12(4): 120–125.

Castaldi, S.; Rutigliano, F. A.; Virzo de Santo, A. 2004. Suitability of soil microbial parameters as indicators of heavy metal pollution, *Water, Air and Soil Pollution* 158(1): 21–35. https://doi.org/10.1023/B:WATE.0000044824.88079.d9

Chaperon, S.; Sauvé, S. 2007. Toxicity interaction of metals (Ag, Cu, Hg, Zn) to urease and dehydrogenase activities in soils, *Soil Biology & Biochemistry* 39(9): 2329–2338. https://doi.org/10.1016/j.soilbio.2007.04.004

Chibuike, G. U.; Obiora, S. C. 2014. Heavy metal polluted soils: effect on plants and bioremediation methods, *Applied and Environmental Soil Science.* Article ID 752708. https://doi.org/10.1155/2014/752708

Clemens, S. 2006. Evolution and function of phytochelatin synthases, *Journal of Plant Physiology* 163(3): 319–332. https://doi.org/10.1016/j.jplph.2005.11.010

De Leij, F. A. A. M.; Whipps, J. M.; Lynch, J. M. 1993. The use of colony development for the characterization of bacterial communities in soil and on roots, *Microbial Ecology* 27: 81–97. https://doi.org/10.1007/BF00170116

Duruibe, J. O.; Ogwuegbu, M. O. C.; Egwurugwu, J. N. 2007. Heavy metal pollution and human biotoxic effects, *International Journal of Physical Sciences* 2(5): 112–118.

Jiang, Z.; Zeng, Q.; Tie, B.; Liao, B.; Pi, H.; Feng, X.; Sun, Y. 2012. Ammonia volatilization and availability of Cu, Zn induced by applications of urea with and without coating in soils, *Journal of Environmental Sciences* 24(1): 177–181. https://doi.org/10.1016/S1001-0742(11)60742-0

Jiao, X-G.; Gao, C-S.; Lü, G-H.; Sui, Y-Y. 2011. Effect of long-term fertilization on soil enzyme activities under different hydrothermal conditions in Northeast China, *Agricultural Sciences in China* 10(3): 412–422. https://doi.org/10.1016/S1671-2927(11)60020-5

Kabata-Pendias, A. 2004. Soil–plant transfer of trace elements – an environmental issue, *Geoderma* 122(2–4): 143–149. https://doi.org/10.1016/j.geoderma.2004.01.004

Kızılkaya, R.; Aşkın, T.; Bayraklı, B.; Sağlam, M. 2004. Microbiological characteristics of soils contaminated with heavy metals, *European Journal of Soil Biology* 40(2): 95–102. https://doi.org/10.1016/j.ejsobi.2004.10.002

Klimek, B. 2012. Effect of long-term zinc pollution on soil microbial community resistance to repeated contamination, *Bulletin of Environmental Contamination and Toxicology* 88(4): 617–622. https://doi.org/10.1007/s00128-012-0523-0

Kucharski, J.; Wieczorek, K.; Wyszkowska, J. 2011. Changes in the enzymatic activity in sandy loam soil exposed to zinc pressure, *Journal of Elementology* 16(4): 577–589. https://doi.org/10.5601/jelem.2011.16.4.07

Kunito, T.; Saeki, K.; Goto, S.; Hayashi, H.; Oyaizu, H.; Matsumoto, S. 2001. Copper and zinc fractions affecting microorganisms in long-term sludge-amended soils, *Bioresource Technology* 79(2): 135–146. https://doi.org/10.1016/S0960-8524(01)00047-5

Lock, K.; Janssen, C. R. 2005. Influence of soil zinc concentrations on zinc sensitivity and functional diversity of microbial communities, *Environmental Pollution* 136(2): 275–281. https://doi.org/10.1016/j.envpol.2004.12.038

Luo, X-S.; Yu, S.; Zhu, Y-G.; Li, X-D. 2012. Trace metal contamination in urban soils of China, *Science of The Total Environment* 421–422: 17–30. https://doi.org/10.1016/j.scitotenv.2011.04.020

Martin, J. 1950. Use of acid rose bengal and streptomycin in the plate method for estimating soil fungi, *Soil Science* 69: 215–233. https://doi.org/10.1097/00010694-195003000-00006

McBride, M. B. 2003. Toxic metals in sewage sludge-amended soils: has promotion of beneficial use discounted the risks?, *Advances in Environmental Research* 8(1): 5–19. https://doi.org/10.1016/S1093-0191(02)00141-7

Meng, L.; Ding, W.; Cai, Z. 2005. Long-term application of organic manure and nitrogen fertilizer on N2O emissions, soil quality and crop production in a sandy loam soil, *Soil Biology and Biochemistry* 37(11): 2037–2045. https://doi.org/10.1016/j.soilbio.2005.03.007

Mikanova, O. 2006. Effects of heavy metals on some biological parameters, *Journal of Geochemical Exploration* 88(1–3): 220–223. https://doi.org/10.1016/j.gexplo.2005.08.043

Moreno, L.; Bastida, F.; Ros, M.; Hernández, T.; García, C. 2009. Soil organic carbon buffers heavy metal contamination on semiarid soils: effect of different metal thresholds levels on soil microbial activity, *European Journal of Soil Biology* 45(3): 220–228. https://doi.org/10.1016/j.ejsobi.2009.02.004

Öhlinger, R. 1996. Dehydrogenase Activity with the Substrate TTC, in F. Schinner, R. Öhlinger, E. Kandeler, R. Mrgesin (Eds.). *Methods in soil biology*. Berlin Heidelberg: Springer Verlag, 241–243.

Oliveira, A.; Pampulha, M. E. 2006. Effects of long-term heavy metal contamination on soil microbial characteristics, *Journal of Bioscience and Bioengineering* 102(3): 157–161. https://doi.org/10.1263/jbb.102.157

Onta, H.; Hattori, T. 1983. Oligotrophic bacteria on organic debris and plant roots in paddy field, *Soil Biology & Biochemistry* 1: 1–8.

Orwin, K. H.; Wardle, D. A. 2004. New indices for quantifying the resistance and resilience of soil biota to exogenous disturbances, *Soil Biology & Biochemistry* 36: 1907–1912. https://doi.org/10.1016/j.soilbio.2004.04.036

Parkinson, D.; Gray, F. R. G.; Williams, S. T. 1971. *Methods for studying the ecology of soil microorganisms*. IBP Handbook, 19. Oxford, Einburg: Blackweel Scientific Publications. 116 p.

Romero, E.; Fernández-Bayo, J.; Díaz, J. M. C.; Nogales, R. 2010. Enzyme activities and diuron persistence in soil amended with vermicompost derived from spent grape marc and treated with urea, *Applied Soil Ecology* 44(3): 198–204. https://doi.org/10.1016/j.apsoil.2009.12.006

Sarathchandra, S. U.; Burch G.; Cox, N. R. 1997. Growth patterns of bacterial communities in the rhizoplane and rhizosphere in white clover (*Trifolium repens* L.) and perennial ryegrass (*Lolium perenne* L.) in long-term pasture, *Applied Soil Ecology* 6: 293–299. https://doi.org/10.1016/S0929-1393(97)00015-2

StatSoft Inc. 2012. *Statistica (data analysis software system), version 10.0* [online], [cited 02 July 2015]. Available from Internet: www.statsoft.com

World Reference Base for Soil Resources (WRB). 2014. *International soil classification system for naming soils and creating legends for soil maps.* World Soil Resources Reports No. 106. FAO, Rome. 181 p.

Wyszkowska, J.; Boros, E.; Kucharski, J. 2007. Effect of interactions between nickel and other heavy metals on the soil microbiological properties, *Plant, Soil and Environment* 53: 544–552.

Wyszkowska, J.; Borowik, A.; Kucharski, M.; Kucharski, J. 2013. Effect of cadmium, copper and zinc on plants, soil microorganisms and soil enzymes, *Journal of Elementology* 18(4): 769–796. https://doi.org/10.5601/jelem.2013.18.4.455

Wyszkowska, J.; Kucharski, J.; Lajszner, W. 2006. The effects of copper on soil biochemical properties and its interaction with other heavy metals, *Polish Journal of Environmental Studies* 15(6): 927–934.

Yang, Z. X.; Liu, S. Q.; Zheng, D. W.; Feng, S. D. 2006. Effects of cadium, zinc and lead on soil enzyme activities, *Journal of Environmental Sciences* 18(6): 1135–1141. https://doi.org/10.1016/S1001-0742(06)60051-X

Yu, C.; Hu, X. M.; Deng, W.; Li, Y.; Xiong, C.; Ye, C. H.; Han, G. M.; Li, X. 2015. Changes in soil microbial community structure and functional diversity in the rhizosphere surrounding mulberry subjected to long-term fertilization, *Applied Soil Ecology* 86: 30–40. https://doi.org/10.1016/j.apsoil.2014.09.013

Zheng, N.; Liu, J.; Wang, Q.; Liang, Z. 2010. Health risk assessment of heavy metal exposure to street dust in the zinc smelting district, Northeast of China, *Science of the Total Environment* 408(4): 726–733. https://doi.org/10.1016/j.scitotenv.2009.10.075

Rafał STRACHEL. MSc Eng. He is a doctoral candidate. He is preparing his doctoral thesis in the Department of Microbiology in the Faculty of Environmental Management and Agriculture of University of Warmia and Mazury in Olsztyn. His scientific discipline is protection and management of the environment.

Jadwiga WYSZKOWSKA. Prof. Dr Hab. She is a full professor at UWM in Olsztyn. She obtained her PhD in 1996 and attained degrees for assistant professor in 2002 and professor in 2006. In 2005–2008, she was Associate Dean of the Faculty of Environmental Management and Agriculture, and in 2008–2012 she was Prorector of UWM in Olsztyn. Within the framework of scientific discipline of protection and management of the environment, she is specialised in soil biochemistry and environmental microbiology. In 2012–2015 she was the Chairman of the Section of Agricultural Microbiology in the Committee of Soil Science and Agricultural Chemistry of Polish Academy of Sciences (PAN).

Małgorzata BAĆMAGA. Dr Eng. She was granted her PhD from the Faculty of Environmental Management and Agriculture of University of Warmia and Mazury in Olsztyn in 2008. She is an associate professor in the Department of Microbiology in the Faculty of Environmental Management and Agriculture of University of Warmia and Mazury in Olsztyn. Her scientific discipline is protection and management of the environment, with the specialisation of soil microbiology.

3

REPRESENTATIVENESS OF EUROPEAN BIOCHAR RESEARCH

Ruben SAKRABANI[a], Jürgen KERN[b], Utra MANKASINGH[c], Costanza ZAVALLONI[d], Giulia ZANCHETTIN[e], Ana Catarina BASTOS[f], Priit TAMMEORG[g], Simon JEFFERY[h], Bruno GLASER[i], Frank G. A. VERHEIJEN[j]

[a]*Cranfield Soil and Agrifood Institute, School of Water, Energy and Environment, Cranfield University, MK43 0AL United Kingdom*

[b]*Leibniz Institute for Agricultural Engineering and Bioeconomy, Max-Eyth Allee 100, 14469 Potsdam, Germany*

[c]*Department of Life and Environmental Sciences, University of Iceland, Sturlugata 7, Reykjavik 107, Iceland*

[d]*California State University Stanislaus, Agricultural Studies Program, One University Circle, Turlock, CA 95382, USA*

[e]*University of Udine, Ecosystems Ecology, Via delle Scienze 206, Udine, 33100, Italy*

[f]*Department of Biology and Centre for Environmental and Marine Studies (CESAM), University of Aveiro, 3810-193 Aveiro, Portugal*

[g]*Department of Agricultural Sciences, P.O. Box 27 (Latokartanonkaari 5, Plant Production Sciences), FIN-00014 University of Helsinki, Finland*

[h]*Crop and Environment Sciences Department, Harper Adams University, Newport, Shropshire, TF10 8NB, United Kingdom*

[i]*Institute of Agronomy and Nutritional Sciences, Martin Luther University Halle-Wittenberg, von-Seckendorff-Platz 3, 06120 Halle/Saale, Germany*

[j]*Department of Environment and Planning and Centre for Environmental and Marine Studies (CESAM), University of Aveiro, 3810-193 Aveiro, Portugal*

Abstract. Biochar research is extensive and there are many pot and laboratory studies carried out in Europe to investigate the mechanistic understanding that govern its impact on soil processes. A survey was conducted in order to find out how representative these studies under controlled experimental conditions are of actual environmental conditions in Europe and biomass availability and conversion technologies. The survey consisted of various key questions related to types of soil and biochar used, experimental conditions and effects of biochar additions on soil chemical, biological and physical properties. This representativeness study showed that soil texture and soil organic carbon contents used by researchers are well reflected in the current biochar research in Europe (through comparison with published literature), but less so for soil pH and soil type. This study provides scope for future work to complement existing research findings, avoiding unnecessary repetitions and highlighting existing research gaps.

Keywords: biochar, soil, Europe, pot experiments, laboratory scale, representativeness.

Introduction

Biochar is a black carbon solid produced by pyrolysis of organic materials (Lehmann *et al.* 2006) that is often distinguished from charcoal by its intended use as soil amendment (Sohi *et al.* 2009). During pyrolysis of biomass, carbon (C) atoms form polycondensed aromatic moieties that slow down biological decomposition (Downie *et al.* 2009). This recalcitrance confers long-term stability of biochar in soil, potentially lasting for centuries (Kuzyakov *et al.* 2014; Lehmann, Joseph 2015).

Corresponding author: Ruben Sakrabani
E-mail: r.sakrabani@cranfield.ac.uk

There has been increasing interest in the potential that biochar has for enhancing C sequestration and improving soil fertility (Spokas, Reikosky 2009), restoring degraded lands (Beesley, Marmiroli 2011), reducing greenhouse gas (GHG) emissions in agriculture (Cayuela *et al.* 2013; Harter *et al.* 2014), influencing soil moisture retention (Ulyett *et al.* 2014) and soil biological activity (Lehmann, Joseph 2015). Recently, Zhang *et al.* (2016) reviewed 798 publications to provide a quantitative assessment of field and laboratory based experiments from across the globe. They found that the majority of biochars used were produced using kiln or lab batch scale rather than commercial scale pyrolysis, using primarily wood and municipal waste feedstock and that more studies involved laboratory and greenhouse pot experiments rather than field studies. Further, in a recent meta-analyses reported that yield response are dependent on crop type and soil/climatic conditions with more positive effects in the tropics (Jeffery *et al.* 2011, 2017).

Previous research activities and presentations of results have been uncoordinated, providing a general picture on the topic but with little in-depth investigation of key themes (Tammeorg *et al.* 2017, this issue). This was one of the stimuli for the initiation of the COST Action TD1107 "Biochar as Option for Sustainable Resource Management" funded by the European Union to bring together key scientists and technologists working in this area. The COST Action TD1107 was a network that aimed to improve data sharing and communication between scientists and stakeholders that developed into a highly useful platform supporting biochar research in Europe. As one of the main tasks of this COST Action was to minimise unnecessary repetition in biochar research, a representativeness survey of European biochar research was conducted consisting of both field-scale (Verheijen *et al.* 2017, this issue) and pot and laboratory scale studies carried out in Europe. Particularly in the last two years (i.e. 2013–2015), many European researchers joined the COST Action, voluntarily submitting data on their work to a database created to gather information about the type of research and field of application, providing meta-data concerning experimental set up and so producing quantitative, and some qualitative information concerning their studies. This analysis does not qualify for a meta-analysis but it aims to investigate how representative experimental conditions were within biochar research of the COST Action members, regarding biochar properties and environmental application, i.e. feedstock type, pyrolysis conditions, soils, land use and land management. The utility of a representativeness survey is that it gathers and combines multiple aspects of biochar research, providing an overview compared with the published dataset. Such comparison can aid identification of aspects that have not been considered or covered within Europe, so that the missing knowledge gaps can be addressed with further research.

This representativeness survey collates the information from various researchers to provide an overview of their activities and clusters the key outcomes of their research in order to direct further work. The representativeness survey is divided into two categories – part I – field trials (Verheijen *et al.* 2017, this issue) and part II – pot experiments (the current paper). As such, representativeness analysis is a valuable tool to provide a general overview that is currently missing for biochar studies in Europe.

The objective of this paper is to identify and quantify the observations of biochar effects in pot and laboratory studies and assess how representative the experimental conditions are compared to likely European scale conditions..

1. Materials and methods

The information provided by the Representativeness Survey was obtained by using an online questionnaire (http://webform.cost.european-biochar.org/) formulated by consulting researchers from COST Action TD1107 countries to prioritise key information. The survey was sent to researchers working with biochar in pots, lysimeters and incubation studies and covered aspects related to methods of the experiment, substrates used, processes analysed, type of biochar used and the effects that were observed. As such, the questionnaire was divided into sections that required data related to type of soil used, biochar type feedstock and pyrolysis technology, and effect of biochar on various soil physical, chemical and biological properties. Following submission by respondents within the EU COST Action on a voluntary basis, results from the database were downloaded as a Microsoft Office Excel spreadsheet.

The database was divided into four main parts: (i) General; (ii) Experimental setup; (iii) Biochar treatments and properties; (iv) Measured effects. The first step was to summarise all the sections of the MS Office Excel document to have a visual starting point to facilitate analysis. All entries were grouped following the categories offered by the survey, to obtain a schematic view about the experiments, such as country where the experiment took place, type of experiment, description of soil, description of biochar applied and responses observed. Data for each category were explored by calculating mean, standard deviation, minimum and maximum values using MS Office Excel. Representativeness was deduced from the graphs produced after the data synthesis by comparing with published literature. This consists of infographics, pie and column charts, depending on the type of data considered with the aim of providing a clear visualisation of the trends in the experiments. The final step was the interpretation of data considering the present status of knowledge on biochar.

2. Results

2.1. Data collection

A total of 20 countries contributed to the Representativeness Survey, which resulted in 356 web form submissions. A respondent who submits an online form containing multiple sets of information categorised into four parts as explained above is considered to submit one submission. Table 1 shows the number of submissions for the various countries involved in the COST Action. The submissions were not uniformly distributed with almost half (149) from Spain, which contributed mostly with experiments focusing on ecotoxicological impact, followed by Belgium (37) and Germany (33).

Table 1. Number of submissions per country involved in the Representiveness Survey

Austria	2	Poland	5
Belgium	37	Portugal	19
Denmark	13	Russia	8
Estonia	2	Slovakia	9
Finland	19	Spain	149
France	11	Sweden	3
Germany	33	Turkey	2
Greece	1	United Kingdom	5
Italy	19	Czech Republic	1
Netherlands	4	Norway	14

Total = 356

The climatic zone represented in current biochar research is mainly dependent on the distribution and contribution of different countries: according to the Köppen classification, most of the observations come from experiments under Mediterranean climate (150 submissions), maritime temperate climate (60 submissions) and humid continental climate (31 submissions). This survey did not seek for specific questions about environmental conditions (e.g. annual average temperature, relative humidity and precipitation), the climatic zone only refers to the country where the experiment took place since it was all carried out under controlled conditions.

2.2. Experimental setup

A quarter of the experiments were conducted as incubation (25.0%) or greenhouse scale (20.5%) experiments, with 43% carried out in climate chambers (Fig. 1). No experiments performed on undisturbed soil cores were submitted.

2.3. Feedstock for production of biochars

The feedstocks used for biochar production were grouped into three main categories: (i) lignocellulosic (including woody); (ii) herbaceous (crop residue, green waste, silage); (iii) biosolids (manure, sewage, liquid organic waste). Almost 70% of the biochar was derived from lignocellulosic feedstock. Biochar produced from herbaceous feedstock accounted for 16% and biosolids for 9% of feedstock (Fig. 2).

2.4. Biochar processing conditions

The main parameters that are generally modified during char production are carbonisation temperature, type of feedstock and residence time in the heating unit. Changes in these variables influence the proportions of the end products of pyrolysis (biochar, biooil, gas, water) and quality of char produced.

Considering the type of thermochemical conversion of biomass to char, slow pyrolysis covered more than half (64%) of the experiments, while 17% of the biochar used in the experiment came from fast pyrolysis. (Fig. 3). The remaining 19% refers to other methods of heating as gasification, stove, retort kiln and hydrothermal carbonisation (HTC), which are separated from biochars due to their different composition (Schimmelpfennig, Glaser 2012). Nevertheless, these products are defined here as chars. The preference for the slow pyrolysis method, due

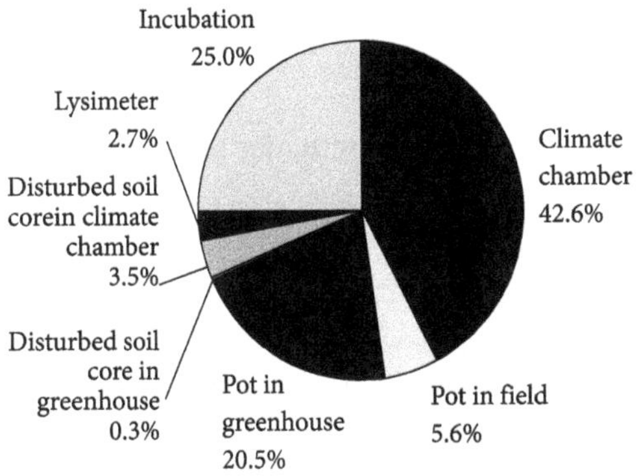

Fig. 1. Survey response of types of experiment (n = 356)

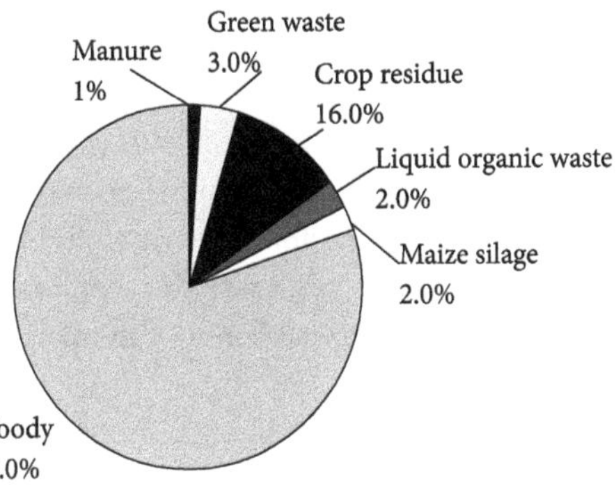

Fig. 2. Survey response of biochar feedstock types (n = 370)

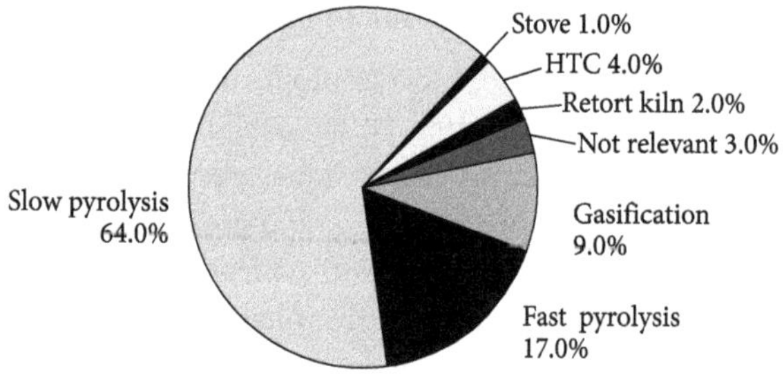

Fig. 3. Survey response of biochar producing technologies (n = 362), HTC – hydrothermal carbonization

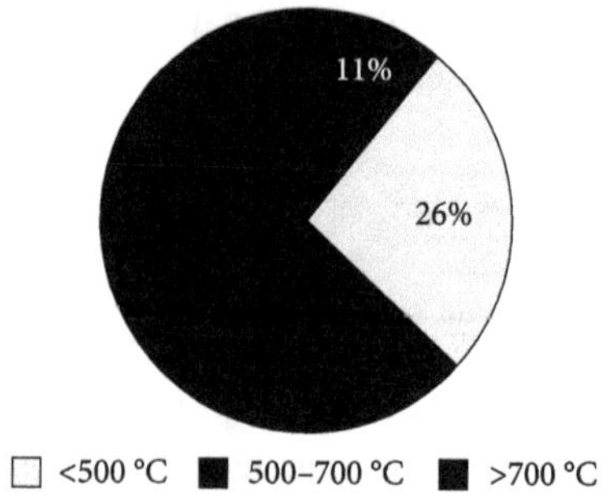

Fig. 4. Survey response of pyrolysis temperature (n = 328)

to the maximisation of the portion of char produced, reflects the purpose of using biochar as soil amendment and as a possible tool for climate change mitigation (Lehmann, Joseph 2015).

The comparison between the type of feedstock with the temperature of pyrolysis is useful to show how the two aspects, in combination with residence time, are strictly correlated in determining biochar physical and chemical properties.

Biochar was described by the carbonization temperature and residence time, as well as by the method of pretreatment (when applicable) and of its application to pots. Mainly high temperatures (>500 °C) were used for biochar production, covering almost 75% of the reported experiments (Fig. 4). Most of the carbonization processes (63%) occurred between 500 and 700 °C, while 11% took place at over 700 °C. The retention time was generally less than an hour (61%), but longer periods of carbonization were also present (Fig. 5).

2.5. Soil type, texture and length of experiments

All soil types shown in Figure 6 were represented by at least one entry. The presence of Calcisols, Fluvisols and Luvisols was dominant. Calcisols and Fluvisols were mostly associated with responses from Spain indicating dominance of this response. The frequency of these fertile soil types is reflected in the soil texture class, which was mostly loamy (77%) and details of the breakdown is shown in Figure 7. Calcisols, Fluvisols and Luvisols cover 5%, 5% and 6% of Europe (Soil Atlas of Europe 2005). According to the Soil Atlas Map of Europe, Albeluvisol is the most common soil in Europe (15%), whilst Podzol is a dominant soil of the northern latitude (14%). The Cambisols, Chernozems and Leptosols cover 12%, 9% and 9% respectively of European soil. Albeluvisols, Chernozems, and Leptosols were not represented in our database. This shows that in this pot study the representativeness is not well covered in comparison to dominant soil types in Europe.

Soil organic carbon (SOC) content was between 1 and 2% in about half of reported studies (52%); 31% of studies had SOC less than 1% whilst 17% had more than

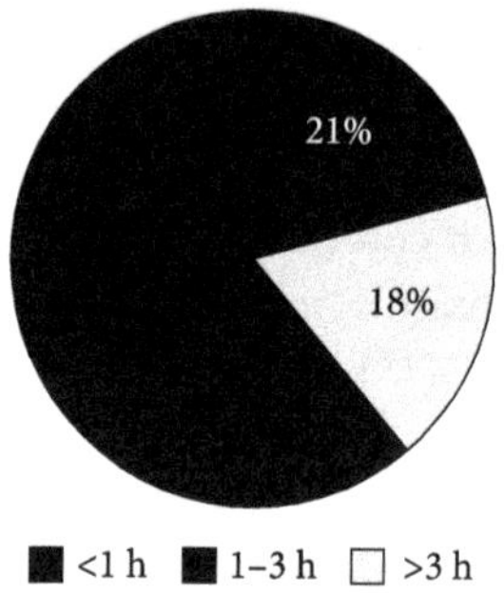

Fig. 5. Survey response of residence time in pyrolysis reactor (n = 203) application

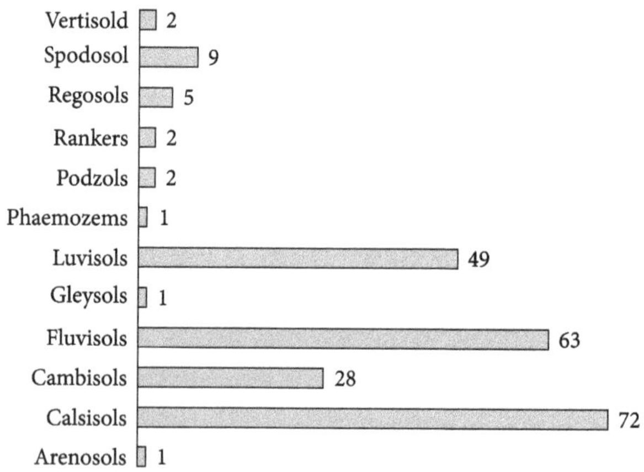

Fig. 6. Survey response of soil types used in biochar application

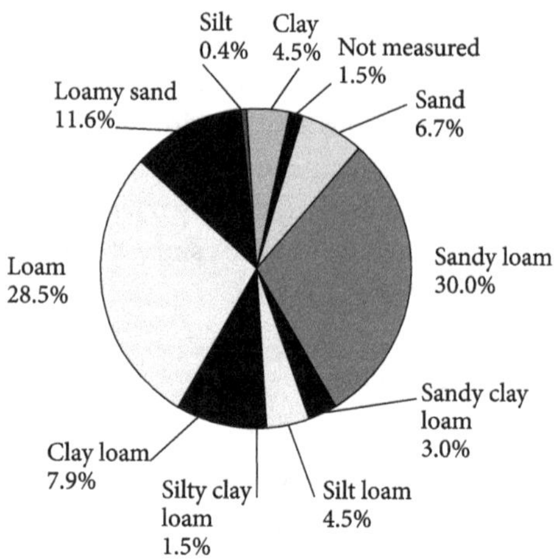

Fig. 7. Survey response of soil texture classes (n = 267)

2%. Therefore, SOC content was reasonably representative of actual SOC contents in European cropland soils, which mostly have SOC contents between 1–2% (Toth *et al.* 2013). The overrepresentation of smaller SOC contents in the laboratory studies is most likely a consequence of the interest to improve soils with low quality, and therefore low SOC contents. Soil pH was represented by mostly alkaline and neutral soils, while acidic soils accounted for 15% of studies.

Biochars had an even distribution of pH and ash content. 60% of biochar samples contained less than 10% ash. Seventy percent of all biochars had a pH between 8 and 10 in line with the findings by Verheijen *et al.* (2017, this issue).

The length of the experiments (i.e. the incubation period) was measured in days. Most of the experiments (53%) lasted one month or less, while 20% lasted more than three months. In terms of the incubation temperature, 64% of the experiments were carried out above 20 °C (data not shown).

3. Discussion

Data were collected using an online survey as this allowed the maximum reach for obtaining data across a range of European Countries. However, such an approach is not without its weaknesses. For example, the timing at which the survey arrived with some institutes may have been more favourable than others in terms of having the time and data availability to allow contribution. Furthermore, the questions posed and format of the questionnaire may have been inhibitory to providing data from some groups than others. These potential biases should be borne in mind while interpreting these data. Nevertheless, 356 web form submissions from 20 countries is a reasonable achievement given the resources and time that was allocated to this task.

The representativeness was judged by comparing the survey response with published literature such as Tammeorg *et al.* (2017, this issue), Toth *et al.* (2013) and Zhang *et al.* (2016) which were used as bench mark in addition to other peer reviewed manuscripts on biochar.

While 20 countries reported on biochar research at the pot scale of experimentation, there is a potential bias driven by the large number of pot scale studies conducted in Spain. However, the majority of these entries were ecotoxicological studies carried out in climate-controlled chambers. As such, the conditions reported from these studies are not "Mediterranean" per se, but representative of standardized conditions based on ISO or OECD guidelines for ecotoxicological assessment of soil and soil materials. In addition to this during a COST Action meeting, biochar effects on soil toxicology was identified as a knowledge gap which encouraged these large submissions. The authors of this manuscript did not want to remove these submissions in order not to add any bias on the representativeness of this study.

The majority of north-western cropland soil in Europe is acidic; soil in the Mediterranean is generally alkaline. In the partly arid sub-continental zone and temperate mountainous regions, neutral to alkaline soils dominate croplands, and these types of soils are present in considerable areas of the sub-oceanic regions as well (Toth *et al.* 2013). So the fact that 41% and 44% respectively of pot studies used soils from pH range 6–8 and 8 shows that pot studies are only partly representative.

Around 45% of the mineral soils in Europe have low or very low organic carbon contents (0–2%) and 45% have a medium content (2–6%) (Louwagie *et al.* 2009). Low contents are particularly evident in cropland soils and in southern Europe where 74% of the land is covered by soils that have less than 2% of organic carbon in the topsoil (0–30 cm) (Zdruli *et al.* 2004). However, areas of low organic carbon can be found almost everywhere, including in cropland soils of northern countries such as Belgium, France, Germany, Norway and the United Kingdom. This study is representative of SOC contents of mineral soils in Europe. However, more than 50% of EU soil organic carbon stocks are in peatlands (Schils *et al.* 2008). This study does not list any study using peatland soils, and therefore is not representative of organic soils in Europe. Biochar experiments with a range of peatland soils are therefore recommended to increase European representativeness. A first example has been reported by Kern *et al.* (2017) in the context of blending peat with biochars in growing media.

Figure 6 shows that Fluvisols and Calcisols are dominant in this study but upon detailed evaluation, these are associated with responses of the online survey from Spain which has skewed the representativeness. However when the information on soil types from Spanish submissions were ignored, the dominant ones are Cambisols and Luvisols. This resonates well with the field scale representativeness reported by Verheijen *et al.* (2017, this issue).

Uzoma *et al.* (2011) concluded that the biochar prepared at higher temperature (500 °C) had a greater effect on movement of water in sandy soil as compared to biochar made at lower temperatures (300 and 400 °C). In that experiment, it was found that the pyrolysis temperature and the mixing rate of biochar both affected the hydraulic properties of sandy soil. Existing research (Kinney *et al.* 2012; Ulyett *et al.* 2014; Ahmed *et al.* 2016) shows that most of the biochar has been applied to course textured soils and its positive effect on soil physical aspects is reflected in this representativeness survey (Fig. 7).

A section of the online survey also contained questions related to effects of biochar on various soil physical, chemical and biological properties which can be mapped to the thematic areas described by Tammeorg *et al.* (2017, this issue). Since the questions posed on the effects were categorised as "positive", "negative" or "no effect" which was deemed as too subjective, further discussions about this results are not elaborated here. Nevertheless it has led to postulation of research questions which can be useful to fulfil knowledge gaps identified by Tammeorg *et al.* (2016). Amongst the aspects that needs further work are as follows:

1. What are the effects of low temperature biochar on soil properties?

This survey indicates that the smallest percentage of biochar used in this study are produced using temperature <500 °C. It will be very valuable if further investigation is carried out using biochar produced at low temperatures. At lower temperatures stability of the carbon in biochar is reduced which can possibly facilitate easier use as substrate for soil microbial proliferation. This can provide new insights where the tradeoff is between carbon sequestration (associated with more biochar produced using high temperature) and carbon mineralisation and lability (associated with biochar produced using low temperatures).

2. What are the effects of livestock manure and biosolids based biochar on soil properties?

This survey shows that woody and crop residue based biochar are dominant as shown in Figure 2. It will be very valuable to explore alternative sources as feedstock for biochar especially biosolids which has usually high contents of phosphorus and how mineralisable this will be in soil to meet crop demands.

3. What are the effects of biochar on undisturbed soil cores?

In this survey, biochar has been added to pots or incubation studies where it is mixed and then the effects studied. However the effects on undisturbed soil cores have not been studied so far. This can entail using biochar of different particle sizes that can be added as top dressing to the undisturbed soil cores.

4. What are the effects of biochar on common soil types in the EU?

The Soil Atlas Map of Europe, shows that Albeluvisol is the most common soil in Europe (15%), whilst Podzol is a dominant soil of the northern latitude (14%). This survey shows that the dominant soils types in these studies are not Albeluvisol or Podzol indicating that further work is needed in this aspect.

The above list of questions is non-exhaustive and only demonstrates the opportunity that this Representativeness Survey can lend itself to. Whilst this Representativeness Survey in itself may not represent all biochar research in the EU, it allows opportunity for future work through knowledge gap analysis.

Since this representativeness survey was focused on providing an overview of the extent of current biochar research in Europe, there are some limitations in this approach. Due to the overwhelming response from Spain, this has potentially skewed the representativeness of this survey. Whilst this has to be acknowledged, one benefit is that this survey has revealed that the responses from Spain are associated with research related to ecotoxicology. This can provide opportunities for collaboration amongst EU COST Actions Members. Whilst this manuscript was targeting to showcase representativeness of biochar research in the EU associated with pots and laboratory scale studies, in reality this was not fully achieved due to lack of response to the online survey. The lack of response is something that we can only postulate as being a combination of, data not being ready for further analysis, unwillingness to participate or survey may seems too onerous despite that the explanation that the expectation is not for the respondent to fill every box but as many as possible and where relevant. It was also challenging to convince colleagues to submit data, which have not been published before and to show that further publishing of original data is still possible after using for the online survey.

This survey did depend on number of respondents so where regions got a good response it enabled to better showcase research from that part of the EU. However, where response was poor from certain regions of the EU, it just shows that response was poor for this survey and does not fully reflect that research is missing in this region. This is a limitation of this survey.

Conclusions

This study provides an overview of generic responses from adding biochar to soil in laboratory scale experiments and can be used as an indicative tool but should not be over-interpreted. The data presented here show that currently the regions in the European continent are not well represented. Further to this, there is a moderate representation of soil pH and weaker representation of soil type, but good representation in terms of SOC content and soil texture in mineral soils. This representiveness survey can also be used to link with the field scale experiment to corroborate any observations that can provide some seed corn ideas that can be further developed into future projects. In terms of the effects of biochar on the Thematic Areas, many aspects of the results were in line with published work and confirmed what was known about biochar until present time. The novelty of this work lies in the approach adopted to provide an overview on soil responses observed when applied with biochar in non-field conditions and as well as starting point for future research on this topic has been brought to light.

Acknowledgements

The authors are grateful for the EU COST Action TD1107 (Biochar as option for sustainable resource management) for funding the meetings that helped to enable this work. We also wish to acknowledge the Portuguese Foundation for Science and Technology (FCT) for the fellowships of A.C. Bastos (SFRH/BPD/98231/2013) and F.G.A. Verheijen (SFRH/BPD/107913/2015), the financial support to CESAM [UID/AMB/50017/2013], to FCT/MEC through national funds and the co-funding by FEDER, within PT2020 Partnership Agreement and Compete 2020. The authors are very indebted to Jan Sorensen who was extremely helpful in converting the Excel version of the survey into an online version that could be easily edited, stored and used to retrieve information. Johannes Harter was also very instrumental in retrieving information for the online version of the survey and collating it into a format that authors could use for this publication. We would like to thank all contributors from the various Member States as without your information this paper would not have been feasible.

References

Ahmed, A.; Kurian, J.; Raghavan, V. 2016. Biochar influences on agricultural soils, crop production, and the environment: a review, *Environmental Reviews* 24(4): 495–502. https://doi.org/10.1139/er-2016-0008

Beesley, L.; Marmiroli, M. 2011. The immobilisation and retention of soluble arsenic, cadmium and zinc by biochar, *Environmental Pollution* 159: 474–480. https://doi.org/10.1016/j.envpol.2010.10.016

Cayuela, M. L.; Sánchez-Monedero, M. A.; Roig, A.; Hanley, K.; Enders, A; Lehmann, J. 2013. Biochar and denitrification in soils: when, how much and why does biochar reduce N_2O emissions?, *Scientific Reports* 3. Article No. 1732. https://doi.org/10.1038/srep01732

Downie, A.; Crosky, A.; Munroe, P. 2009. Physical properties of biochar, in J. Lehmann, S. Joseph (Eds.). *Biochar for environmental management: science and technology.* London, UK: Earthscan, 13–32.

Harter, J.; Krause, H-M.; Schuetter, S.; Ruser, R.; Fromme, M.; Scholten, T.; Kappler, A.; Behrens, S. 2014. Linking N_2O emissions from biochar-amended soil to the structure and function of the N-cycling microbial community, *International Society for Microbial Ecology Journal* 8(3): 660–674. https://doi.org/10.1038/ismej.2013.160

Jeffery, S.; Abalos, D.; Prodana, M.; Bastos, A. C.; van Groenigen, J. W.; Hungate, B.; Verheijen, F. G. A. 2017. Biochar boosts tropical but not temperate yields, *Environmental Research Letters* 12(5). https://doi.org/10.1088/1748-9326/aa67bd

Jeffery, S.; Verheijen, F. G. A.; Van Der Velde, M.; Bastos, A. C. 2011. A quantitative review of the effects of biochar application to soils on crop productivity using meta-analysis, *Agriculture, Ecosystems & Environment* 144(1): 175–187. https://doi.org/10.1016/j.agee.2011.08.015

Kern, J.; Tammeorg, P.; Shanskiy, M.; Sakrabani, R.; Knicker, H.; Kammann, C.; Tuhkanen, E.-M.; Smidt, G.; Prasad, M.; Tiilikkala, K.; Sohi, S.; Gasco, G.; Steiner, C.; Glaser, B. 2017. Synergistic use of peat and charred material in growing media – an option to reduce the pressure on peatlands?, *Journal of Environmental Engineering and Landscape Management* (this issue).

Kinney, T. J.; Masiello, C. A.; Dugan, B.; Hockaday, W. C.; Dean, M. R.; Zygourakis, K. 2012. Hydrologic properties of biochars produced at different temperatures, *Biomass Bioenergy* 41: 34–43. https://doi.org/10.1016/j.biombioe.2012.01.033

Kuzyakov, Y.; Bogomolova. I.; Glaser, B. 2014. Biochar stability in soil: decomposition during eight years and transformation as assessed by compound-specific ^{14}C analysis, *Soil Biology and Biochemistry* 70: 229–236. https://doi.org/10.1016/j.soilbio.2013.12.021

Lehmann, J.; Joseph, S. 2015. *Biochar for environmental management.* London and New York: Routledge.

Lehmann, J.; Gaunt, J.; Rondon, M. 2006. Bio-char sequestration in terrestrial ecosystems – a review, *Mitigation & Adaptation Strategies for Global Change* 11: 395–419. https://doi.org/10.1007/s11027-005-9006-5

Louwagie, G.; Gay, S. H.; Burrell, A. 2009. *Final report on the project 'Sustainable Agriculture and Soil Conservation (SoCo)' JRC Scientific and Technical Reports.* European Commission, Joint Research Centre, Luxembourg.

Schils, R.; Kuikman, P.; Liski, J.; van Oijen, M.; Smith, P.; Webb, J.; Alm, J.; Somogyi, Z.; van den Akker, J.; Billett, M.; Emmett, B.; Evan, C.; Lindner, M.; Palosuo, T.; Bellamy, P.; Jandl, R.; Hiederer, R. 2008. *Final report on review of existing information on the interrelations between soil and climate change (Climsoil).* Final Report for the European Commission's Directorate-General for the Environment. European Commission, Brussels.

Schimmelpfennig, S.; Glaser, B. 2012. One step forward toward characterization: some important material properties to distinguish biochars, *Journal of Environmental Quality* 41: 1001–1013. https://doi.org/10.2134/jeq2011.0146

Sohi, S.; Lopez-capel, E.; Krull, E; Bol, R. 2009. *Biochar, climate change and soil a review to guide future research.* CSIRO, Glen Osmond.

Soil Atlas of Europe. 2005. European Soil Bureau Network, European Commission. Office for Official Publications of the European Communities, Luxembourg. 128 p.

Spokas, K. A.; Reikosky, D. C. 2009. Impacts of sixteen different biochars on soil greenhouse gas production, *Annals of Environmental Science* 3: 179–193.

Tammeorg, P.; Bastos, A. C.; Jeffery, S.; Rees, F.; Kern, J.; Graber, E.; Ventura, M.; Kibblewhite, M.; Amaro, A.; Budai, A.; Cordovil, C. M. d. S.; Domene, X.; Gardin, G.; Gasco, G.; Horak, J.; Kammann, C.; Kondrlova, E.; Laird, D.; Loureiro, S.; Martins, M. A. S.; Panzacchi, P.; Prasad, M.; Prodana, M.; Puga, A. P.; Ruysschaert, G.; Sas-Paszt, L.; Silva, F. C.; Teixeira, W. G.; Tonon, G.; Vedove, G. D.; Zavalloni, C.; Glaser, B.; Verheijen, F. G. A. 2017. Biochars in soils: towards the required level of scientific understanding, *Journal of Environmental Engineering and Landscape Management* (this issue). https://doi.org/10.3846/16486897.2016.1239582

Tóth, G.; Jones, A.; Montanarella, L. 2013. The LUCAS topsoil database and derived information on the regional variability of cropland topsoil properties in the European Union,

Environmental Monitoring and Assessment 185(9): 7409–7425. https://doi.org/10.1007/s10661-013-3109-3

Ulyett, J.; Sakrabani, R.; Hann, M.; Kibblewhite, M. 2014. Impact of biochar addition on water retention, nitrification and carbon dioxide evolution from two landy loam soils, *European Journal of Soil Science* (Special Issue on Biochar) 65(1): 96–104. https://doi.org/10.1111/ejss.12081

Uzoma, K. C.; Inoue, M.; Andry, H.; Fujimaki, H.; Zahoor, A.; Nishihara, E. 2011. Effect of cow manure biochar on maize productivity under sandy soil condition, *Soil Use Manage* 27: 205–212. https://doi.org/10.1111/j.1475-2743.2011.00340.x

Verheijen, F. G. A.; Mankasingh, U.; Penizek, V.; Panzacchi, P.; Glaser, B.; Jeffery, S.; Bastos, A. C.; Harter, J.; Tammeorg, P.; Kern, J.; Zavalloni, C.; Zanchettin, G.; Sakrabani, R. 2017. Representativeness of European biochar research: Part I – field experiments, *Journal of Environmental Engineering and Landscape Management* (this issue).

Zdruli, P.; Jones, R. J. A.; Montanarella, L. 2004. *Organic matter in the soils of Southern Europe European soil bureau technical report*. Publications Office of the European Union, Luxembourg.

Zhang, D.; Yan, M.; Niu, Y.; Liu, X.; van Zwieten, L.; Chen, D.; Bian, R.; Cheng, K.; Li, L.; Joseph, S.; Zheng, J.; Zhang, X.; Zheng, J.; Crowley, D.; Fillet, T. R.; Pan, G. 2016. Is current biochar research addressing global soil constraints for sustainable agriculture?, *Agriculture, Ecosystems and Environment* 226: 25–32. https://doi.org/10.1016/j.agee.2016.04.010

Ruben SAKRABANI. PhD, Senior Lecturer in Soil Chemistry, Cranfield Soil and Agrifood Institute, Cranfield University, 30 peer-reviewed articles. H-index 12 (Google Scholar).

Jürgen KERN. PhD, Senior Scientist of Biogeochemistry, Leibniz Institute for Agricultural Engineering and Bioeconomy, Germany, 1211 citations. H-index 16 (Google Scholar).

Utra MANKASINGH. PhD, Research Fellow in the Faculty of Life and Environmental Sciences at the University of Iceland. 6 publications in international peer reviewed journals, 3 major reports, 159 citation; H-index 4 (Google Scholar).

Costanza ZAVALLONI. PhD, Assistant Professor in plant science in the Agricultural Studies program, California State University Stanislaus, Turlock, 19 peer-reviewed articles in international scientific journals, 1 book chapter, 514 citations, H-index 11 (Google Scholar).

Giulia ZANCHETTIN. MSc graduate student in Ecology and Soil Science and was funded under the Short Term Scientific Mission (within the Biochar COST Action) to analyse representativeness datasets.

Ana Catarina BASTOS. PhD, research fellow of the Portuguese Science Foundation (FCT), at Department of Biology and Centre for Environmental and Marine Studies (CESAM), University of Aveiro, Portugal. 22 publications in international peer-reviewed journals, 1 European Commission report, 4 book chapters. 1745 citations; H-index 13 (Google Scholar).

Priit TAMMEORG. DSc, Postdoctoral Researcher in Agroecology, Dept of Agricultural Sciences, University of Helsinki, 11 peer-reviewed articles in international scientific journals, H-index 6 (TR WoS).

Simon JEFFERY. PhD, Senior Lecturer at Harper Adams University, Department of Crop and Environment Sciences, 2031 citations, H-index 16 (Google Scholar).

Bruno GLASER. PhD, Professor of Soil Biogeochemistry, Institute of Agronomy and Nutritional Sciences, Martin Luther University Halle-Wittenberg, Germany, 140 peer-reviewed articles in scientific journals. H-index 39 (WoS).

Frank G. A. VERHEIJEN. PhD, research fellow of the Portuguese Science Foundation (FCT), Centre for Environmental and Marine Studies (CESAM), University of Aveiro, Portugal. 25 publications in international peer-reviewed journals, 7 major European reports, 5 book chapters. 2348 citations; H-index 16 (Google Scholar).

4

NUMERICAL MODELING OF UNSATURATED LAYERED SOIL FOR RAINFALL-INDUCED SHALLOW LANDSLIDES

Chih-Yu LIU[a], Cheng-Yu KU[a], Jing-En XIAO[a], Chi-Chao HUANG[b], Shih-Meng HSU[a]

[a] *Department of Harbor and River Engineering, National Taiwan Ocean University, Keelung, Taiwan*
[b] *Central Geological Survey, Ministry of Economic Affairs, Taiwan*

Abstract. In this paper, a pioneer study on numerical modeling of rainfall-induced shallow landslides in unsaturated layered soil using the variably saturated flow equation is presented. To model the shallow landslides, the infinite slope stability analysis coupled with the hydrological model with the consideration of the fluctuation of time-dependent pore water pressure and Gardner equation for soil water characteristic curve was developed. A linearization process for the nonlinear Richards equation to deal with groundwater flow in unsaturated layered soil is derived using the Gardner model. To solve one-dimensional flow in the unsaturated zone of layered soil profiles, flux conservation and the continuity of pressure potential at the interface between two consecutive layers are considered in the numerical discretization of the finite difference method. The validity of the proposed model is established in three numerical problems by comparing the results with the analytical and other numerical solutions. Application examples have also been conducted. Obtained results demonstrate that the fluctuation of pore water pressure in unsaturated layered soil dominates slope stability of landslides and the lowest factor of safety may occur at the interface between two consecutive layers. The findings observed in this study are a fundamental contribution to environmental protection engineering for landslides in areas with higher occurrence and vulnerability to extreme precipitation.

Keywords: Variably saturated flow equation, Layered soil, Shallow Landslides, Groundwater flow.

Introduction

Due to global climate change, severe weather phenomena such as extreme precipitation events are becoming much more frequent around the world (Pereira *et al.* 2010; Csete, Buzasi 2016). These affect the stability of slopes and have consequences on landslides. Since global warming is expected to increase the frequency and intensity of severe rainfall events, the number of people exposed to landslide risk may dramatically be increased (Gariano, Guzzetti 2016). Accordingly, it is of importance to understand and measure how climate variables and their variability affect landslide hazards. Landslides commonly observed in the foothill areas involve sliding surfaces which is typically called a shallow landslide with an indication of the high diffusion of shallow landslides phenomena in different environmental and geological contexts, all over the world. In shallow landslides, the sliding surface is located in the soil mantle or regolith typically to depths of a few centimeters to a few meters. Shallow landslides may often occur within the vadose zone under unsaturated soil conditions (Lu, Godt 2008; Yeh, Lee 2013). The assessment of rainfall-induced shallow landslides has long been an interest of research topic in many branches of science and engineering such as soil science, hydrology, and geotechnical engineering (Montgomery, Dietrich 1994; Baum *et al.* 2002; Frattini *et al.* 2004; Tsai, Yang 2006; Tsai, Chen 2010).

In the past, theoretical models have been developed to predict slope stability assuming a steady or quasi-steady water table associated with an infinite-slope stability analysis (Das 2010). With regard to the influence of rainfall infiltration on slope stability in unsaturated zones, Iverson (2000) assumed that this occurred when soil was nearly saturated, further solving the analytical solution of the Richards equation in the simplified form of a one-dimensional linear diffusion equation, and allowing the development of an analytical model of rainfall-induced shallow landslides. Baum *et al.* (2002) expanded Iverson's work and developed a transient rainfall infiltration and grid-based regional slope-stability analysis (TRIGRS) model. Because of its practicability, the TRIGRS model became popular for assessing shallow landslides (Crosta, Frattini

Corresponding author: Cheng-Yu Ku
E-mail: chkst26@mail.ntou.edu.tw

2003; Keim, Skaugset 2003; Lan *et al.* 2005; D'Odorico *et al.* 2005; Liu *et al.* 2008; Baum *et al.* 2010; Bordoni *et al.* 2015; Schilirò *et al.* 2015; Alvioli, Baum 2016).

In these numerical models, the hillslope soil is typically assumed to be homogeneous. Besides, Iverson's method assumed that nearly saturated soil was a simplified form of the linear diffusion equation of the Richards equation, in which the influence of matric suction of soil and the soil water characteristic curve (SWCC) cannot be comprehensively considered during analysis. A complete theory of groundwater flow when rainfall infiltrates unsaturated zones can be described using either the variably saturated flow equation or the generalized Richards equation. The Richards equation is highly nonlinear and cannot directly provide an analytical solution. The soil water content is a major factor that influences the nonlinear physical relationship of the hydraulic conductivity in unsaturated zones (Van Genuchten 1980; Fredlund *et al.* 1994).

Since the appearance of layered soil is much more common than homogeneous soil in nature, the hydrological process in heterogeneous porous media has drawn much attention and been studied (Fok 1970; Aylor, Parlange 1973; Hachum, Alfaro 1980; Samani *et al.* 1989), using analytical or numerical methods (Hanks Bowers 1962; Whisler, Klute 1966; Romano *et al.* 1988; Moldrup *et al.* 1989; Srivastava, Yeh 1991; Corradini *et al.* 2000; Ku, Tsai 2013). Because the numerical modeling of rainfall-induced shallow landslides in unsaturated layered soil using the variably saturated flow equation has hardly been reported, the study proposed a pioneer work on numerical modeling of rainfall-induced shallow landslides in unsaturated layered soil.

To model the shallow landslides, the infinite slope stability analysis coupled with the hydrological model with the consideration of the fluctuation of time-dependent pore water pressure and the SWCC proposed by Gardner was developed. A linearization process for the nonlinear Richards equation to deal with groundwater flow in unsaturated layered soil is derived using the Gardner model. To solve one-dimensional flow in the unsaturated zone of layered soil profiles, flux conservation and the continuity of pressure potential at the interface between two consecutive layers are considered in the numerical discretization of the finite difference method. In addition, model verification and comparison were performed, and a numerical model of rainfall-induced shallow landslides in unsaturated zones was subsequently developed. The formulation of the proposed method is described as follows.

1. Methods

1.1. Formulation of the variably saturated flow equation

Unsaturated soil is composed through a three-phase mixing process that involves a gas phase (air), a liquid phase (pore water) and a solid phase (soil particles). Soil layers above the water table are termed as the unsaturated zone. In geotechnical engineering, pore water pressure fluctuation reduces soil suction which may cause soil failure. Generally, continuous rainfall reduces soil suction, affecting the stability of slope. The complete theory of groundwater flow and rainfall infiltration of the unsaturated zone can be expressed using the Richards equation:

$$K\nabla^2 H = \left[S_s S_w + C(h)\right]\frac{\partial H}{\partial t}, \tag{1}$$

where H represents the total groundwater head, K represents hydraulic conductivity, S_s represents the specific storage, S_w represents saturation, $C(h)$ represents specific capacity, and t represents time. If porous media are heterogeneous and anisotropic, Eq. (1) becomes:

$$\nabla K \nabla H = \left[S_s S_w + C(h)\right]\frac{\partial H}{\partial t}. \tag{2}$$

This is known as the generalized Richards equation or variably saturated flow equation, which can be used to describe the groundwater flow in saturated and unsaturated zones simultaneously (Tóthová *et al.* 2007). Eq. (2) can be rewritten as follows for a complete three-dimensional problem:

$$S_s S_w \frac{\partial H}{\partial t} + C(h)\frac{\partial H}{\partial t} = \frac{\partial}{\partial x}\left(K_x(h)\frac{\partial H}{\partial x}\right) + \frac{\partial}{\partial y}\left(K_y(h)\frac{\partial H}{\partial y}\right) + \frac{\partial}{\partial z}\left(K_z(h)\frac{\partial H}{\partial z}\right), \tag{3}$$

where $K_x(h)$, $K_y(h)$, and $K_z(h)$ represent hydraulic conductivity in the *x*, *y* and *z* axis, which are all functions of the pore water pressure head h when unsaturated. To convert Eq. (3) into an equation that considers infiltration, we assume that the *x* and *y* planes represent slope vertical sections, the *z* axis is perpendicular to the *xy* plan, and the total water head H can be expressed using the pressure head h and elevation head E as follows:

$$H = h + E. \tag{4}$$

When the slope infiltration problem is considered, the elevation head E varies with the slope angle α, as shown in Figure 1. The elevation head can be presented as follows:

$$E = x\sin(\alpha) + z\cos(\alpha). \tag{5}$$

The governing equation can be obtained by substituting Eqs (4) and (5) into Eq. (3):

$$\left(S_s S_w + C(h)\right)\frac{\partial h}{\partial t} = \frac{\partial}{\partial x}\left(K_x(h)\left(\frac{\partial h}{\partial x} - \sin\alpha\right)\right) + \frac{\partial}{\partial y}\left(K_y(h)\frac{\partial h}{\partial y}\right) + \frac{\partial}{\partial z}\left(K_z(h)\left(\frac{\partial h}{\partial z} + \cos\alpha\right)\right). \tag{6}$$

Thus, the majority of the variably saturated flow problem was simplified using Eq. (6). Considering a

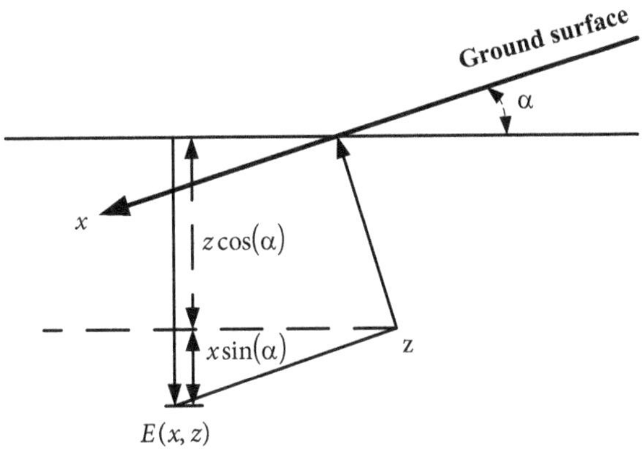

Fig. 1. The definition of the coordinate system in this study

one-dimensional problem for example, rainfall duration was far shorter than the transmission time of pore water pressure in the x and y axis. This consideration can be realistic for slopes, especially steep slopes usually covered by a thin soil mantle. Thus, only vertical water flow variations were considered, which can be presented as a one-dimensional vertical infiltration equation:

$$\left(S_s S_w + C(h)\right)\frac{\partial h}{\partial t} = \frac{\partial}{\partial z}\left(K_z(h)\left(\frac{\partial h}{\partial z} + \cos\alpha\right)\right). \quad (7)$$

When soil is saturated, $S_w = 1$ and $C(h) = 1$. Therefore, the saturated flow equation is a commonly used governing equation for groundwater in the saturated zone, where:

$$S_s \frac{\partial h}{\partial t} = \frac{\partial}{\partial z}\left(K_z\left(\frac{\partial h}{\partial z} + \cos\alpha\right)\right). \quad (8)$$

When soil is not saturated, the groundwater flow equation in unsaturated zone can be expressed as:

$$C(h)\frac{\partial h}{\partial t} = \frac{\partial}{\partial z}\left(K_z(h)\left(\frac{\partial h}{\partial z} + \cos\alpha\right)\right). \quad (9)$$

When the soil layer is not saturated, Eq. (9) can be used to calculate fluctuations in pore water pressure. When the soil layer is gradually saturated by rainfall, Eq. (8) is used for analysis because Eq. (7) primarily describes the pore water pressure changes in saturated and unsaturated zones. This study adopted Eq. (7) as the core governing equation of rainfall-induced pore water pressure fluctuations in unsaturated zones. In 1958, Gardner (Gardner 1958) proposed a simple-one parameter model as follows:

$$S_e = e^{\alpha_g h}, \quad (10)$$

where α_g is the soil pore-size distribution parameter which is related to the pore size distribution of soil. S_e is the effective saturation defined by normalizing volumetric water content with its saturated and residual values as:

$$S_e = \frac{(\theta - \theta_r)}{(\theta_s - \theta_r)}, \quad (11)$$

where θ represents the volumetric water content function, θ_r represents the residual water content and θ_s represents the saturated water content. Substituting Eq. (11) into Eq. (10), we have

$$\theta = \theta_r + (\theta_s - \theta_r)e^{\alpha_g h}. \quad (12)$$

It is common to normalize the hydraulic conductivity of unsaturated soil with respect to their maximum value. The normalized value, referred to as the relative hydraulic conductivity, K_r can be expressed as:

$$K_r(h) = \frac{K_z(h)}{K_s}, \quad (13)$$

where K_s is the saturated hydraulic conductivity. $K_r(h)$ is the relative hydraulic conductivity and it is a function of the pressure head which makes the Richards equation highly nonlinear. Rather than using the van Genuchten expression for S_e, a simpler version is used as follows:

$$S_e = K_r. \quad (14)$$

Therefore, $K_r = e^{\alpha_g h}$. Using the Gardner model, we can derive a linearized Richards equation. First of all, we define a new parameter h^* which can be expressed as

$$h^* = e^{\alpha_g h} - \chi, \quad (15)$$

where h^* is the pressure head of the linearized Richards equation and χ is a constant which is a key transformation allowing the solution of the linearized Richards equation using the transform methods (Duffy 1994; Tracy 2006, 2011; Liu *et al.* 2015). The parameter χ has no physical meaning and can be expressed as $\chi = e^{\alpha_g h_d}$. h_d is the pressure head when the soil is dry. Taking the derivative of Eq. (15) with respect to z, we obtain

$$\frac{\partial h^*}{\partial z} = \alpha_g e^{\alpha_g h}\frac{\partial h}{\partial z}. \quad (16)$$

Rearranging Eq. (16), we have

$$\frac{\partial h}{\partial z} = \frac{1}{\alpha_g}e^{-\alpha_g h}\frac{\partial h^*}{\partial z}. \quad (17)$$

Again, substituting Eq. (15) into Eq. (17), we have

$$K_r\frac{\partial h}{\partial z} = e^{\alpha_g h}\left(\frac{1}{\alpha_g}e^{-\alpha_g h}\frac{\partial h^*}{\partial z}\right). \quad (18)$$

Eq. (18) can be written as

$$K_r\frac{\partial h}{\partial z} = \frac{1}{\alpha_g}\frac{\partial h^*}{\partial z}. \quad (19)$$

Taking the derivative of $K_r = e^{\alpha_g h}$ with respect to z, we have

$$\frac{\partial K_r}{\partial z} = \alpha_g e^{\alpha_g h}\frac{\partial h}{\partial z}. \quad (20)$$

Substituting Eq. (17) into Eq. (20), we have

$$\frac{\partial K_r}{\partial z} = \frac{\partial h^*}{\partial z}. \quad (21)$$

Recall Eq. (12), we have

$$\frac{\partial\theta}{\partial t}=\left(\theta_s-\theta_r\right)\frac{\partial S_e}{\partial t}. \quad (22)$$

Because $S_e = K_r$, $K_r = e^{\alpha_g h}$, and $h^* = e^{\alpha_g h} - \chi$, we can rewrite the above equation as

$$\frac{\partial\theta}{\partial t}=\left(\theta_s-\theta_r\right)\frac{\partial h^*}{\partial t}. \quad (23)$$

Finally, we substitute Eqs. (17), (12), and (23) to Eq. (19). The linearized Richards equation can be obtained as

$$\frac{\partial^2 h^*}{\partial z^2}+\alpha_g\frac{\partial h^*}{\partial z}=\frac{\alpha_g\left(\theta_s-\theta_r\right)}{K_s}\cos\alpha\frac{\partial h^*}{\partial t}. \quad (24)$$

For simplicity, the equation can be expressed as

$$K_{\alpha\theta}\frac{\partial^2 h^*}{\partial z^2}+K_\theta\cos\alpha\frac{\partial h^*}{\partial z}=\frac{\partial h^*}{\partial t}, \quad (25)$$

where

$$K_\theta=\frac{K_s}{\left(\theta_s-\theta_r\right)} \text{ and } K_{\alpha\theta}=\frac{K_\theta}{\alpha_g}. \quad (26)$$

1.2. The finite difference method

The finite difference method (FDM) is adopted for the numerical discretization. The reason for choosing the FDM is that the FDM can be easily used to integrate the formulation for the discontinuities (or soil interface) which characterize the layers for unsaturated layered soil. In addition, several studies of numerical modeling of shallow landslides have been carried out using the FDM such as Yoshimatsu, Abe (2006), Wang *et al.* (2006), Singh *et al.* (2010), Sarkar *et al.* (2012). The main advantage of the FDM is that it is probably the most straight forward numerical method to formulate complicated partial different equations such as the Richards equation. In this study, we need to derive the sophisticated mathematical formulation for the unsaturated layered soil. Accordingly, we adopted the FDM as the numerical method. The linearized Richards equation based on the Gardner model is depicted in Eq. (25). To apply the FDM, we have to consider the discretization in spatial and temporal domains. For the spatial discretization, we divide the domain into sections, each of length Δz along the z axis and approximated the first and second derivatives in the linearized Richards equation for each grid point by the central difference formulas. For the time discretization, the implicit scheme is adopted. Accordingly, we obtain the finite difference equation as:

$$K_{\alpha\theta}\left(\frac{h^*_{i+1}-2h^*_i+h^*_{i-1}}{\Delta z^2}\right)+K_\theta\cos\alpha\left(\frac{h^*_{i+1}-h^*_{i-1}}{2\Delta z}\right)=\left(\frac{h^{*n}_i-h^{*n-1}_i}{\Delta t}\right), \quad (27)$$

where i is the nodal point and Δz is the interval. Δt is the time interval. K_s, α_g, θ_s and θ_r are the same definition as described in the previous section. Considering the steady-state situation, we obtain the finite difference equation of the linearized Richards equation as:

$$K_{\alpha\theta}\left(\frac{h^*_{i+1}-2h^*_i+h^*_{i-1}}{\Delta z^2}\right)+K_\theta\cos\alpha\left(\frac{h^*_{i+1}-h^*_{i-1}}{2\Delta z}\right)=0, \quad (28)$$

Eq. (28) can be used to deal with homogenous soil. For unsaturated layered soil, it is necessary to account for the discontinuities of the parameters which characterize the layers. This gives rise to the so-called interface problem for the linearized Richards equation. For the interface between layers, the conservation of water flux and the Darcy's law have to be considered, as shown in Figure 2. The finite difference equation of the first term, $K_{\alpha\theta}\frac{\partial^2 h^*}{\partial z^2}$ in Eq. (25), can be derived as follows:

$$K_{\alpha\theta}\frac{\partial^2 h^*}{\partial z^2}=K_{\alpha\theta}\frac{2\left(\frac{K_{\alpha\theta A}}{K_{\alpha\theta A}+K_{\alpha\theta B}}\right)h^*_{i-1}-2h^*_i+2\left(\frac{K_{\alpha\theta B}}{K_{\alpha\theta A}+K_{\alpha\theta B}}\right)h^*_{i+1}}{\Delta z^2}, \quad (29)$$

where $K_{\alpha\theta A}$ and $K_{\alpha\theta B}$ represent the parameter, $K_{\alpha\theta}$, in layer A and layer B respectively. The finite difference equation of the second term, $K_\theta\frac{\partial h^*}{\partial z}$ in Eq. (25), can be derived as follows:

$$K_\theta\frac{\partial h^*}{\partial z}=K_\theta\frac{\left[\frac{K_{\theta A}}{K_{\theta B}}\left(h^*_i-h^*_{i-1}\right)+\frac{K_{\theta B}}{K_{\theta A}}\left(h^*_{i+1}-h^*_i\right)\right]}{2\Delta z}, \quad (30)$$

where $K_{\theta A}$ and $K_{\theta B}$ represent the parameter, K_θ, in layer A and layer B respectively. Substituting Eqs. (29) and (30) into Eq. (27) we obtain:

$$K_{\alpha\theta}\frac{\left[2\alpha_A h^{*n}_{i-1}-2h^{*n}_i+2\alpha_B h^{*n}_{i+1}\right]}{\Delta z^2}+K_\theta\cos\alpha\frac{\left[\beta_A\left(h^{*n}_i-h^{*n}_{i-1}\right)+\beta_B\left(h^{*n}_{i+1}-h^{*n}_i\right)\right]}{2\Delta z}=\left(\frac{h^{*n}_i-h^{*n-1}_i}{\Delta t}\right), \quad (31)$$

where $\alpha_A=\frac{K_{\alpha\theta A}}{K_{\alpha\theta A}+K_{\alpha\theta B}}$, $\alpha_B=\frac{K_{\alpha\theta B}}{K_{\alpha\theta A}+K_{\alpha\theta B}}$, $\beta_A=\frac{K_{\theta A}}{K_{\theta B}}$ and $\beta_B=\frac{K_{\theta B}}{K_{\theta A}}$.

In Eqs. (28) and (31), $K_{\alpha\theta}$ and K_θ at the interface have to be determined. In a stratified soil deposit where the hydraulic conductivity for flow in a given direction changes from layer to layer, an equivalent hydraulic conductivity can be obtained by the following equation:

$$K_{(eqv)} = \frac{H}{\left(\frac{H_1}{K_1}\right)+\left(\frac{H_2}{K_2}\right)+\cdots\left(\frac{H_n}{K_n}\right)}, \tag{32}$$

where $K_{(eqv)}$ is the equivalent hydraulic conductivity, H is the height of the soil, H_i is the thickness of each soil layer, and K_i is the hydraulic conductivity of each soil layer. i is the number of the soil layer. Because we considered only two consecutive layers at the interface in the finite difference discretization. Using Eq. (32), $K_{\alpha\theta}$ and K_θ at the interface can be obtained as:

$$K_{\alpha\theta(i)} = \frac{2K_{\alpha\theta A}K_{\alpha\theta B}}{K_{\alpha\theta A}+K_{\alpha\theta B}} \text{ and } K_{\theta(i)} = \frac{2K_{\theta A}K_{\theta B}}{K_{\theta A}+K_{\theta B}}. \tag{33}$$

Accordingly, the finite difference equations of the linearized Richards equation for steady-state and transient conditions can be expressed as:

$$K_{\alpha\theta(i)}\left(\frac{h^*_{i+1}-2h^*_i+h^*_{i-1}}{\Delta z^2}\right)+ K_{\theta(i)}\cos\alpha\left(\frac{h^*_{i+1}-h^*_{i-1}}{2\Delta z}\right)=0; \tag{34}$$

$$K_{\alpha\theta(i)}\frac{\left[2\alpha_{-}h^{*n}_{-1}-2h^{*n}+2\alpha\ h^{*n}_{+1}\right]}{\Delta z^2}+ K_{\theta(i)}\cos\alpha\frac{\left[\beta_A\left(h_i^{*n}-h_{i-1}^{*n}\right)+\beta_B\left(h_{i+1}^{*n}-h_i^{*n}\right)\right]}{2\Delta z}= \left(\frac{h^{*n}_i-h^{*n-1}_i}{\Delta t}\right). \tag{35}$$

Furthermore, Eqs (34) and (35) can be expressed as the matrix form and can be written as

$$\mathbf{Ax}=\mathbf{b}, \tag{36}$$

where $\mathbf{A}$ is a $n \times n$ square matrix, $\mathbf{x}$ is a $n \times 1$ vector in which $\mathbf{x}=[h_1^* \cdots h_n^*]^T$, and $\mathbf{b}$ is a $n \times 1$ vector. n is the number of unknowns to be solved. Eq. (36) can then be simply solved by $\mathbf{x}=\mathbf{A}^{-1}\mathbf{b}$.

1.3. Slope stability theory of unsaturated soil

The failure mechanism of infinite slopes primarily involves colluvium, weathered rock formations, or shallow failure and planar slides in bedrock alternations located at thin depth below the ground level. The sliding soil thickness is far less than the slope height. In most studies, infinite slopes are used to analyze shallow landslides. The factor of safety (FS) for the infinite slope theory can be expressed as follows:

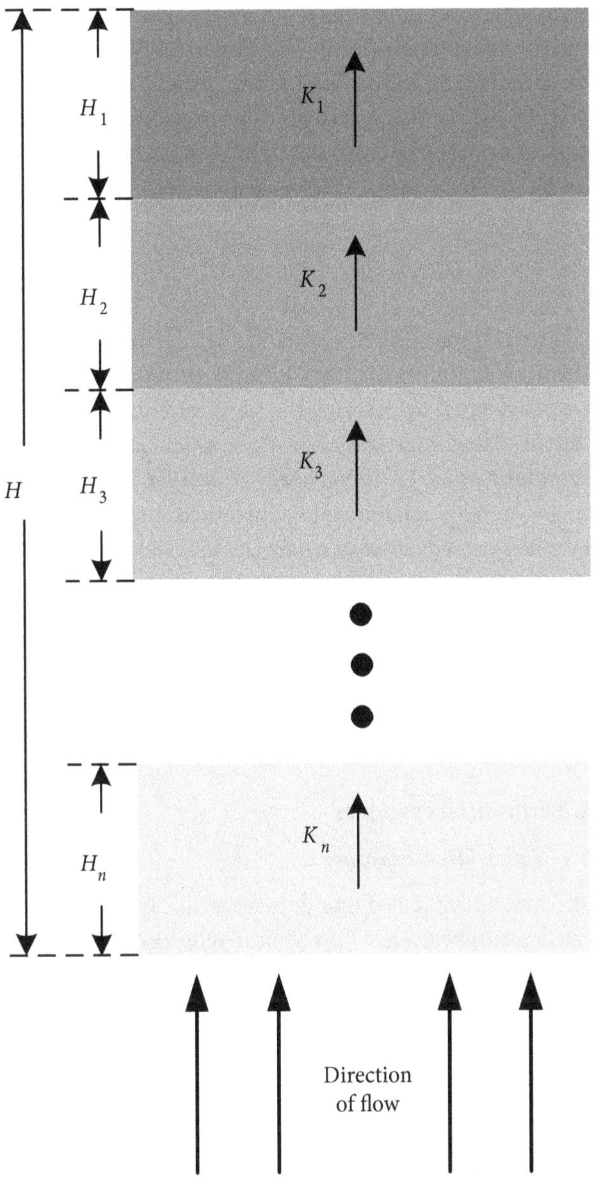

Fig. 2. The configuration of water flow through layered soils in the FDM formulation

$$\text{FS} = \frac{c' + (z\gamma_t\cos^2\alpha - h\gamma_w\cos^2\alpha)\tan\varphi'}{z\gamma_t\cos\alpha\sin\alpha}, \tag{37}$$

where c' represents the effective cohesiveness, φ' represents the effective friction angle, z represents the soil thickness, γ_t represents the unit weight of soil and γ_w represents the unit weight of water. Fredlund adopted the perspectives of continuum mechanics and demonstrated that the shear strength equation can be applied to unsaturated soil (Fredlund, Morgenstern 1977). The reason of choosing Fredlund's approach for modeling the FS in unsaturated condition is that its description of the stress state of the unsaturated soil examine the context of multiphase continuum mechanics. Unlike saturated condition, which are two-phase mixing process involved a solid phase and a gas phase or a liquid phase, unsaturated soil is composed through a three-phase mixing process that comprises of a gas phase, a liquid phase and a solid phase. Therefore, the corresponding pressure of a gas and a liquid phase in unsaturated condition have a direct impact on physical behavior of the shear strength (Lu, Likos 2004). Vanapalli *et al.* proposed the shear strength equation based on the SWCC, which is suitable for unsaturated zones (Vanapalli *et al.* 1996):

$$\tau_f = c' + \sigma \tan\varphi' - h\gamma_w \left(\frac{\theta - \theta_r}{\theta_s - \theta_r} \right) \tan\varphi', \quad (38)$$

where σ represents the total stress, τ_f represents the shear strength. The shear strength of unsaturated soil was developed to describe the physical consideration to the air-water interphase and these stress state variables (Fredlund *et al.* 1978; Vanapalli *et al.* 1996). The revised shear strength equation was substituted into the original infinite slope equation to obtain the following FS equation for analyzing slope stability in shallow unsaturated zones:

$$\mathrm{FS} = \frac{c'}{z\gamma_t \cos\alpha \sin\alpha} + \frac{\tan\varphi'}{\tan\alpha} - \frac{h\gamma_w \left(\frac{\theta - \theta_r}{\theta_s - \theta_r} \right) \tan\varphi'}{z\gamma_t \cos\alpha \sin\alpha}. \quad (39)$$

2. Verification examples

2.1. Verification example 1

The first example is a one-dimensional transient unsaturated groundwater flow problem in a homogenous soil. A column of soil is initially dry until water begins to infiltrate the soil. A pool of water at the ground surface is then maintained holding the pressure head to zero. The thickness of the soil L is 10 m. The initial condition is described as follows:

$$h(z,\ t=0) = -10^5. \quad (40)$$

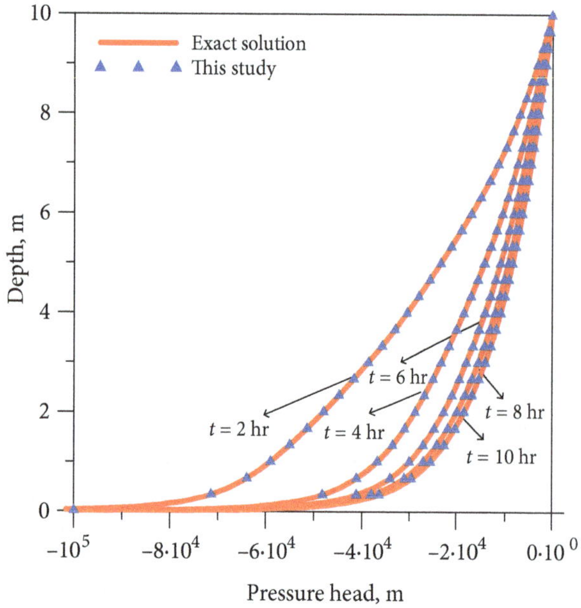

Fig. 3. Result comparison of verification example 1 with the exact solution

The boundary conditions of the top and bottom boundaries are as follows:

$$h(z=0,t) = -10^5, \quad (41)$$

$$h(z=L,t) = 0. \quad (42)$$

Using Eq. (15), the suction head can be converted to h^* as follows:

$$h^*(z=0,t) = 0, \quad (43)$$

$$h^*(z=L,t) = 1. \quad (44)$$

In this example, the soil is assumed to be silt which the α_g of the Gardner model is 10^{-4}. The saturated hydraulic conductivity (K_s), saturated water content (θ_s), and residual water content (θ_r) for this example are 9×10^{-5} m/s, 0.50, and 0.11, respectively, as shown in Table 1. For one-dimensional transient unsaturated groundwater flow in a homogenous soil, Tracy proposed the analytical transient solution as follows (Tracy 2011):

$$h^*(z,t) = (1-\chi)e^{\frac{\alpha_g}{2}(L-z)} \times \left[\frac{\sinh\left(\frac{\alpha_g}{2}z\right)}{\sinh\left(\frac{\alpha_g}{2}L\right)} + \frac{2}{Lc}\sum_{k=1}^{\infty}(-1)^k \frac{\lambda_k}{\mu_k}\sin\lambda_k z e^{-\mu_k t} \right], \quad (45)$$

where $\lambda_k = k\pi/L$, $c = \alpha_g(\theta_s - \theta_r)/K_s$, $\mu_k = (\alpha_g^2/4 + \lambda_k^2)/c$, t is time. L is the thickness of the soil. Figure 3 demonstrates that the numerical results agreed well with the analytical solution for the one-dimensional transient unsaturated groundwater flow in a homogenous soil. It is found that the mean absolute error (MAE) was 1.18×10^{-4} m. The above verification example demonstrates that the proposed numerical scheme for the Richards equation to solve one-dimensional flow processes in the unsaturated zone of soil is validated.

Table 1. Parameters used in the verification example 1

saturated hydraulic conductivity (m/s)	9×10^{-5}
α_g of the Gardner model	10^{-4}
saturated water content	0.50
residual water content	0.11

2.2. Verification example 2

The second example under investigation is also a one-dimensional, transient, unsaturated groundwater flow problem in homogeneous soil. A soil column is initially dry until water begins to infiltrate the soil. A pool of water at the ground surface is maintained, holding the infiltration rate to zero. This is known as one-dimensional Green-Ampt problem (Green, Ampt 1911). The thickness of the soil L is 10 m. The initial condition is described as follows:

$$h(z,\ t=0)=-100\,. \tag{46}$$

The boundary conditions of the top and bottom boundaries are as follows.

$$h(z=0,t)=-100\,, \tag{47}$$

$$h(z=L,t)=0\,. \tag{48}$$

The soil was assumed to be sand, which has saturated hydraulic conductivity (K_s) of 3×10^{-3} m/s. The saturated water content (θ_s) and residual water content (θ_r) are 0.50 and 0.11, respectively. In the Gardner model, there is only one fitting parameter, α_g. The SWCC obtained from the laboratory is shown in Figure 4. Using the Gardner model, α_g can be obtained by fitting the SWCC with the Gardner model as shown in Figure 4. It is found that the $\alpha_g=1.6\times10^{-2}$ and the MAE was 1.78×10^{-2}.

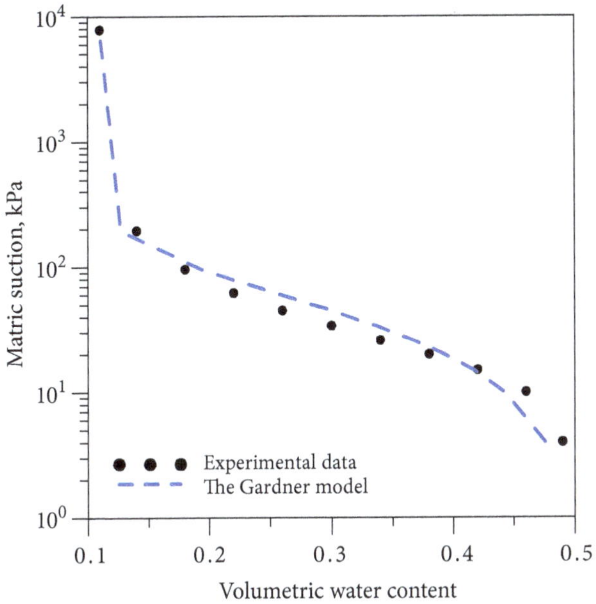

Fig. 4. The fitting curve of the SWCC using the Gardner model

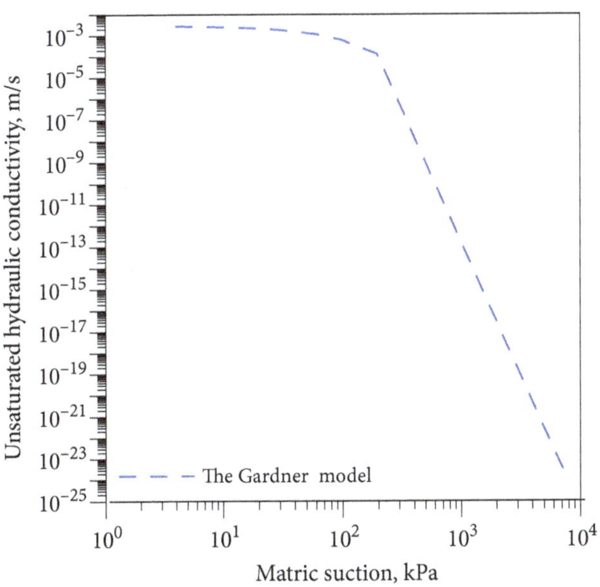

Fig. 5. The relationship between matric suction and unsaturated hydraulic conductivity for the Gardner model

Using the Gardner model, the nonlinear relationship between the matric suction and unsaturated hydraulic conductivity can be obtained as shown in Figure 5. In this example, the total simulation time is one hour. Figure 6 depicted the comparison of the solution with the numerical solution using the Skeel and Berzins' method (Skeel, Berzins 1990). It is found that the computed results agreed well with those obtained from the Skeel and Berzins' method and the MAE was 2.74×10^{-4} m.

2.3. Verification example 3

The governing equation for unsaturated flow in response to infiltration at ground surface is known as the Richards equation. For one-dimension, it can be written as Eq. (9). Iverson assumed that the Richards equation as a linear diffusion equation and derived the analytical solution (Iverson 2000). Baum *et al.* revised the analytical solution and implemented in TRIGRS (Baum *et al.* 2002). The analytical solution proposed by Baum *et al.* can be written as follows:

$$h(Z,t)=\left[Z-d\right]\beta+2\sum_{n=1}^{N}\frac{I_{nZ}}{K_Z}\times$$
$$\left[H\left(t-t_n\right)\left[D_1\left(t-t_n\right)\right]^{\frac{1}{2}} ierfc\left[\frac{Z}{2\left[D_1\left(t-t_n\right)\right]^{\frac{1}{2}}}\right]\right]-2\sum_{n=1}^{N}\frac{I_{nZ}}{K_Z}\times$$
$$\left[H\left(t-t_{n+1}\right)\left[D_1\left(t-t_{n+1}\right)\right]^{\frac{1}{2}} ierfc\left[\frac{Z}{2\left[D_1\left(t-t_{n+1}\right)\right]^{\frac{1}{2}}}\right]\right], \tag{49}$$

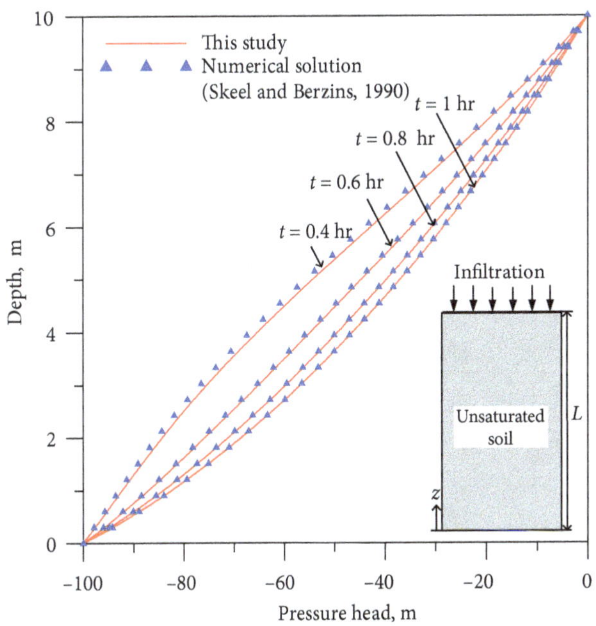

Fig. 6. Result comparison of verification example 2 with the numerical solution

where $Z = z/\cos\alpha$, d is the depth of steady-state water table measured in z direction, $\beta = \lambda\cos\alpha$ and $\lambda = \cos\delta - \left[I_z / K_z\right]_{LT}$, where I_z is the long term rainfall flux at ground surface. K_z is the hydraulic conductivity, I_{nz} is the surface flux for the nth time interval, n is the total number of time intervals, D_1 is related to the hydraulic diffusivity D_0, $H(t - t_{n+1})$ is the Heaviside step function. The ierfc is the complementary error function and can be expressed as follows:

$$ierfc(\eta) = \frac{1}{\sqrt{\pi}}\exp\left(-\eta^2\right) - \eta erfc(\eta). \tag{50}$$

We conducted a comparison example with the consideration of unsaturated flow in response to infiltration at ground surface. The input parameters were listed in Table 2. The numerical results of this example were compared to those obtained from the analytical solution of equation. In a scenario set in this study, a soil layer thickness is assumed to be 100 cm and a water table at the junction between the soil base and rock bed. Figure 7 and Figure 8 depict that the results from the proposed model and the analytical solution. It is interested that the analytical solution from the simplified linear diffusion equation cannot obtain the appropriate result when the value of the hydraulic diffusivity is large as revealed in Figure 8. The

Table 2. Parameters used in the verification example 3

	This study	Baum *et al.* (2002)
Slope angle (degree)	30	
Thickness (m)	1	
Infiltration rate (m/hr)	0.2	
Infiltration time (hr)	5	
Saturated hydraulic conductivity (m/hr)	2.5	
Other parameters	$\theta_s = 0.43$ $\theta_r = 0.08$ $\alpha_g = 0.2$	$D_0 = (10 \sim 100) \times K_s$ (m/hr)

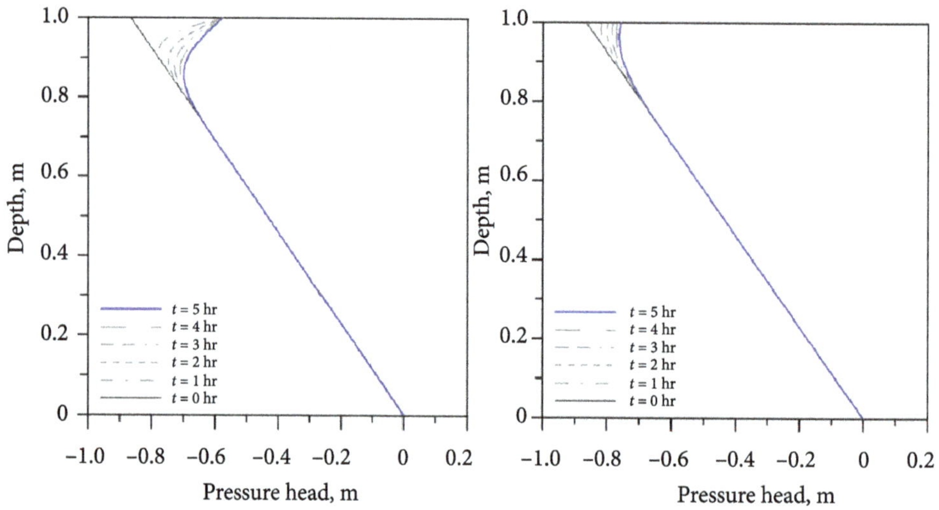

Fig. 7. Comparison of the results for $D_0 = 10 \cdot K_s$ (left: this study, right: TRIGRS model)

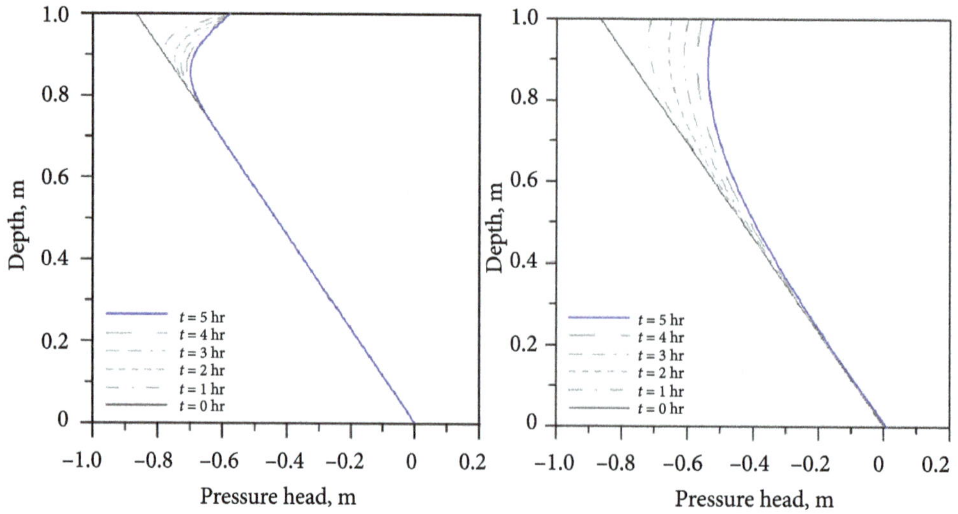

Fig. 8. Comparison of the results for $D_0 = 100 \cdot K_s$ (left: this study, right: TRIGRS model)

results from this example demonstrate that the analytical solution of equation can only be applied on the problems within certain condition and should be used with caution.

3. Application examples

3.1. Application example 1

The first application example under investigation is a one-dimensional, transient, unsaturated groundwater flow problem in a homogeneous soil. The thickness of the soil L is 150 cm and the slope angle is 31 degrees. Since the Richards equation is an initial value problem, an initial condition must be provided to solve the equation. The initial pressure head is given as follows:

$$h(z,t=0)=(-z-z_w)\cos\alpha\,, \tag{51}$$

where z_w represents the height of water table. The pore pressure head distribution in unsaturated zones is congruent with the water table function. In this example, we considered $z_w=50$ cm. The total simulation time is five hours. Considering that a pool of water at the ground surface is maintained holding the infiltration rate, the boundary condition can be defined according to Darcy's law $q=-K_s\nabla H$, and assumes that infiltration rate facilitates complete infiltration and $q=R_f$ as follows:

$$R_f=-K_s\frac{\partial H}{\partial z}\,. \tag{52}$$

The definition of K_s and H are the same as defined in the previous section. Using Eqs. (4) and (5), the above equation can be written as:

$$R_f=-K_s\left(\frac{\partial h}{\partial z}+\cos\alpha\right). \tag{53}$$

The above equation can then be rewritten as the following equation which is also known as the Neumann boundary condition:

$$\frac{\partial h}{\partial z}=-\cos\alpha-\frac{R_f}{K_s}\,. \tag{54}$$

If the infiltration rate is greater than saturated hydraulic conductivity, the maximal soil infiltration equals the saturated hydraulic conductivity. The bottom boundary may be assumed to be a no-flow boundary, and is inferred as follows:

$$0=-K_s\frac{\partial H}{\partial z}\,. \tag{55}$$

The above equation can be written as:

$$0=-K_s\left(\frac{\partial h}{\partial z}+\cos\alpha\right). \tag{56}$$

Then, we have

$$\frac{\partial h}{\partial z}=-\cos\alpha\,. \tag{57}$$

To solve the Richards equation, the boundary conditions must be considered. We adopted $R_f=5\times10^{-5}$ cm/hr. The soil was assumed to be sand, which has saturated hydraulic conductivity (K_s) of 9×10^{-2} cm/hr. The saturated water content (θ_s) and residual water content (θ_r) are 0.43 and 0.08, respectively. The α_g of the Gardner model is 1×10^{-2}. In this example, the total simulation time is five hours. The effective cohesiveness, effective friction angle and unit weight of soil are 4.6 kPa, 30 degrees and 21.5 kN/m³, respectively. Figure 9 and Figure 10 indicate the computed results of pressure head and FS for application example 1.

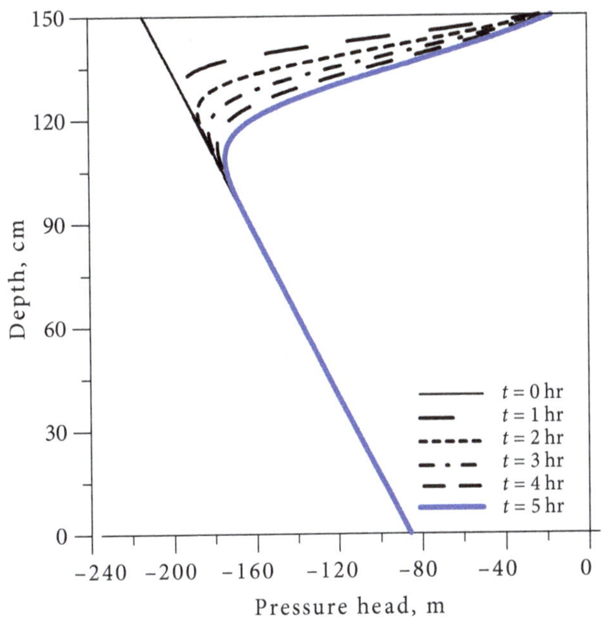

Fig. 9. Computed pressure head of the soil profile for application example 1

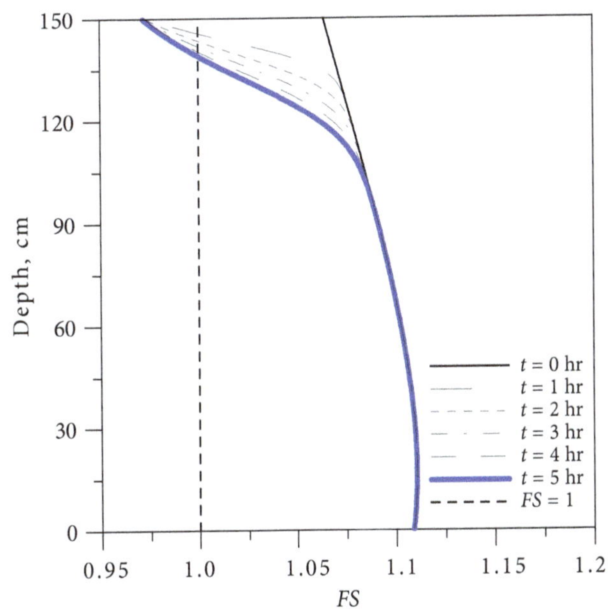

Fig. 10. Computed FS of the soil profile for application example 1

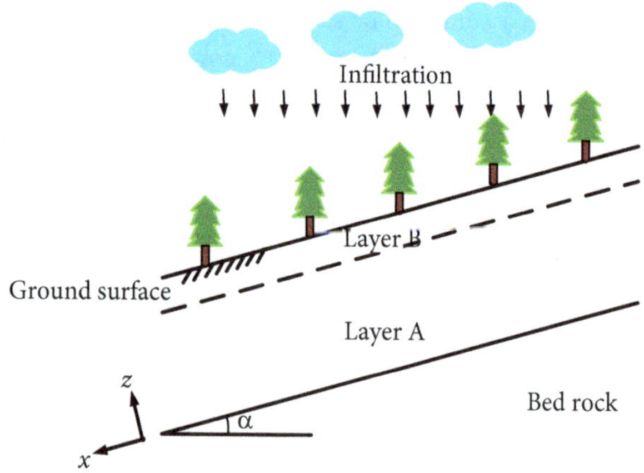

Fig. 11. Schematic illustration of application example 2

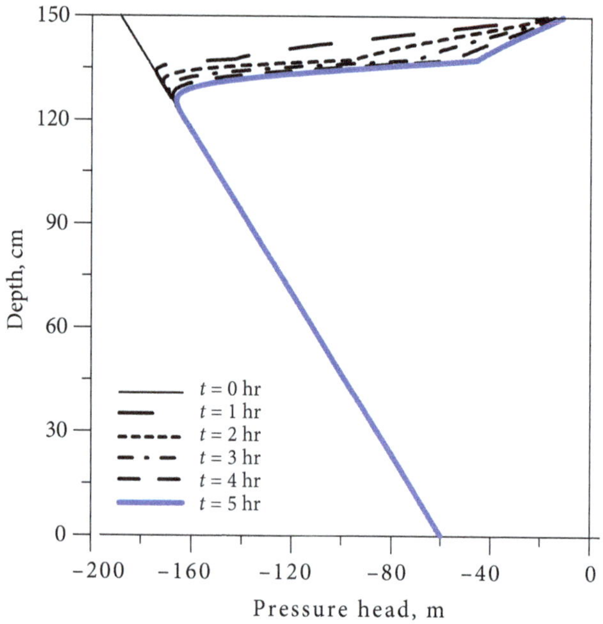

Fig. 12. Computed pressure head of the soil profile for application example 2

3.2. Application example 2

The second application example under investigation is a one-dimensional, transient, unsaturated groundwater flow problem in two-layered soil. The thickness of the soil L is 150 cm. In the case of two layered soils, layer A and layer B have thickness of 130 cm and 20 cm, respectively, as shown in Figure 11. Table 3 shows the soil parameters for two soil layers. In this example, layer B is more permeable than layer A. The hillslope has a slope angle of 31 degrees. The initial pressure head can be described as Eq. (53). The boundary conditions of the top and bottom boundaries are showed as Eq. (56) and Eq. (59). In this example, we considered z_w = 80 cm and $R_f = 5\times10^{-5}$ cm/hr. The total simulation time is five hours. The computed pressure heads versus elapsed time for this example are depicted in Figure 12. Figure 12 demonstrates that the fluctuation of pore water pressure in unsaturated layered soil dominates slope stability behavior. Besides, the pressure head at the interface increases quickly and the lowest FS may occur at the interface between two consecutive soil layers during a rainfall even if the top soil layer has the lower hydraulic conductivity. Figure 13 shows the computed FS with respect to time. It shows that the lowest FS occurs at the interface between two consecutive soil layers during a rainfall event. Figure 14 indicates the results of the FS at the depth of 135 cm for this example. It is found that the FS is strongly affected by the rainfall with respect to time. Accordingly, we may conclude that the variation of the pressure head caused by the rainfall event is strongly associated with soil layers and the interface between two consecutive soil layers may play a crucial role for the slope stability.

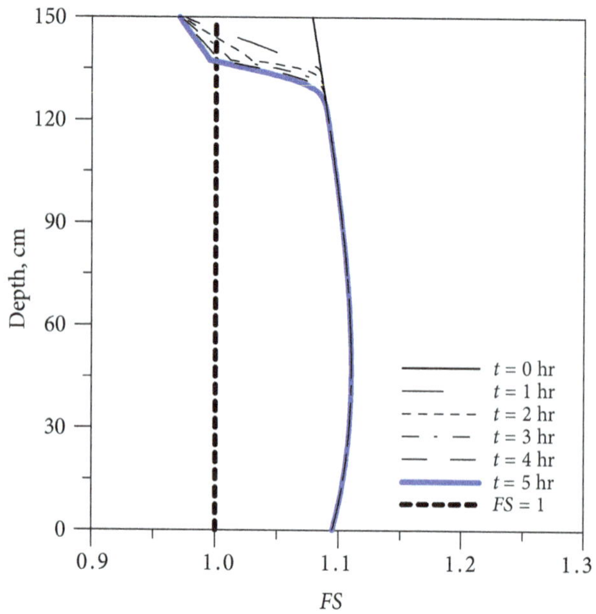

Fig. 13. Computed FS of the soil profile for application example 2

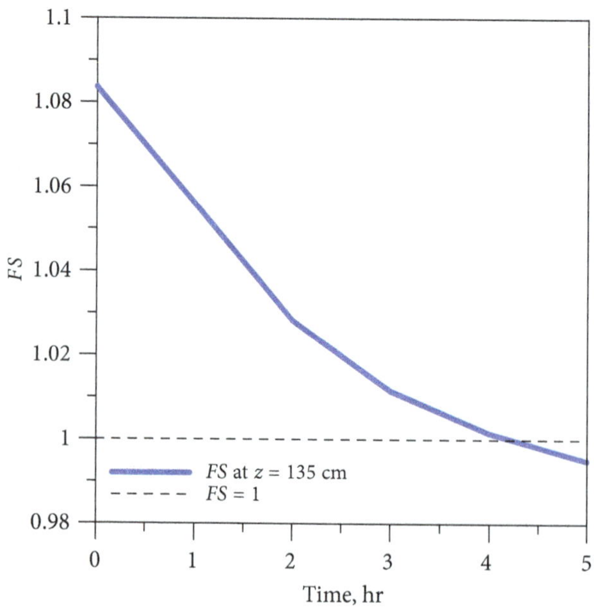

Fig. 14. The computed results of FS at z = 135 cm for application example 2

Table 3. Parameters used in the application example 2

Soil	Layer A	Layer B
Hydraulic conductivity K_s (cm/hr)	1×10^{-2}	1×10^{-1}
Saturated water content θ_s	0.43	0.50
Residual water content θ_r	0.08	0.11
α_g of the Gardner model	1×10^{-2}	1×10^{-2}

4. Discussions

In this paper, we carried out three verification examples and two application examples. From the results of verification examples, it is found that the computed numerical results agreed well with the analytical transient solution and those from the Skeel and Berzins' method. Comparison between the proposed numerical model and TRIGRS model (a simplified analytical-based model by assuming the Richards equation as a linear diffusion equation; Iverson 2000; Baum *et al.* 2002) was also conducted. We conducted a comprehensive study in this example with the consideration of several scenarios. It is found that the proposed model in this study can obtain more appropriate results when the value of the hydraulic diffusivity is larger than those from the TRIGRS model.

In the application examples, we further adopted the proposed model to study the slope stability of a landslide with the consideration of the infiltration from the ground surface for homogenous and layered soils. Since the numerical modeling of rainfall-induced shallow landslides in unsaturated layered soil using the variably saturated flow equation is still hardly found, this study demonstrates the advantages of using the proposed model to study the slope stability of landslides for layered soils. From the results of application examples, we also found that the fluctuation of pore water pressure in unsaturated layered soil dominates slope stability of landslides. Accordingly, it can be concluded that the proposed model in this study may have great potential to deal with problems of rainfall-induced shallow landslides in unsaturated layered soil.

The nature of the Richards equation is highly nonlinear and cannot directly provide an analytical solution. The SWCC is a major factor that influences the nonlinear physical relationship. To derive the proposed numerical model of rainfall-induced shallow landslides in unsaturated layered soil, we adopted the Gardner model to formulate the linearized Richards equation. It should be noted that the Gardner model is based on the one-parameter equation with an indication of the rate of desaturation of a soil. Parameters used in mathematical models for the SWCC include fixed points and two or more fitting constants that are usually selected to capture the general shape of the SWCC. Nevertheless, the Gardner model has only one parameter which may have limitations for fitting the SWCC with complicated shape.

Conclusions

In this study, a numerical model of unsaturated layered soil for rainfall-induced shallow landslides was developed. The infinite slope stability analysis coupled with the hydrological model with the consideration of the SWCC proposed by Gardner was developed to model the shallow landslides triggered by rainfall. The fundamental concepts and the construct of the proposed method are clearly addressed. Findings from this study are drawn as follows:

1. The FDM for the numerical discretization of the variably saturated flow equation based on the Gardner model to deal with groundwater flow in unsaturated layered soil is derived. The proposed method can be used to solve slope stability problems with the appearance of layered soil.

2. The validity of the model is established for a number of test problems by comparing numerical results with the analytical solutions. It is found that the proposed method can be used to model groundwater flow in unsaturated layered soil with high accuracy.

3. The results obtained from this study demonstrate that the slope stability of landslides is strongly dependent to the hydraulic conductivity. It is found that the variation of pore water pressure in unsaturated layered influences the stability of a slope. Besides, the pressure head at the interface increases quickly and the lowest FS may occur at the interface between two consecutive soil layers during a rainfall even if the top soil layer is less permeable.

Acknowledgements

This study was partially supported by the Central Geological Survey, Ministry of Economic Affairs, R.O.C. The authors would like to thank the Central Geological Survey for the generous financial support.

Disclosure statement

The authors declare that they don't have any competing financial, professional, or personal interests from any other parties.

References

Alvioli, M.; Baum, R. L. 2016. Parallelization of the TRIGRS model for rainfall-induced landslides using the message passing interface, *Environmental Modelling & Software* 81: 122–135.

Aylor, D. E.; Parlange, J. Y. 1973. Vertical infiltration into a layered soil, *Soil Science Society of America Journal* 37(5): 673–676. https://doi.org/10.2136/sssaj1973.03615995003700050015x

Baum, R. L.; Godt, J. W.; Savage, W. Z. 2010. Estimating the timing and location of shallow rainfall-induced landslides using a model for transient, unsaturated infiltration, *Journal of Geophysical Research* 115(F03013). https://doi.org/10.1029/2009JF001321

Baum, R. L.; Savage, W. Z.; Godt, J. W. 2002. *TRIGRS – A FORTRAN program for transient rainfall infiltration and grid-based regional slope-stability analysis*, U.S. Geological Survey Open-File Report 02-0424.

Bordoni, M.; Meisina, C.; Valentino, R.; Bittelli, M.; Chersich, S. 2015. Site-specific to local-scale shallow landslides triggering zones assessment using TRIGRS, *Natural Hazards and Earth System Science* 15(5): 1025–1050. https://doi.org/10.5194/nhess-15-1025-2015

Corradini, C.; Melone, F.; Smith, R. E. 2000. Modeling local infiltration for a two-layered soil under complex rainfall patterns, *Journal of Hydrology* 237: 58–73. https://doi.org/10.1016/S0022-1694(00)00298-5

Crosta, G. B.; Frattini, P. 2003. Distributed modeling of shallow landslides triggered by intense rainfall, *Natural Hazards and Earth System Science* 3: 81–93. https://doi.org/10.5194/nhess-3-81-2003

Csete, M.; Buzasi, A. 2016. Climate-oriented assessment of main street design and development in Budapest, *Journal of Environmental Engineering and Landscape Management* 24(4): 258–268. https://doi.org/10.3846/16486897.2016.1185431

D'Odorico, P.; Fagherazzi, S.; Rigon, R. 2005. Potential for landsliding: dependence on hyetograph characteristics, *Journal of Geophysical Research* 110(F01007). https://doi.org/10.1029/2004JF000127

Das, B. M. 2010. *Principles of geotechnical engineering*. 7th ed. Cengage Learning. USA.

Duffy, D. G. 1994. *Transform methods for solving partial differential equations*. CRC Press, Boca Raton, Fla.

Fok, Y. S. 1970. One-dimensional infiltration into layered soils, *Journal of the Irrigation and Drainage Division* 96(2): 121–129.

Frattini, P.; Crosta, G. B.; Fusi, N.; Negro, P. D. 2004. Shallow landslides in pyroclastic soil: a distributed modeling approach for hazard assessment, *Engineering Geology* 73: 277–295. https://doi.org/10.1016/j.enggeo.2004.01.009

Fredlund, D. G.; Morgenstern, N. R. 1977. Stress state variables for unsaturated soils, *Journal of the Geotechnical Engineering Division* 103: 447–466.

Fredlund, D. G.; Morgenstern, N. R.; Widger, R. A. 1978. The shear strength of unsaturated soils, *Canadian Geotechnical Journal* 15(3): 313–321. https://doi.org/10.1139/t78-029

Fredlund, D. G.; Xing, A.; Huang, S. 1994. Predicting the permeability function for unsaturated soils using the soil-water characteristic curve, *Canadian Geotechnical Journal* 31(4): 533–546. https://doi.org/10.1139/t94-062

Gardner, W. R. 1958. Some steady-state solutions of the unsaturated moisture flow equation with application to evaporation from a water table, *Soil science* 85(4): 228–232. https://doi.org/10.1097/00010694-195804000-00006

Gariano, S. L.; Guzzetti, F. 2016. Landslides in a changing climate, *Earth-Science Reviews* 162: 227–252. https://doi.org/10.1016/j.earscirev.2016.08.011

Green, W. H.; Ampt, G. A. 1911. Studies on soil physics I. The flow of air and water through soils, *Journal of Agricultural Science* 4: 1–24.

Hachum, A. Y.; Alfaro, J. F. 1980. Rain infiltration into layered soils: prediction, *Journal of the Irrigation and Drainage Division. American Society of Civil Engineers* 106: 311–319.

Hanks, R.; Bowers, S. 1962. Numerical solution of the moisture flow equation for infiltration into layered soils, *Soil Science Society of America Journal* 26(6): 530–534. https://doi.org/10.2136/sssaj1962.03615995002600060007x

Iverson, R. M. 2000. Landslide triggering by rain infiltration, *Water Resources Research* 36(7): 1897–1910. https://doi.org/10.1029/2000WR900090

Keim, R. F.; Skaugset, A. E. 2003. Modelling effects of forest canopies on slope stability, *Hydrological Processes* 17: 1457–1467. https://doi.org/10.1002/hyp.5121

Ku, C. Y.; Tsai, Y. H. 2013. Solving nonlinear problems with singular initial conditions using a perturbed scalar homotopy method. *International Journal of Nonlinear Sciences and Numerical Simulation* 14(6): 367–375. https://doi.org/10.1515/ijnsns-2013-0029

Lan, H. X.; Lee, C. F.; Zhou, C. H.; Martin, C. D. 2005. Dynamic characteristics analysis of shallow landslides in response to rainfall event using GIS, *Environmental Geology* 47: 254–267. https://doi.org/10.1007/s00254-004-1151-8

Liu, C. N.; Wu, C. C. 2008. Mapping susceptibility of rainfall-triggered shallow landslides using a probabilistic approach, *Environmental Geology* 55(4): 907–915. https://doi.org/10.1007/s00254-007-1042-x

Liu, C. Y.; Ku, C. Y.; Huang, C. C.; Lin, D. G.; Yeih, W. 2015. Numerical solutions of groundwater flow in unsaturated layered soil with extreme physical property contrasts, *International Journal of Nonlinear Sciences and Numerical Simulation* 16: 325–335. https://doi.org/10.1515/ijnsns-2015-0060

Lu, N.; Godt, J. 2008. Infinite slope stability under steady unsaturated seepage conditions, *Water Resources Research* 44(W11404). https://doi.org/10.1029/2008WR006976

Lu, N.; Likos, W. J. 2004. *Unsaturated soil mechanics*. John Wiley and Sons, Inc.

Moldrup, P.; Rolston, D. E.; Hansen, L. A. 1989. Rapid and numerically stable simulation of one dimensional, transient water flow in unsaturated, layered soils, *Soil Science* 1483(3): 219–226. https://doi.org/10.1097/00010694-198909000-00009

Montgomery, D. R.; Dietrich, W. E. 1994. A physically based model for the topographic control on shallow landslide, *Water Resources Research* 30(4): 1153–1171. https://doi.org/10.1029/93WR02979

Pereira, P.; Oliva, M.; Baltrenaite, E. 2010. Modelling extreme precipitation in hazardous mountainous areas, *Journal of Environmental Engineering and Landscape Management* 18(4): 329-342. https://doi.org/10.3846/jeelm.2010.38

Romano, N.; Brunoneb, B.; Santini, A. 1988. Numerical analysis of one-dimensional unsaturated flow in layered soils, *Advances in Water Resources* 21(4): 315–324. https://doi.org/10.1016/S0309-1708(96)00059-0

Samani, Z.; Cheraghi, A.; Willardson, L. 1989. Water movement in horizontally layered soils, *Journal of irrigation and drainage engineering* 115(3): 449–456. https://doi.org/10.1061/(ASCE)0733-9437(1989)115:3(449)

Sarkar, K.; Singh, T. N.; Verma, A. K. 2012. A numerical simulation of landslide-prone slope in Himalayan region – a case study, *Arabian Journal of Geosciences* 5(1): 73–81. https://doi.org/10.1007/s12517-010-0148-8

Schilirò, L.; Esposito, C.; Mugnozza, G. S. 2015. Evaluation of shallow landslide-triggering scenarios through a physically based approach: an example of application in the southern Messina area (northeastern Sicily, Italy), *Natural Hazards and Earth System Science* 15(9): 2091–2109. https://doi.org/10.5194/nhess-15-2091-2015

Singh, T. N.; Verma, A. K.; Sarkar, K. 2010. Static and dynamic analysis of a landslide, *Geomatics, Natural Hazards and Risk* 1(4): 323–328. https://doi.org/10.1080/19475705.2010.521354

Skeel, R. D.; Berzins, M. 1990. A method for the spatial discretization of parabolic equations in one space variable, *SIAM Journal on Scientific and Statistical Computing* 11(1): 1–32. https://doi.org/10.1137/0911001

Srivastava, R.; Yeh, T. C. 1991. Analytical solutions for one-dimensional, transient infiltration toward the water table in homogeneous and layered soils, *Water Resources Research* 27(5): 753–762. https://doi.org/10.1029/90WR02772

Tóthová, I.; Igaz, D.; Antal, J. 2007. Soil water movement modelling in hapllicc luvisols (HMm) and albi hapllic luvisols (HMl) under Slovak climatic conditions, *Journal of Environmental Engineering and Landscape Management* 15(2): 69–75.

Tracy, F. T. 2006. Clean two- and three-dimensional analytical solutions of Richards' equation for testing numerical solvers, *Water Resources Research* 42: W08503. https://doi.org/10.1029/2005WR004638

Tracy, F. T. 2011. *Analytical and numerical solutions of Richards' equation with discussions on relative hydraulic conductivity.* INTECH Open Access Publisher, Janeza Trdine 9, 51000 Rijeka, Croatia.

Tsai, T. L.; Chen, H. F. 2010. Effects of degree of saturation on shallow landslides triggered by rainfall, *Environmental Earth Sciences* 59(6): 1285–1295. https://doi.org/10.1007/s12665-009-0116-3

Tsai, T. L.; Yang, J. C. 2006. Modeling of rainfall-triggered shallow landslide, *Environmental Geology* 50(4): 525–534. https://doi.org/10.1007/s00254-006-0229-x

Van Genuchten, M. T. 1980. A closed form equation for predicting the hydraulic conductivity of unsaturated soils, *Soil Science Society of American Journal* 44(5): 892–898. https://doi.org/10.2136/sssaj1980.03615995004400050002x

Vanapalli, S. K.; Fredlund, D. G.; Pufahl, D. E.; Clifton, A. W. 1996. Model for the prediction of shear strength with respect to soil suction, *Canadian Geotechnical Journal* 33(3): 379–392. https://doi.org/10.1139/t96-060

Wang, C.; Esaki, T.; Xie, M.; Qiu, C. 2006. Landslide and debris-flow hazard analysis and prediction using GIS in Minamata–Hougawachi area, Japan, *Environmental Geology* 51: 91–102. https://doi.org/10.1007/s00254-006-0307-0

Whisler, F.; Klute, A. 1966. Analysis of infiltration into stratified soil columns, in *Proc Wageningen Symp, Proc. IAHS Symposium on Water in the Unsaturated Flow.* Wageningen, The Netherlands, 451–470.

Yeh, H. F.; Lee, C. H. 2013. Soil water balance model for precipitation-induced shallow landslides, *Environmental Earth Sciences* 70(6): 2691–2701. https://doi.org/10.1007/s12665-013-2326-y

Yoshimatsu, H.; Abe, S. 2006. A review of landslide hazards in Japan and assessment of their susceptibility using an analytical hierarchic process (AHP) method, *Landslides* 3: 149–158. https://doi.org/10.1007/s10346-005-0031-y

Chih-Yu LIU. Doctoral Candidate, Department of Harbor and River Engineering, National Taiwan Ocean University, Taiwan, 2 peer-reviewed articles in scientific journals, 14 conference papers.

Cheng-Yu KU. PhD, Professor, Department of Harbor and River Engineering, National Taiwan Ocean University, Taiwan, 58 peer-reviewed articles in scientific journals, 4 book chapters, 4254 citations, H-index 18 (Google Scholar).

Jing-En XIAO. Doctoral Candidate, Department of Harbor and River Engineering, National Taiwan Ocean University, Taiwan, 1 peer-reviewed articles in scientific journals, 13 conference papers.

Chi-Chao HUANG. Doctoral Candidate, Department of Resources Engineering, National Cheng Kung University, Taiwan, Chief Secretary, Central Geological Survey, Ministry of Economic Affairs, Taiwan.

Shih-Meng HSU. PhD, Assistant Professor, Department of Harbor and River Engineering, National Taiwan Ocean University, Taiwan, 32 peer-reviewed articles in scientific journals, 3 book chapters, 254 citations, H-index 8 (Google Scholar).

5

EXPERIMENTAL INSIGHTS OF USING WASTE MARBLE FINES TO MODIFY THE GEOTECHNICAL PROPERTIES OF A LATERITIC SOIL

Isaac I. AKINWUMI[a], Colin A. BOOTH[b]

[a]*Department of Civil Engineering, College of Engineering, Covenant University, Ota, Nigeria*
[b]*Construction and Property Research Centre, Faculty of Environment and Technology, University of the West of England, BS16 1QY Bristol, UK*

Abstract. Marble spoil waste is an environmental nuisance. The effects of adding waste marble fines (WMF) on the plasticity, strength and permeability of a lateritic clay have been investigated for its potential use as a soil modifier or stabilizer of road pavement layer materials or earth-building materials. The chemical compositions of the WMF and soil were determined using X-ray fluorescence and atomic absorption spectrometry, respectively. The specific gravity, Atterberg limits, compaction, strength and permeability characteristics of the soil were determined for varying proportions of the soil-WMF blends. The properties of the natural soil – classified as clay of low plasticity (CL) and A-7-6(7), according to unified soil classification and AASHTO classification systems, respectively – were improved after the addition of 10% WMF such that it behaves like a silt of low plasticity. Therefore, WMF is recommended as a low-cost soil modifier or stabilizer for lateritic soil and well-suited for road construction applications.

Keywords: marble dust, waste reuse, soil modification, soil plasticity, soil strength, soil permeability, soil stabilization.

Introduction

Marble is a versatile material, principally used for construction, interior decoration and in sculptures. Its use in construction of monumental buildings, such as the Taj Mahal in India, attests to its durability, ease of maintenance, lustrous colour, glossy appearance and its ability to take a very high polish (Liguori *et al.* 2008; El-Gammal *et al.* 2011).

While extracting or quarrying, cutting and polishing to obtain finished marble products, a significant amount of marble waste is usually generated. During the quarrying process, the extracted stones are cut into large blocks, which are typically 3 m long, 1.5 m wide and 1.5 m deep. The blocks are then transported to processing plants and cut into slabs (2–3 cm thick) and tiles. The quarrying, cutting and polishing to obtain finished marble products generate about 40% of the volume of the extracted marble blocks as waste (Saboya Jr. *et al.* 2007). The cutting process alone generates marble dust of about 20–30% of the weight of the marble blocks (Agarwal, Gulati 2006; Alyamac, Ince 2009).

Marble waste is a source of concern to marble manufacturers and environmentalists (Segadaes *et al.* 2005; Acchar *et al.* 2006a, 2006b). The finest wastes are readily transported by wind, which can cause air and visual pollution in marble quarries and processing plants. Those wastes dumped on land usually alter the landscape, where the wastes are placed as overburden, and can greatly reduce surface water percolation into the ground. Moreover, they can increase the alkalinity of a soil and reduce its fertility, making the soil unsuitable for cultivation (Montero *et al.* 2009).

Given the nuisance of the spoil, attempts have been made to incorporate marble waste into many construction materials, which include modifying cement mortar properties (Hwang *et al.* 2008; Corinaldesi *et al.* 2010), application as a recycled aggregate in concrete production (Binici *et al.* 2008; Gencel *et al.* 2012; Topcu *et al.* 2009; Ergun 2011; Hebhoub *et al.* 2011; Belaidi *et al.* 2012; Bacarji *et al.* 2013; Andre *et al.* 2014), as an aggregate in asphaltic concrete (Akbulut, Cahit 2007; Karasahin, Terzi 2007), as a raw material in brick production (Bilgin *et al.* 2012; Eliche-Quesada *et al.* 2012) and as raw material in cement production (Aruntas *et al.* 2010; Aliabdo *et al.* 2014).

Few works have investigated the effective use of marble wastes for geotechnical engineering purposes such as

Corresponding author: Isaac I. Akinwumi
E-mail: isaac.akinwumi@covenantuniversity.edu.ng

for improving the workability, modifying the drainage characteristics, strengthening and improving the deformation characteristics of soils. Marble wastes also have the potential of being used to remediate contaminated soils. The workability of a soil is improved by reducing its plasticity, therefore, allowing construction engineers to easily manipulate the physical properties of the soil. The drainage characteristics of a soil can be obtained from tests that determine the hydraulic conductivity or permeability of the soil. For soils used as the core of earth-fill dams and base (pavement) layer materials, it is necessary that they exhibit low permeability. California bearing ratio (CBR) and unconfined compressive strength (UCS) of a soil are measures of its strength and deformation characteristics. The higher the CBR and UCS of a soil the greater its capacity to bear structural load and the lesser its deformation. Okagbue, Onyeobi (1999) and Baser (2009) evaluated the possibility of using marble dust to stabilize a red tropical soil and an expansive soil, respectively. While Okagbue, Onyeobi (1999) focused on assessing the plasticity, compressive strength and California bearing ratio (CBR) of red tropical soil and its mixture with varying proportions of marble dust, Baser (2009) focused on evaluating the plasticity and swell potential of the expansive soil and its mixture with varying percentages of marble dust.

The purpose of this study is to investigate the effects of adding waste marble fines (WMF) on the plasticity, strength and permeability of the most widely available soil type (lateritic soil) in tropical and sub-tropical countries. The geotechnical properties of some of the lateritic soils found in these countries do not satisfy standard requirements for use as road construction materials. Portland cement and lime are well-known and effective stabilizers or modifiers for the improvement of the geotechnical properties of lateritic soils, but these stabilizing agents are expensive. In recent years, research works on the use of low-cost alternative materials and particularly waste materials in the construction industry are becoming attractive and have been encouraged (Marzouk *et al.* 2013; Malaiskiene *et al.* 2014; Marciukaitis, Jurkenas 2013). The use of WMF to improve the geotechnical properties of lateritic soils seem, therefore, appealing because of their low cost and potential to rid the environment of the nuisance and human health impact associated with their indiscriminate disposal.

1. Methodology

1.1. Materials

The WMF used in this experimental study were sourced from a marble manufacturing company in Lagos, Southwestern Nigeria. The particles of the WMF easily passed through a 0.425 mm sieve, while 87% of it passed through a 0.075 mm sieve. This is similar to the gradation of marble dust used by Karasahin, Terzi (2007).

A clay soil obtained from a borrow pit at Agbara, Ogun State, Southwestern Nigeria, with a deep lateritic soil profile, was used in this study. The soil was collected from the sidewall of the soil profile at a depth of about 10 m from the top of the pit. This soil profile had been exposed by borrowing activities of several months. Some of the soil samples were collected and stored in a water-tight container for laboratory determination of their natural moisture contents before they were transported to the soil laboratory. Prior to conducting tests on the soil samples for other laboratory determinations, they were air-dried, passed through a 4.75 mm sieve (AASHTO T 99-10 2010) and thoroughly mixed.

1.2. Experimental Work

Geochemical characterization of the soil and WMF were determined using atomic absorption spectrophotometer (Pelkin Elmer 3300) and X-ray fluorescence (Philips PW 1400), respectively. Representative samples of the air-dried soil were obtained by the process of quartering into parts of equal mass. WMF were added to each of the four parts in 0%, 2%, 6% and 10% proportions by dry weight of the soil sample, respectively.

Gradation tests were carried out on the soil sample. Specific gravity, Atterberg limits, compaction, unsoaked and soaked CBR, unconfined compression and permeability tests were conducted on the natural (0% WMF content) and treated (2%, 6% and 10% WMF contents) soil samples, respectively. Each of these tests was carried out in triplicate (Akinwumi 2014c) and their mean values determined. The tests were conducted according to the procedures outlined in BSI (1990a; 1990b).

Sieve analysis was carried out on samples of soil retained on a sieve with 0.075 mm opening while hydrometer analysis, using sodium hexametaphosphate, was conducted on the soil sample passing the sieve. Specific gravity of the soil and WMF were determined using a pycnometer. Liquid and plastic limits were determined using the Casagrande apparatus and glass plate.

The British Standard heavy compaction energy was used for preparing the specimens for compaction and CBR tests. Unsoaked CBR test specimens were cured for 7 days under controlled temperature (25±2 °C) and relative humidity (100%). For the determination of soaked CBR, specimens were cured for 6 days under the controlled temperature and the relative humidity before being immersed in water for 24 hours. Unconfined compressive strength (UCS) test specimens (50 mm × 100 mm), prepared and extruded from a cylindrical mould, were cured in sealed plastic bags. The UCS for each test batch was determined after 28 days of curing. Falling head permeameter was used to determine the coefficient of permeability of the samples.

Gradation tests, comprising of sieve analysis and hydrometer analysis, were carried out on the soil sample to determine the particle size distribution of the soil. Soil moisture content determinations were carried out using the oven drying method. Determinations of the specific gravity of the soil samples were necessary because it gives a measure of the density of the soil solids and can also be helpful in calculating their phase relationships. Liquid and plastic limits tests were the Atterberg limits tests conducted. They were performed to characterize the state or condition of the soil sample based on its water content. The CBR values of the natural and treated soil samples were used to evaluate their potential strength, when used as road pavement layer materials. Unconfined compression test on the natural and treated soil samples were carried out to determine their unconsolidated undrained strength. Falling head permeability test was conducted on the natural and treated soil samples to determine their coefficients of permeability.

2. Results

2.1. Chemical Composition of Soil and WMF

The chemical constituents of the WMF and soil sample are presented in Figure 1. The WMF consists predominantly of lime (CaO) while the soil sample is rich in silica (SiO_2). The ratio of silica-sesquioxides for the soil was determined to be 1.66, indicating that the soil is lateritic. A ternary of the composition of silica and sesquioxides of the soil sample with Schellmann (1986) scheme of classification of weathering products is shown in Figure 2.

The plot for this soil sample fell within the area of the chart classified as kaolinization, indicating that the soil sample was taken from a kaolinized profile.

2.2. Natural soil

The physical and mechanical properties of the natural soil are summarily presented in Table 1. The soil sample, classified as A-7-6 (7) in accordance with American Association of State Highway and Transportation Officials

Table 1. Physical and mechanical soil property characteristics

	Properties	Mean Value / Description
Gradation / Classification	Gravel (>4.75 mm), %	0.5 (±0.05)
	Sand (0.075 – 4.75 mm), %	44.0 (±1.70)
	Silt and Clay (<0.075 mm), %	55.5 (±1.20)
	AASHTO Soil Classification System	A-7-6 (7)
	Unified Soil Classification System	CL – Sandy clay
Physical	Colour	Brown
	Natural Moisture Content (%)	15.6 (±1.28)
	Specific Gravity	2.51 (±0.03)
	Liquid Limit (%)	41.0 (±2.59)
	Plastic Limit (%)	23.0 (±1.06)
	Plasticity Index (%)	18.0 (±0.35)
	Maximum Dry Unit weight (kN/m^3)	18.2 (±0.10)
	Optimum Moisture Content (%)	15.3 (±0.36)
	Coefficient of Permeability (cm/s)	8.24×10^{-6} ($\pm 0.31 \times 10^{-6}$)
Strength	Unsoaked CBR (%)	18 (±1.15)
	Soaked CBR (%)	10 (±0.35)
	Unconfined Compressive Strength (kN/m^2)	560 (±22.3)

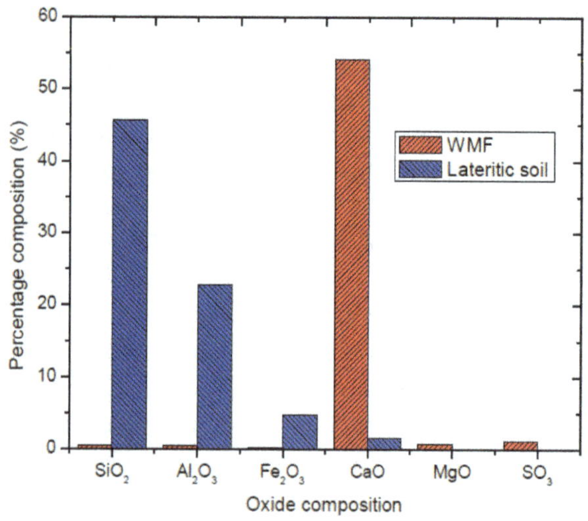

Fig. 1. Chemical compositions of the WMF and soil sample

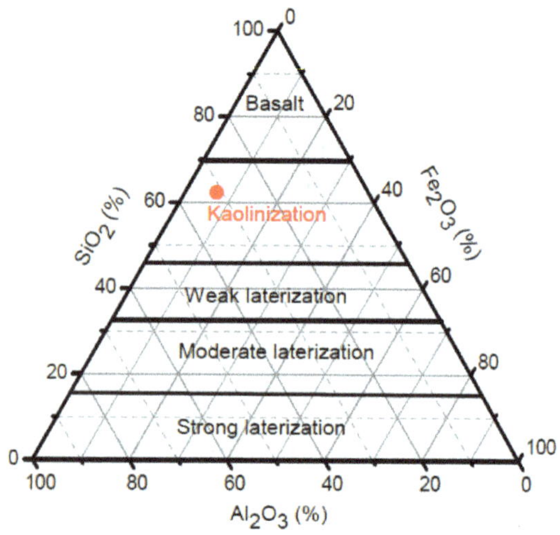

Fig. 2. Al_2O_3-SiO_2-Fe_2O_3 ternary plot for the soil sample

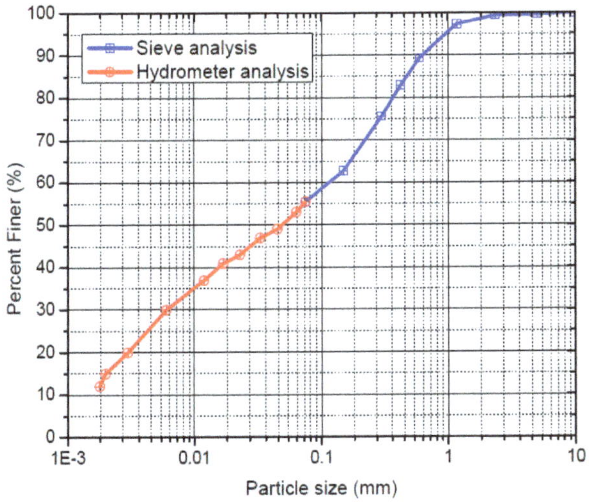

Fig. 3. Particle size distribution of tested soils

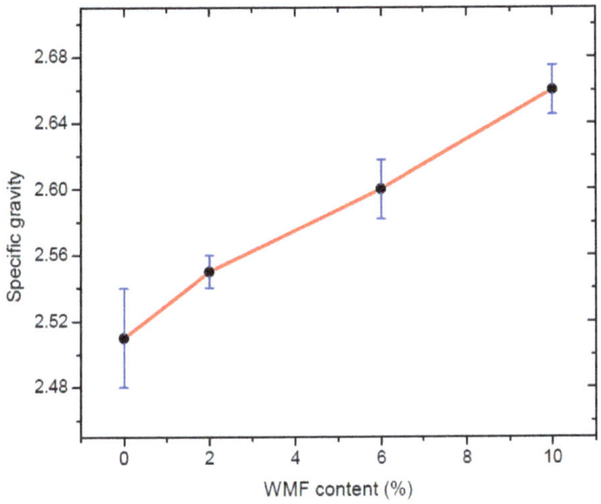

Fig. 4. Variation of specific gravity of the treated soil samples with their WMF contents

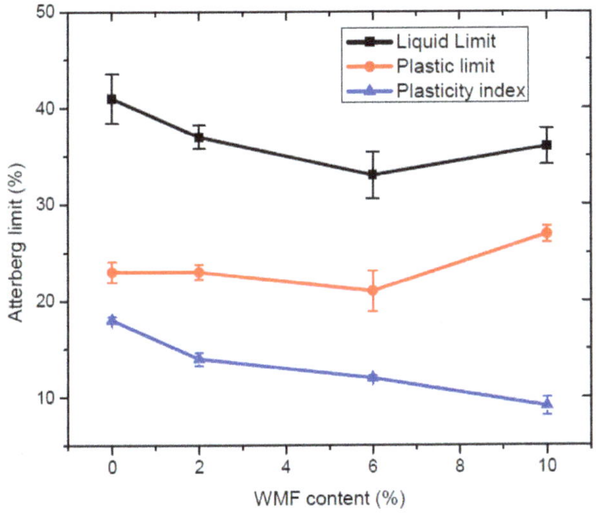

Fig. 5. Variation of Atterberg limits of the treated soil samples with their WMF contents

(AASHTO) system of soil classification and sandy lean clay (CL) in accordance with the Unified soil classification (USC) system, has an average natural moisture content of 15.6% and a specific gravity of 2.51. Its unconfined compressive strength, unsoaked and soaked CBR values are generally low. The particle size distribution of the soil is shown in Figure 3. The shape of the particle size distribution curve indicates that the soil is well-graded. About fifty six percent (55.5%) of the soil's particle sizes passed the 0.075 mm sieve, indicating that this portion significantly influenced the plasticity, strength and permeability characteristics of the soil.

2.3. Effects of Adding WMF to the Soil

2.3.1. Specific gravity

The specific gravity of the WMF is 2.8. The result of the specific gravity tests on the treated soil samples is graphically presented in Figure 4. The specific gravity of the treated soil samples increased as their WMF content increased. After adding 10% WMF, the specific gravity of the treated soil (when compared with that of the natural soil) increased by 4.4%. There is a strong correlation ($r = 0.997$) between the specific gravities of the treated soil samples and their WMF contents.

2.3.2. Atterberg limits

The result of the Atterberg limits tests on the treated soil samples is graphically presented in Figure 5. There is a slight negative correlation between the liquid limits and plasticity indices of the treated soil samples and their WMF contents. There is, however, no correlation between the plastic limits of the treated soil samples and their WMF contents. This implies that while the water contents at which the natural soil and the treated soil samples changed from being plastic to being semi-solid remains fairly constant, the range of water contents at which the treated soil samples behaved like a plastic decreased with increasing WMF content.

This may be attributed to cation exchange and dissociation. The cation exchange occurred between the lime-rich WMF and the water in the soil and this led to the coating of some of the clay particles of the soil by the clay-size particles of the WMF. This coating minimized the interaction between the water and the clay particles of the soil, thereby resulting in the reduction of the soil's moisture-holding capacity and making it more workable. This is a consequence of the decrease in liquid limits and plasticity indices of the treated soil samples as their WMF contents increased.

$$CaO + H_2O \rightarrow Ca(OH)_2 \text{ (Cation exchange);}$$

$$Ca(OH)_2 \rightarrow Ca^{2+} + 2OH^- \text{ (Dissociation).}$$

Addition of 10% WMF to the soil reduced its plasticity index by 49.4%. The decrease in the plasticity indices of the treated soil samples as their WMF contents increased is strongly correlated, r = −0.964.

2.3.3. Compaction characteristics

Optimum moisture content (OMC) and maximum dry unit weight of the natural and treated soil samples were obtained from compaction tests. Figure 6 shows a graphical illustration of the maximum dry unit weight and OMC of the natural and treated soil samples. The maximum dry unit weight of the treated soil samples increased as their WMF contents increased. This increase is strongly correlated, r = 0.961. The increase in maximum dry unit weight of the treated soil samples as their WMF content increased can be attributed to the higher specific gravity or density of the WMF (2.8), when compared with that of the soil (2.51).

The OMC of the treated soil samples decreased as their WMF content increased. This decrease is strongly correlated, r = −0.979.

Figure 7 shows the plasticity charts for AASHTO and USC systems with the plasticity plots for the natural and treated soil samples. It shows that the lateritic clay soil (classified as A-7-6 and CL) became silty (A-4 and ML) after 10% WMF was mixed with the soil.

The decrease in the OMC can, thus, be attributed to the agglomeration of the clay particles of the soil and clay-size particles of the WMF. This makes them behave like silt-size particles (Fig. 7). The coarser a soil becomes the lesser water it requires to reach optimum (Akinwumi *et al.* 2012; Akinwumi 2014b).

2.3.4. Strength characteristics

CBR tests were conducted for both unsoaked and soaked treated soil samples. The variation of unsoaked and soaked CBR values of the treated soil samples with their WMF content is presented in Figure 8. The unsoaked CBR of the treated soil samples increased as their WMF contents increased. After adding 10% WMF, the unsoaked CBR of the treated soil sample, when compared with that of the natural soil, increased by 205.6%. The increase in the unsoaked CBR values of the treated soil samples, as their WMF contents increased, is strongly correlated (r = 0.854). The soaked CBR value of the treated soil samples also increased as their WMF contents increased. Addition of 10% WMF increased the soaked CBR value of the treated soil sample by 360%. The increase in the soaked CBR values of the treated soil samples, as their WMF contents increased, is strongly correlated (r = 0.901).

Figure 8 also shows the variation of unconfined compressive strength (UCS) of the treated soil samples

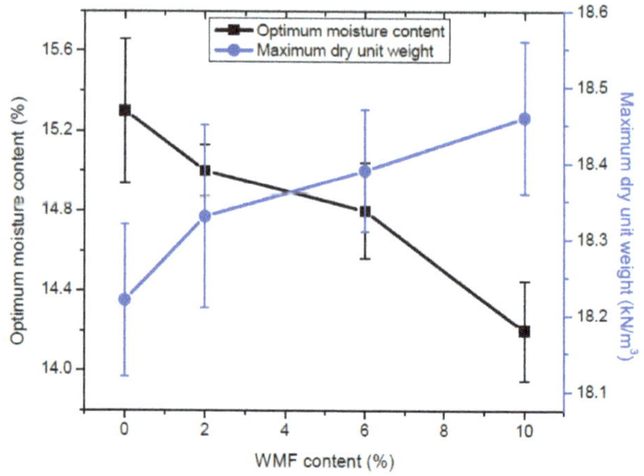

Fig. 6. Variation of compaction characteristics of the treated soil samples with their WMF contents

(a)

(b)

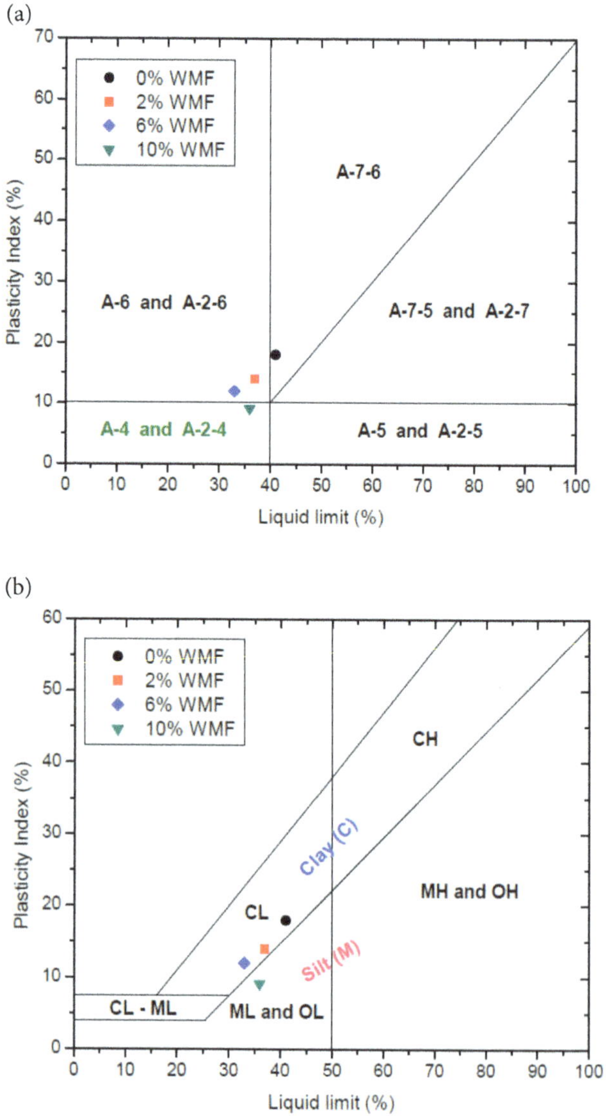

Fig. 7. Plasticity charts (a) AASHTO (b) USC showing the plots for the mean values for the natural and treated soil samples

with their WMF contents. The UCS of the treated soil samples increased as their WMF contents increased. This increase is strongly correlated, r = 0.976. After adding 10% WMF, the UCS of the treated soil sample increased by 311.3%.

The reactions between the lime-rich WMF and the silica- and alumina-rich soil sample, in the presence of water, forms calcium-silicate-hydrates and calcium-aluminate-hydrates, respectively. These cementitious products are thought to be responsible for the increase in the unsoaked and soaked CBR values and unconfined compressive strength of the treated soil samples, as their WMF contents increased.

$$Ca^{2+} + 2OH^- + SiO_2 \rightarrow CaO.SiO_2.H_2O;$$
(Calcium-silicate-hydrate)

$$Ca^{2+} + 2OH^- + Al_2O_3 \rightarrow CaO.Al_2O_3.H_2O.$$
(Calcium-aluminate-hydrate)

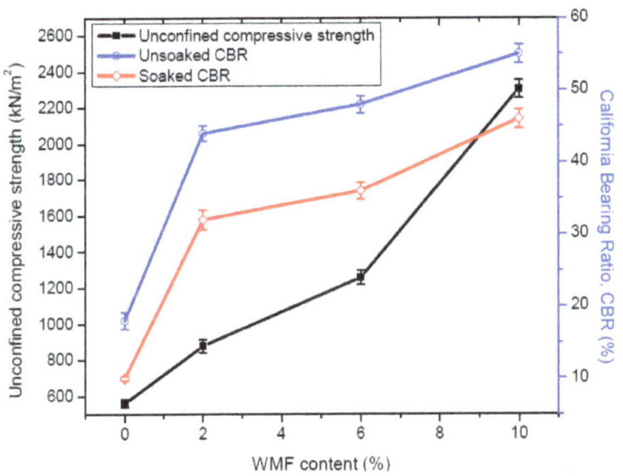

Fig. 8. Variation of strength characteristics of the treated soil samples with their WMF contents

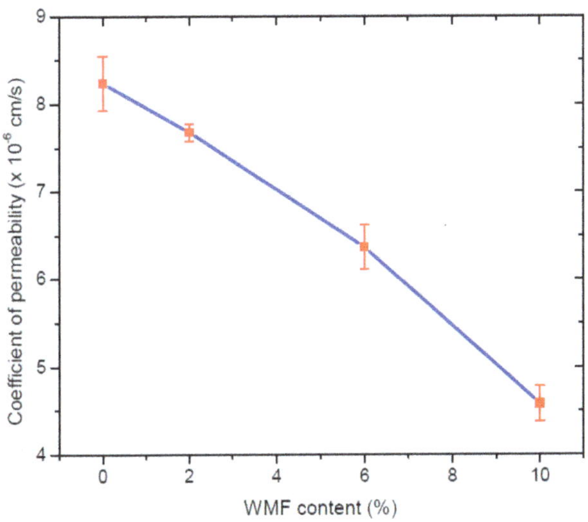

Fig. 9. Variation of permeability characteristics of the treated soil samples with their WMF contents

2.3.5. Permeability

The result of permeability tests on the treated soil samples is graphically presented in Figure 9. The coefficient of permeability of the treated soil samples decreased as their WMF contents increased. After adding 10% WMF, the coefficient of permeability of the treated soil decreased by 44.4%. The decrease in the coefficient of permeability of the treated soil samples, as their WMF contents increased, is strongly correlated (r = −0.995).

The WMF occupied part of the pore space within the soil, thereby reducing the volume of voids within it. This reduction in the void space is responsible for the reduction in the ease with which water flows through the soil.

3. Discussion

The increase in specific gravities of the treated lateritic clay samples as their WMF contents increased did not deviate from expectation reasoned by deduction from the knowledge that the specific gravity of the WMF is higher than that of the soil sample. Similar results were obtained for expansive soils by Sivrikaya *et al.* (2014) and Baser (2009).

The liquid limits and plasticity indices of the treated soil samples decreased with their WMF contents, in alignment with the results of Sivrikaya *et al.* (2014) and Baser (2009) on expansive soils. Although Sivrikaya *et al.* (2014) and Baser (2009) reported an increase in the plastic limits of the expansive soil with increasing marble dust, the plastic limits of the treated lateritic clay samples were not significantly affected by increasing their WMF contents.

The maximum dry unit weights and OMCs of the treated soil samples increased and decreased, respectively, with an increase in their WMF contents. Although the soil considered by Sivrikaya *et al.* (2014) is not lateritic, the effect of adding marble dust to the expansive soil on its compaction characteristics is similar to that obtained for this study.

The unsoaked and soaked CBR values and UCS of the treated lateritic clay samples increased with increasing percentages of WMF. Okagbue, Onyeobi (1999) also reported an increase in the unsoaked CBR of red tropical soils with increasing marble dust content. The effects of treating lateritic clay with WMF on its soaked CBR and UCS have not been reported in literature.

The coefficients of permeability decreased as the percentage of WMF in the treated lateritic clay sample increased. The effect of treating lateritic clay with WMF on its permeability has not been reported in literature.

Conclusions

Addition of WMF to the soil notably altered its geotechnical properties, causing: (i) a reduction of its plasticity, thereby making the soil more workable; (ii) an improvement in its strength; and (iii) a reduction in the soil permeability.

These findings are important, especially when the treated soil is used as a road pavement layer material or when it directly receives the structural foundation of buildings, as the amount of water that can pass through it becomes reduced–thereby reducing the possibility of water-induced failures and extending the life of the pavement or building structures founded on such soil layers. Flow of water through earth materials used as pavement layer materials weakens the pavement structure and this accounts for the failure or reduction in the lifespan of many flexible pavements.

Evaluating the properties of the soil, relative to a benchmark standard such as TRL (1993), shows that the natural soil–which barely met the requirements for use as a subgrade material–became improved after the addition of 10% WMF, such that it satisfies the requirement for use as a sub-base material for road pavement construction.

The potential implications of these findings indicate that the environmental nuisance and human health impact associated with the indiscriminate disposal of WMF can be minimized by using it to improve the plasticity, strength and permeability of an in situ lateritic soil, which is otherwise considered unsuitable in its natural state for use as an earth foundation material to bear the structural foundation of proposed civil engineering infrastructures, such as buildings and roads. Furthermore, it may be useful for improving the engineering properties of traditional earth-building construction materials such as those described by Akinwumi (2014a).

Finally, WMF is recommended as a low-cost additive for improving the geotechnical properties of lateritic clay soils. There is need for further research works, especially, on the performance of constructed roads and buildings using WMF. The possibility of using WMF as total or partial replacement for sand in Portland cement mortar and concrete may also be investigated.

Acknowledgements

The authors would like to acknowledge the contribution of Mr. Bobby Ekanem towards the samples collection and for conducting some of the laboratory tests. Professor Solomon A. Isiorho assisted in proofreading the earlier draft of this paper. The authors hereby appreciate the managements of Julius Berger Nigeria Plc., Euro 65 laboratory, Lagos and Covenant University, whose laboratory facilities were used for this research work.

References

AASHTO T 99-10 2010. *Standard method of test for moisture-density relations of soils using a 2.5-kg (5.5-lb) rammer and a 305-mm (12-in.) drop*. American Association of State Highway and Transportation Officials, Washington, DC.

Acchar, W.; Vieira, F. A.; Hotza, D. 2006a. Effect of marble and granite sludge in clay materials, *Materials Science and Engineering A* 419: 306–309. http://dx.doi.org/10.1016/j.msea.2006.01.021

Acchar, W.; Vieira, F. A.; Segadaes, A. M. 2006b. Using ornamental stone cutting rejects as raw materials for red clay ceramic products: Properties and microstructure development, *Materials Science and Engineering A* 435–436: 606–610. http://dx.doi.org/10.1016/j.msea.2006.07.091

Agarwal, S. K.; Gulati, D. 2006. Utilization of industrial wastes and unprocessed micro-fillers for making cost effective mortars, *Construction and Building Materials* 20: 999–1004. http://dx.doi.org/10.1016/j.conbuildmat.2005.06.009

Akbulut, H.; Cahit, G. 2007. Use of aggregates produced from marble quarry waste in asphalt pavements, *Building and Environment* 42: 1921–1930. http://dx.doi.org/10.1016/j.buildenv.2006.03.012

Akinwumi, I. I.; Adeyeri, J. B.; Ejohwomu, O. A. 2012. Effects of steel slag addition on the plasticity, strength and permeability of a lateritic soil, in *2nd Internation Conference of Sustainable Design, Engineering and Construction proceedings*, 7–9 November 2012, Fort Worth, Texas, United States. ASCE, 457–464. http://dx.doi.org/10.1061/9780784412688.055

Akinwumi, I. I. 2014a. Earth building construction processes in Benin City, Nigeria and engineering classification of earth materials used, *Indian Journal of Traditional Knowledge* 13(4): 686–690.

Akinwumi, I. I. 2014b. Plasticity, strength and permeability of reclaimed asphalt pavement and lateritic soil blends, *International Journal of Scientific and Engineering Research* 5(6): 631–636.

Akinwumi, I. 2014c. Soil modification by the application of steel slag, *Periodica Polytechnica Civil Engineering* 58(4): 371–377. http://dx.doi.org/10.3311/PPci.7239

Aliabdo, A. A.; Elmoaty, A. E. M. A.; Auda, E. M. 2014. Re-use of waste marble dust in the production of cement and concrete, *Construction and Building Materials* 50: 28–41. http://dx.doi.org/10.1016/j.conbuildmat.2013.09.005

Alyamac, K. E.; Ince, R. 2009. A preliminary concrete mix design for SCC with marble powders, *Construction and Building Materials* 23: 1201–1210. http://dx.doi.org/10.1016/j.conbuildmat.2008.08.012

Andre, A.; de Brito, J.; Rosa, A.; Pedro, D. 2014. Durability performance of concrete incorporating coarse aggregates from marble industry waste, *Journal of Cleaner Production* 65: 389–396. http://dx.doi.org/10.1016/j.jclepro.2013.09.037

Aruntas, H. Y.; Guru, M.; Dayi, M.; Tekin, I. 2010. Utilization of waste marble dust as an additive in cement production, *Materials and Design* 31: 4039–4042. http://dx.doi.org/10.1016/j.matdes.2010.03.036

Bacarji, E.; Filho, R. D. T.; Koenders, E. A. B.; Figueiredo, E. P.; Lopes, J. L. M. P. 2013. Sustainability perspective of marble and granite residues as concrete fillers, *Construction and Building Materials* 45: 1–10. http://dx.doi.org/10.1016/j.conbuildmat.2013.03.032

Baser, O. 2009. *Stabilization of expansive soils using waste marble dust*: MSc Dissertation. Middle East Technical University, Ankara.

Belaidi, A. S. E.; Azzouz, L.; Kadri, E.; Kenai, S. 2012. Effect of natural pozzolana and marble powder on the properties of self-compacting concrete, *Construction and Building Materials* 31: 251–257. http://dx.doi.org/10.1016/j.conbuildmat.2011.12.109

Bilgin, N.; Yeprem, H. A.; Arslan, S.; Bilgin, A.; Gunay, E.; Marsoglu, M. 2012. Use of waste marble powder in brick industry, *Construction and Building Materials* 29: 449–457. http://dx.doi.org/10.1016/j.conbuildmat.2011.10.011

Binici, H.; Shah, T.; Aksogan, O.; Kaplan, H. 2008. Durability of concrete made with granite and marble as recycle aggregates, *Journal of Materials Processing Technology* 208: 299–308. http://dx.doi.org/10.1016/j.jmatprotec.2007.12.120

BSI 1990a. *Methods of test for soils for civil engineering purposes*, British Standards Institution, BS 1377, London.

BSI 1990b. *Stabilized materials for civil engineering purposes. General requirements, sampling, sample preparation and tests on materials before stabilization*, British Standards Institution, BS 1924-1, London.

Corinaldesi, V.; Moriconi, G.; Naik, T. R. 2010. Characterization of marble powder for its use in mortar and concrete, *Construction and Building Materials* 24: 113–117. http://dx.doi.org/10.1016/j.conbuildmat.2009.08.013

El-Gammal, M. I.; Ibrahim, M. S.; Badr, E. A.; Asker, S. A.; El-Galad, N. M. 2011. Health risk assessment of marble dust at marble workshops, *Nature and Science* 9(11): 144–154.

Eliche-Quesada, D.; Corpas-Iglesias, F. A.; Perez-Villarejo, L.; Iglesias-Godino, F. J. 2012. Recycling of sawdust, spent earth from oil filtration, compost and marble residues for brick manufacturing, *Construction and Building Materials* 34: 275–284. http://dx.doi.org/10.1016/j.conbuildmat.2012.02.079

Ergun, A. 2011. Effects of the usage of diatomite and waste marble powder as partial replacement of cement on the mechanical properties of concrete, *Construction and Building Materials* 25: 806–812. http://dx.doi.org/10.1016/j.conbuildmat.2010.07.002

Gencel, O.; Ozel, C.; Koksal, F.; Erdogmus, E.; Martinez-Barrera, G.; Browstow, W. 2012. Properties of concrete paving blocks made with waste marble, *Journal of Cleaner Production* 21: 62–70. http://dx.doi.org/10.1016/j.jclepro.2011.08.023

Hebhoub, H.; Aoun, H.; Belachia, M.; Houari, H.; Ghorbel, E. 2011. Use of waste marble aggregates in concrete, *Construction and Building Materials* 25: 1167–1171. http://dx.doi.org/10.1016/j.conbuildmat.2010.09.037

Hwang, E.; Ko, Y. S.; Jeon, J. 2008. Effect of polymer cement modifiers on mechanical and physical properties of polymer-modified mortar using recycled artificial marble waste fine aggregate, *Journal of Industrial and Engineering Chemistry* 14: 265–271. http://dx.doi.org/10.1016/j.jiec.2008.02.009

Karasahin, M.; Terzi, S. 2007. Evaluation of marble waste dust in the mixture of asphaltic concrete, *Construction and Building Materials* 21: 616–620. http://dx.doi.org/10.1016/j.conbuildmat.2005.12.001

Liguori, V.; Rizzo, G.; Traverso, M. 2008. Marble quarrying: an energy and waste intensive activity in the production of building materials, *Environmental Economics and Investment Assessment II, WIT Transactions on Ecology and the Environment* 108: 197–207. http://dx.doi.org/10.2495/EEIA080201

Malaiskiene, J.; Maciulaitis, R. J.; Mikalauskaite, R. 2014. Possibilities to recycle auto glass waste in building ceramics, *Journal of Environmental Engineering and Landscape Management* 22(1): 21–29. http://dx.doi.org/10.3846/16486897.2013.867863

Marciukaitis, G.; Jurkenas, G. 2013. Possibilities of using sulphur and plant origin waste for light-weight and thermal insulation construction composites, *Journal of Environmental Engineering and Landscape Management* 21(2): 106–113. http://dx.doi.org/10.3846/16486897.2013.767805

Marzouk, M.; Abdelhamid, M.; Elsheikh, M. 2013. Selecting sustainable building materials using system dynamics and ant colony optimization, *Journal of Environmental Engineering and Landscape Management* 21(4): 237–247. http://dx.doi.org/10.3846/16486897.2013.788506

Montero, M. A.; Jordan, M. M.; Almendro-Candel, M. B.; Sanfeliu, T.; Hernandez-Crespo, M. S. 2009. The use of a calcium carbonate residue from the stone industry in manufacturing of ceramic tile bodies, *Applied Clay Science* 43: 186–189. http://dx.doi.org/10.1016/j.clay.2008.08.003

Okagbue, C. O.; Onyeobi, T. U. S. 1999. Potential of marble dust to stabilise red tropical soils for road construction, *Engineering Geology* 53: 371–380. http://dx.doi.org/10.1016/S0013-7952(99)00036-8

Saboya Jr., F.; Xavier, G. C.; Alexandre, J. 2007. The use of the powder marble by-product to enhance the properties of brick ceramic, *Construction and Building Materials* 21: 1950–1960. http://dx.doi.org/10.1016/j.conbuildmat.2006.05.029

Schellmann, W. 1986. A new definition of laterite, *Geological survey of India Memoir* 120: 1–7.

Segadaes, A. M.; Carvalho, M. A.; Acchar, W. 2005. Using marble and granite rejects to enhance the processing of clay products, *Applied Clay Science* 30: 42–52. http://dx.doi.org/10.1016/j.clay.2005.03.004

Sivrikaya, O.; Kiyildi, K. R.; Karaca, Z. 2014. Recycling waste from natural stone processing plants to stabilise clayey soil, *Environmental Earth Sciences* 71: 4397–4407. http://dx.doi.org/10.1007/s12665-013-2833-x

Topcu, I. B.; Bilir, T.; Uygunoglu, T. 2009. Effect of waste marble dust content as filler on properties of self-compacting concrete, *Construction and Building Materials* 23: 1947–1953. http://dx.doi.org/10.1016/j.conbuildmat.2008.09.007

TRL 1993. *Overseas Road Note 31: a guide to the structural design of bitumen-surfaced roads in tropical and sub-tropical countries*. Transport Research Laboratory (TRL), Berkshire.

Isaac I. AKINWUMI. Lecturer of Civil Engineering at the School of Applied Engineering, College of Engineering, Covenant University, Ota, Ogun State, Nigeria and doctoral researcher in Civil Engineering at Covenant University. Publications: author/co-author of ~15 scientific papers. Research interests: soil stabilization/improvement, remediation of contaminated soils, removal of heavy metals from wastewater and soil, waste management, civil engineering materials, sustainability, indigenous construction technologies, and engineering education.

Colin A. BOOTH. Dr, Associate Professor of Sustainability, is Associate Head of Research and Scholarship in the Faculty of Environment and Technology at the University of the West of England, Bristol, UK. Publications: author/co-author of ~130 scientific papers and chapters. Research interests: sustainability, environmental management, climate change mitigation and adaptation strategies, water resources management, and built environment studies.

6

MAPPING OF HEAVY METAL CONTAMINATION IN ALLUVIAL SOILS OF THE MIDDLE NILE DELTA OF EGYPT

Mohamed S. SHOKR[a,b], Ahmed A. EL BAROUDY[a], Michael A. FULLEN[b], Talaat R. EL-BESHBESHY[a], Ramadan R. ALI[c], Abd ELHALIM[a], Antonio J. T. GUERRA[d] and Maria C. O. JORGE[d]

[a]*Soils and Water Department, Faculty of Agriculture, Tanta University, Tanta, Egypt*
[b]*The University of Wolverhampton, Wolverhampton WV1 1LY, UK*
[c]*Soils and Water Use Department, National Research Centre, Giza, Egypt*
[d]*Department of Geography, Federal University of Rio de Janeiro, Brazil*

Abstract. Areas contaminated by heavy metals were identified in the El-Gharbia Governorate (District) of Egypt. Identification used remote sensing and Geographical Information Systems (GIS) as the main research tools. Digital Elevation Models (DEM), Landsat 8 and contour maps were used to map physiographic units. Nine soil profiles were sampled in different physiographic units in the study area. Geochemical analysis of the 33 soil samples was conducted using X-ray fluorescence spectrometry (XRF). Vanadium (V), nickel (Ni), chromium (Cr), copper (Cu) and zinc (Zn) concentrations were measured. V, Ni and Cr concentrations exceeded recommended safety values in all horizons of the soil profiles, while Cu had a variable distribution. Zn concentrations slightly exceeded recommended concentration limits. Concentrations were mapped in each physiographic unit using the inverse distance weighted (IDW) function of Arc-GIS 10.1 software. Pollution levels were closely associated with industry and urban areas.

Keywords: soil contamination, x-ray fluorescence spectrometry, remote sensing, geographical information systems, Middle Nile Delta, Egypt.

Introduction

Sustainable agriculture is mainly related to environmental, agronomic, ethical and socio-economic issues (Abd Elgawad *et al.* 2007). One aspect of sustainability is accumulation of heavy metals in soils, which may cause serious problems, if certain levels are exceeded. In recent years, much concern has been articulated over problems of soil contamination with heavy metals. These metals can accumulate in plants and animals and then in humans through the food chain (Govil *et al.* 2001; Lu, Bai 2010; Romic, M., Romic, D. 2003). Thus, heavy metals may damage human health and the environment (Jankaite, Vasarevičius 2005).

The Nile Delta (area ~20,000 km^2) represents only 2.3% of the area of Egypt, but it has ~46% of the total cultivated area (55,040 km^2) and accommodates ~45% of Egypt's population (Fanos 2002), with densities ≤1600 inhabitants per km^2 (Zeydan 2005). On the Nile Delta ~63% land is agricultural, due to suitable soil properties and the presence of irrigation systems (Dawoud 2004). The River Nile divides into two branches, the Rosetta and Damietta, and the Delta region is located between them (Dumont 2009). The Nile Delta (area 404,686 ha) depends on drainage water for irrigation (Abu 2011).

There are three major layers in the middle Delta aquifer (Atwia *et al.* 2006). The uppermost layer is composed of clay deposits, the second layer is formed from sandy clay deposits and the third layer is composed of saturated sand and gravel. Thus, the thin clay layers and presence of sandy clay lenses facilitate percolation of sewage water to the aquifer. Many activities, including agricultural development and industrial activities and inadequate rural sanitation, have impacts on eutrophication and contamination status, ecological value and environmental conditions in the Nile Delta (Zeydan 2005).

Heavy metal contamination of soil may present risks and hazards to humans and the ecosystems through: direct consumption or contact with contaminated soil, the food chain (soil-plant-human or soil-plant-animal-human), drinking of contaminated ground-water, decreased food quality (safety and marketability) via

Corresponding author: Mohamed Shokr
E-mail: m.s012@yahoo.com

phytotoxicity, reduction in land usability for agricultural production causing food insecurity, and land tenure problems (McLaughlin *et al.* 2000; Ling *et al.* 2007). Huge amounts of fertilizers are regularly added to soils in intensive farming systems to provide sufficient nitrogen (N), phosphorus (P) and potassium (K) for crop growth. The compounds used to supply these elements contain trace amounts of heavy metals which, after continued fertilizer application, may significantly increase soil metal contents (Raven *et al.* 1998). Integration of remote sensing information within a GIS database can quickly provide detailed soil survey information at low cost. GIS databases can also help derive Digital Elevation Models (DEM), which can help derive landscape attributes utilized in landform characterization (Brough 1986; Dobos *et al.* 2000).

It is critical to analyse the distribution and concentration of metals. This will enable identification of contamination levels and assess associated impacts, on both the environment and human health. Soils are a vital sink for these metals, because of their high metal retention capacities. The assessment and mapping of soil heavy metals can assist the development of strategies to promote sustainable use of soil resources, decrease soil degradation and expand crop production. Remediation of soils polluted by heavy metals is a major global ecological issue. Remote sensing is one of the most important methods used for soil survey, mapping and environmental investigations (Lillesand, Kiefer 2003). Geostatistical interpolation is used to survey and interpret the spatial distribution of pollutants in soil (He, Jia 2004; Woo *et al.* 2009). The inverse distance weighted (IDW) function is helpful when the purpose is to investigate overall pollution patterns (Zheng 2006). The Middle Nile Delta is affected by different pollution sources, because of the increasing number and types of industries, urban expansion, increased traffic volumes, use of drain-water and waste deposits (Abu Khatita 2011). The latter may well present a long-term danger. Usually waste deposits just settle within the normal Nile sediments and no special effort is made to construct barriers, which hinder the migration of water from these deposits into ground-water. High concentrations of vanadium (V) can damage human health, while the inhalation of airborne V-compounds can affect eyes, throat and lungs, produce weakness, ringing in the ears, nausea, vomiting, headaches and damage nerve systems (Lagerkvist, Oskarsson 2007). In Egypt, measured chromium (Cr) contents in soils range between 11.6–179 ppm, and depend on soil types and land management (Abdel-Sabour *et al.* 2002). Cr toxicity depends on its oxidation status. While $Cr3^{+}$ is considered relatively harmless, $Cr6^{+}$ is highly toxic. Cr uptake can cause diarrhoea, bleeding in the stomach and intestines, liver and kidney damage and cramp. Nickel (Ni) compounds are relatively non-toxic for plants and animals, but there is an increased risk of respiratory tract cancer, due to exposure to nickel sulphide and oxides (Sundermann, Oskarsson 1991).

Copper (Cu) is an essential element for all life-forms. In plants, Cu is required in small amounts (5–15 ppm) (Bowen 1979). The amount of Cu in soils may affect crop growth and yields. The application of Cu salts to Cu-deficient soils increases crop yields, because it compensates for Cu deficiency in plants (Baker, Senft 1995). Coal fly-ash contains 48 µg/g of Cu (Wong, M. H.; Wong, J. W. C. 1986). In Ohio (USA), measured Cu concentrations in indoor dust were twice that of outdoor dust (Tong 1998). Cu toxicity in humans is relatively rare, because they can tolerate levels ≤12 mg/day (WHO 1996). However, Cu deficiency in humans causes anaemia, bone and cardiovascular disorders, mental and nervous system deterioration and defective keratinization of hair.

Zinc (Zn) is the fourth most used metal in the world, after iron (Fe), aluminium (Al) and Cu (Bradl, Xenidis 2005). Zn uptake can lead to health disorders, including pancreatic diseases. Inhalation of Zn-oxide (particle size 0.2–1 µm) during Zn-processing causes metal fume fever, which is characterized by a sore throat, cough, fever, vomiting and pneumonitis (Ohnesorge, Wilhelm 1991).

The main aim of this research is to identify land contaminated by heavy metals in the El-Gharbia Governorate (District) of Egypt. This was undertaken using remote sensing, Geographical Information Systems (GIS) and X-ray fluorescence (XRF) spectrometry (Fig. 1).

1. Materials and methods

1.1. Study area

The study area occupies the Middle part of the Nile Delta of Egypt. It is bounded by 30°45′20″–31°10′50″E and 30°35′10″–31°10′05″N, and covers an area of 1927.4 km^2 (Fig. 2). Based on the US Soil Taxonomy (USDA 2010) the soil temperature regime of the study area is Thermic and the soil moisture regime is Torric. The mean annual

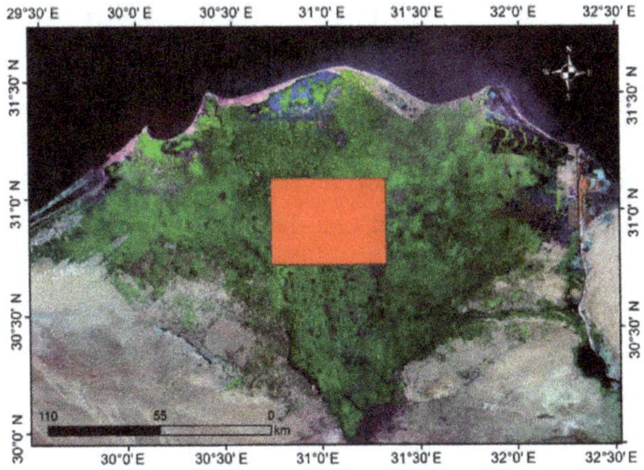

Fig. 1. Location of the study area in the middle Nile Delta of Egypt. – study area

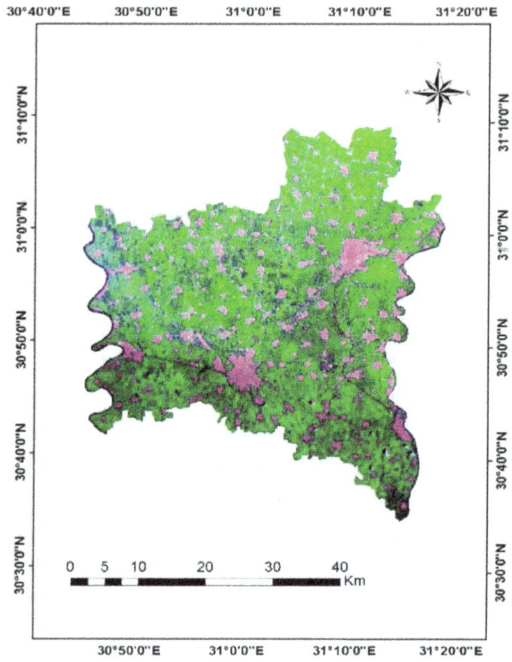

Fig. 2. Landsat 8 mosaic of the study area

temperature reaches its maximum in June, July and August and often exceeds 30 °C. The mean minimum temperature (11.2 °C) usually occurs in January, February or March at Tanta Meteorological Station (Climatologically Normal for Egypt 2011). Precipitation is unequally distributed through the rainy season. Annual rainfall is very low and mostly falls in winter; with a mean 3.8 mm/year. Rain mainly falls in the cold season (November–March) and the minimum amount is in June and September. The area belongs to the late Pleistocene era, which is evidenced by the deposits of the Neonile, which are composed of medium and fine silts (Said 1993).

1.2. Digital image processing and physiographic mapping

Digital image processing was completed for two Landsat 8 satellite images (path 177/row 38 and path 177/row 39), with a spatial resolution of 30 m, acquired in May 2014. The images were pre-processed, including radiometric correction (used to modify digital values of pixels to remove noise). Images were geometrically rectified using the Universal Transverse Mercator (UTM) co-ordinates, with the World Geodetic System datum (WGS 1984) and then maps were constructed. Images were atmospherically corrected using the FLAASH module (ITT 2009). Data were calibrated to radiance using the inputs of image type, acquisition date and time. Images were subject to linear stretching by 2%, smooth-filtered, and their histograms were matched, adopting the procedures of Lillesand and Kiefer (2007) and mosaicked using ENVI 5.1 software. The extraction of landform units used high spatial resolution images, so the spatial resolution of satellite image was enhanced using the data merge function of Envi5.1 software. Merging is performed by using multispectral bands (~30 m) as low spatial resolution, and band 8 (panchromatic band) with ~15 m resolution. Landforms were extracted using contour maps (scale 1:25,000) and enhanced satellite images. Both enhanced satellite images were processed with DEM in ERDAS Imagine 8.7, to extract the landform information (Dobos *et al.* 2002). The initial landform maps were ground-truthed using field observations.

1.3. Spatial distribution of heavy metals

Spatial interpolation is widely used when data are collected at distinct locations (e.g. soil profiles) for producing continuous information (Ali, Moghanm 2013). Inverse distance weighted (IDW) is an interpolation method which uses measured values surrounding the prediction location. The measured values closest to the prediction location have more influence on the predicted value than those farther away, thus giving greater weight to points closest to the prediction location, and the weights decrease as a function of distance (Shepard 1968). Geostatistical relationships among the known points (IDW) of Arc-GIS 10.1 software were used to interpolate heavy metal concentrations in the study area. The spatial interpolation method (IDW) was used with 12 neighbouring samples for estimation of each grid point. A power of two was used to weight the nearest points.

1.4. Assessment of contamination risk

The Geoaccumulation Index (Igeo) was originally used to evaluate bottom sediment contamination. However, it has been successfully used to evaluate soil contamination (Gowd *et al.* 2010). The Igeo Index means the assessment of contamination depends on comparing heavy metal concentrations in soils to background values. The calculation of the Geoaccumulation Index uses the equation:

$$\text{Igeo} = \log 2 \frac{C_n}{1.5 B_n}, \tag{1}$$

where C_n = the measured concentration of the element in soil; B_n = the geochemical background concentration of the heavy metal.

The Geoaccumulation Index (Igeo) is shown in Table 1.

1.5. Soil analysis

Soil samples were collected from nine profiles in El-Gharbia Governorate. The selected profiles represent the different soil units. Pedological descriptions of profiles were conducted using the procedures of FAO (2006) (Table 2). About 1 kg was collected from each horizon of

Table 1. The Geoaccumulation Index (Igeo) for assessing contamination levels in soil (Rahman *et al.* 2012)

Igeo Class	Igeo value	Contamination level
0	Igeo ≤ 0	Uncontaminated
1	0 < Igeo < 1	Uncontaminated/moderately contaminated
2	1< Igeo < 2	Moderately contaminated
3	2 < Igeo < 3	Moderately/strongly contaminated
4	3 < Igeo< 4	Strongly contaminated
5	4 < Igeo< 5	Strongly/extremely contaminated
6	5 < Igeo	Extremely contaminated

each profile. Soil samples were air-dried and large stones and organic debris were removed before sieving. Samples were gently ground, homogenized, sieved through a 2.0 mm sieve and then crushed to a fine (<125 μm) powder. Oven-dry samples were ignited at 375 °C for 16 hours (overnight), adopting the procedures of Ball (1964). Sub-samples of 8.5 g of soil powder were added to 1.5 g of wax (Lico waxc micropowder PM, Hoechst wax)) and then compressed under 12 tonnes pressure by a semi-automatic hydraulic press to make a pellet. The geochemical composition of soil pellets were analysed using an XRF spectrometer model Epsilon3 XLE. XRF analyses were performed at the University of Wolverhampton, UK.

Table 2. Pedological descriptions of soil profiles

Profile Number	Depth (cm)	Colour	Texture	Structure	Soil consistency
1	0–50	5YR 3/4 5YR5/4	Loam	Sub-angular blocky	Sticky, plastic
	50–85	7.5YR 5/6 7.5YR 6/6	Sandy loam	Sub-angular blocky	Slightly sticky, slightly plastic
	85–120	10YR 5/8 10YR 6/8	Loam	Sub-angular blocky	Sticky, plastic
	120–150	10YR 5/8 10YR 7/8	Sandy loam	Sub-angular blocky	Slightly sticky, slightly plastic
2	0–45	5YR 4/6 5YR 5/6	Sandy Loam	Sub-angular blocky	Slightly sticky, slightly plastic
	45–85	10YR 5/8 10YR 6/8	Sandy Clay Loam	Sub-angular blocky	Sticky, plastic
	85–110	10YR 5/8 10YR 7/8	Sandy Loam	Sub-angular blocky	Slightly sticky, slightly plastic
3	0–75	10 YR 5/6 10YR 6/8	Sandy Loam	Sub-angular blocky	Slightly sticky, slightly plastic
	75–100	10YR 5/6 10YR 7/8	Loam	Sub-angular blocky	Sticky, plastic
	100–150	10YR 5/8 10YR 7/8	Silt loam	Sub-angular blocky	Sticky, plastic
4	0–60	5YR 4/8 5YR 4/6	Loam	Sub-angular blocky	Sticky, plastic
	60–100	10YR 5/8	Sandy Loam	Sub-angular blocky	Slightly sticky, slightly plastic
	100–120	10YR 7/6	Loam	Sub-angular blocky	Sticky, plastic
	120–150	5YR 4/8 5YR 4/6	Sandy Loam	Sub-angular blocky	Slightly sticky, slightly plastic
5	0–45	10YR 5/8 10YR 7/8	Sandy Loam	Sub-angular blocky	Slightly sticky, slightly plastic
	45–65	7.5YR 5/6 7.5YR 5/8	Loam	Sub-angular blocky	Sticky, plastic
	65–110	10YR 5/8 10YR 6/8	Silt loam	Sub-angular blocky	Sticky, plastic
	110–150	10YR 5/8 10YR 6/8	Sandy Loam	Sub-angular blocky	Slightly sticky, slightly plastic
6	0–35	10YR 5/6 10YR 6/8	Sandy Loam	Sub-angular blocky	Slightly sticky, slightly plastic

End of Table 2

Profile Number	Depth (cm)	Colour	Texture	Structure	Soil consistency
6	35–65	10YR 6/8 10YR 7/8	Loam	Sub-angular blocky	Sticky, plastic
	65–100	10YR 5/8 10YR 7/8	Loam	Sub-angular blocky	Sticky, plastic
	100–150	5YR 5/6 5YR 4/8	Sandy Loam	Sub-angular blocky	Slightly sticky, slightly plastic
7	0–55	5YR 4/8 5YR 6/8	Sandy Loam	Sub-angular blocky	Slightly sticky, slightly plastic
	55–110	5YR 4/8 5YR 6/8	Loam	Sub-angular blocky	Sticky, plastic
	110–150	7.5YR 5/8 7.5YR 6/4	Sandy Loam	Sub-angular blocky	Slightly sticky, slightly plastic
8	0–30	7.5YR 4/4 7.5YR 5/8	Silt loam	Sub-angular blocky	Sticky, plastic
	30–60	7.5YR 4/6 7.5YR 6/8	Silt loam	Sub-angular blocky	Sticky, plastic
	60–100	7.5YR 5/8 7.5YR 6/8	Silt loam	Sub-angular blocky	Sticky, plastic
	100–150	7.5YR 5/8 7.5YR 5/8	Silt loam	Sub-angular blocky	Sticky, plastic
9	0–45	7.5YR 5/8 7.5YR 6/8	Sandy Loam	Sub-angular blocky	Slightly sticky, slightly plastic
	45–105	10YR 5/8 10YR 6/8	Loam	Sub-angular blocky	Sticky, plastic
	105- 130	10YR 5/8 10YR 6/8	Sandy Loam	Sub-angular blocky	Slightly sticky, slightly plastic
	130–150	7.5YR 6/8 7.5YR 5/8	Loam	Sub-angular blocky	Sticky, plastic

Table 3. Physiographic units on the soil map

Physiographic unit	Landforms	Mapping unit	Soil profile	Profile elevation (masl)	Area (km^2)	Area (%)
Flood plain	High terraces	T1	9	12	232.21	12.05
	Moderately high terraces	T2	4	8	431.99	22.41
	Low Terraces	T3	1	0	417.8	21.68
	High Decantation Basin	D1	3	10	39.53	2.05
	Low Decantation Basin	D2	5	6	236.29	12.26
	High overflow Basin	OB1	6	7	244.461	12.68
	Low Overflow basin	OB2	7	5	206.45	10.71
	Levees	L	8	9	103.82	5.39
	Swales	S	2	8	14.89	0.77
	Total	–	–	–	1927.441	100.00

*masl = metres above sea level.

1: River terraces: these soils represent the late Pleistocene deltaic plain and occur at the edge of decantation basins (these are basins in which sedimentation, particularly of silt and clay, occur during floods). The soils are formed on terraces at various heights above the valley floor.

2: Basins: these are artificially enclosed areas of a river or harbour, designed so that water levels are unaffected by tides.

3: River levees: these are a type of dam that runs along the banks of rivers or canals. Levees reinforce the banks and help prevent flooding. By confining the flow, levees can also increase water velocity.

4: Swales: these are low tracts of land, usually consisting of moist and marshy lands. The term can refer to both natural and artificial landscape features. Artificial swales are often designed to manage water runoff, filter pollutants and increase rainwater infiltration.

2. Results and discussion

2.1. Physiographic map of the study area

The satellite images show that the study area is a floodplain and includes high terraces (12.04% of area), moderately high terraces (22.41%), low terraces (21.67%), high decantation basins (2.05%), low decantation basins (12.26%), high overflow basins (12.68%), low overflow basins (10.71%), river levees (5.38%) and swales (0.77%). The main physiographic soil units of the study area are reported in Table 3 and Figure 3.

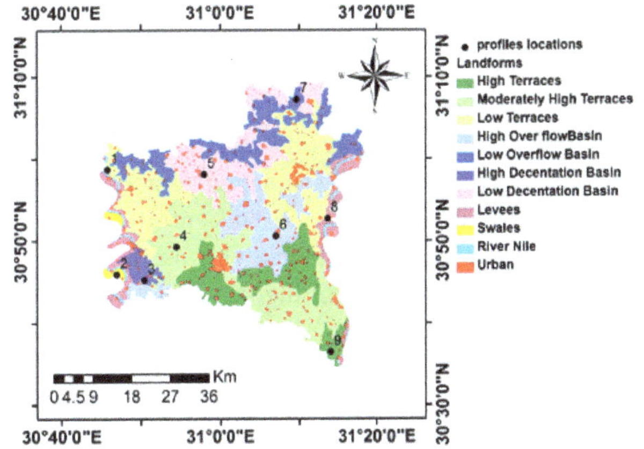

Fig. 3. The main landforms of the study area and profile locations

2.2. Heavy metal contamination

XRF analyses of the soil samples identified the presence of SiO_2, Al_2O_3, P_2O_5, K_2O, CaO, MgO, Na_2O and Fe_2O_3 (major) and Cr, Cu, Zn, Ni, Br, Rb, Sr, Y, Zr, Nb, Sn, Te, Ba, Eu, Yb, Re, Ga, Ir, Mo, As and Pb (minor). Concentrations of the heavy metals Cr, Cu, Ni, V and Zn for each profile are reported in Table 4. For the metals Te, Mo, As and Pb, results are not reported, because their concentrations were below detection limits. Spatial interpolation maps (Figs 4–7) of heavy metal concentrations were prepared using the IDW function (inverse distance weighted) interpolation method in Arc GIS 10.1.

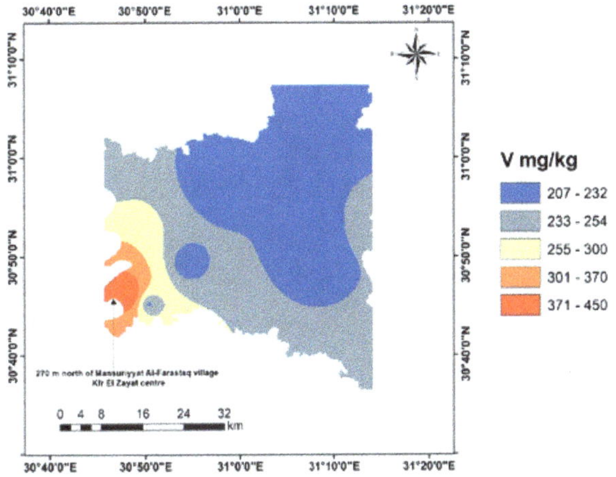

Fig. 4. Spatial interpolation of the weighted mean of vanadium

Vanadium

The concentrations and the interpolation map for V in the soil samples are given in Table 4 and Figure 4. V concentrations ranged from 194.0–744.4 mg/kg with a weighted mean ranging from 206.79–450.58 mg/kg (Table 5). The highest measured concentration of V was in the upper horizon of Profile 2, which represents a swales unit and is located 270 m north of Mansuriyyat Al-Farastaq village, ~6.5 km south-west from the centre of the town of Kfr Elzayat (population in 2015 was 448,965). The Igeo Index showed that all soil samples were in the uncontaminated/moderately contaminated categories, except for first horizon of Profile 2 in the swales mapping unit, which is classified as moderately/strongly contaminated (Table 6). The high deposition of V might be caused by the numerous local factories. V concentrations are higher than the permissible limits (90 mg/kg), recommended by Bowen (1979) in all soil profile horizons (Table 5). The spatial interpolation shows an increasing trend from north-east to south-west. The highest weighted mean (weighting concentration by representative area) (450.58 mg/kg) was found in 0.77% of the study area. From the interpolation map of V in the study area (Fig. 4) we can conclude that the order of concentration ascending in the mapping units is: low decantation basin (D2), high overflow basin (OB1), low overflow basin (OB2), moderately high terraces (T2), high decantation basin (D1), levees (L), high terraces (T1), low terraces (T3) and swales (S).

Chromium

Anthropogenic sources of Cr include alloys, chrome plating, pigments, chemical catalysts, dyes, tanning, wood impregnation and refractory bricks (Reimann, de Caritat 1998). The highest concentration of Cr (519 mg/kg) was in the top-soil of Profile 2, which may be due to the many local factories. The lowest (140.3 mg/kg) was in the top-soil of Profile 6, which represents a high over-flow basin (Table 4). The mean weight of Cr concentrations ranged from 152.84–314.73 mg/kg (Table 5). All concentrations exceeded the recommended values given by Bowen (1979) and, according to the Igeo Index, most soil samples are in the uncontaminated/moderately contaminated category (Table 6). Cr concentrations increased from east to west and south of the study area (Fig. 5). The highest Cr concentrations tended to be in the swales unit and the lowest in the high overflow basin unit.

Nickel

Baghdady, Sippola (1984) reported that the mean total Ni content in Egyptian alluvial soils is 64.4 mg/kg, ranging from 20–74 mg/kg, while the mean NH_4OAC–EDTA extractable Ni is 1.9 mg/kg, ranging from 1.0–2.2 mg/kg. However, in this study, Ni concentrations ranged from 60.60–267.30 mg/kg (Table 4), with mean weight ranging from 69.60–148.39 mg/kg (Table 5). Ni concentrations were higher in alluvial soils than previous studies and exceeded the permissible limit (50 mg/kg) (Bowen 1979). According to the Igeo Index, all samples are in the uncontaminated and moderately contaminated categories (Table 6). Table 4 reports Ni concentrations in soil profiles and Figure 6 shows the spatial trends, which increased from north to south and west. The highest concentrations were in swales, which occupy 14.89 km² of the study area. The interpolation of Ni shows high spatial variability, with the lowest values in the high decantation basin units.

Copper

Table 4 and Figure 7 report Cu concentrations and the spatial interpolation of weighted mean Cu concentrations, respectively. Cu contents in horizons ranged from

Table 4. XRF analysis of soils collected from the study area

Profile no	Mapping unit	Depth (cm)	Metal concentrations (mg/kg)				
			V	Cr	Ni	Cu	Zn
1	T3	0–50	227.1	179.7	68.30	78.60	94.30
		50–85	265.9	158.1	77.80	60.90	85.40
		85–120	250.4	167.3	73.50	81.40	93.30
		120–150	221.8	163.3	61.60	60.10	80.80
2	S	0–45	744.4	519.0	267.30	288.90	377.60
		45–85	244.9	161.4	63.50	50.90	75.50
		85–110	250.8	192.4	70.20	71.40	86.80
3	D1	0–75	221.0	170.7	63.80	94.50	90.20
		75–100	241.4	179.6	63.00	73.00	86.90
		100–150	238.9	166.6	81.60	76.60	85.00
4	T2	0–60	203.1	159.2	72.30	118.70	308.00
		60–100	258.3	166.3	70.90	0.00	88.70
		100–120	222.8	166.2	65.50	75.20	84.90
		120–150	206.1	179.8	70.70	72.30	91.60
5	D2	0–45	194.0	149.1	74.40	95.20	103.10
		45–65	197.5	159.6	69.20	93.30	98.40
		65–110	225.3	143.0	73.20	95.50	95.00
		110–150	205.0	150.5	72.20	94.60	98.80
6	OB1	0–35	210.1	140.3	60.60	131.80	124.50
		35–65	216.0	152.7	76.70	76.40	109.40
		65–100	220.9	164.3	84.50	97.50	103.20
		100–150	219.5	153.7	85.00	0.00	100.60
7	OB2	0–55	228.2	180.8	76.50	93.50	94.70
		55–110	230.3	170.7	73.30	89.10	84.70
		110–150	196.5	168.7	64.50	74.10	84.30
8	L	0–30	218.7	164.1	68.90	0.00	93.20
		30–60	241.3	151.3	73.90	0.00	97.10
		60–100	247.3	155.4	82.10	0.00	89.50
		100–150	231.9	154.9	75.40	85.80	93.80
9	T1	0–45	250.7	156.8	78.20	0.00	112.10
		45–105	232.8	158.1	80.90	74.30	103.30
		105- 130	223.7	168.8	76.40	74.50	89.50
		130–150	231.1	168.4	69.70	74.10	94.00

Table 5. Heavy metal concentrations in soil samples and concentration limits recommended by Bowen (1979)

Profile No	Mapping unit	Mean weight of metals concentrations (mg/kg)				
		V	Cr	Ni	Cu	Zn
1	T3	240.53	168.48	70.39	71.42	89.29
2	S	450.58	314.73	148.39	152.92	201.65
3	D1	230.36	170.81	69.60	84.94	87.91
4	T2	221.04	166.14	70.70	71.96	176.49
5	D2	206.79	149.06	72.76	94.87	98.89
6	OB1	216.93	152.84	77.53	68.79	108.54
7	OB2	220.51	173.87	72.12	86.71	88.26
8	L	235.24	156.15	75.58	28.60	93.19
9	T1	236.42	160.86	83.28	52.01	102.40
Concentration limits (mg/kg)		90	70	50	25	90

0–288.9 mg/kg and mean weight ranged from 28.90–152.92 mg/kg (Table 5). All concentrations exceeded the permissible limit of 25 mg/kg (Bowen 1979), except for the second horizon of Profile 4 and the first, second and the third horizons of Profile 8 (Table 4). In addition, in Profile 9 the deepest horizon exceeded the limit, whereas concentrations in the upper layer were 0. This is probably due to percolation and illuviation of Cu associated with irrigation water. These profiles represent moderately high terraces, levees and high terraces, respectively. The Igeo Index showed that soil samples were in three contamination categories (uncontaminated/moderately contaminated, moderately contaminated and moderately/strongly contaminated) (Table 6). Two important sources of Cu in the Nile Delta are: (i) applications of Cu-based liquid fungicides, and (ii) use of $CuSO_4$ as an algicide in treating and controlling problematic macro-algal blooms in the Nile, especially during summer (Abdel-Moati, El-Sammak 1997). The lowest Cu concentrations were in the river levees units and the highest values were in the swale units.

Zinc

Zn concentrations slightly exceed the permissible concentration limit of 90 mg/kg (Bowen 1979) (Table 5). Exceptions include the upper layers of Profiles 2 and 4, where concentrations greatly exceeded permissible limits (377.6 and 308.0 mg/kg, respectively). The highest concentrations were in the upper horizon, but in Profile 8 the highest concentration was in the subsurface (Table 4). This could be caused by infiltration of irrigation water through the profile. The mean weight of Zn ranged between 88.26–201.65 mg/kg. According to the Igeo Index, the Zn concentrations of all soil samples fell into the uncontaminated category, except for the first horizons of Profiles 2 and 4, which were moderately contaminated (Table 6). The spatial interpolation of Zn is presented in Figure 8. The highest concentration was in the south-west of the study area, which is located 270 m north of Mansuriyyat Al-Farastaq

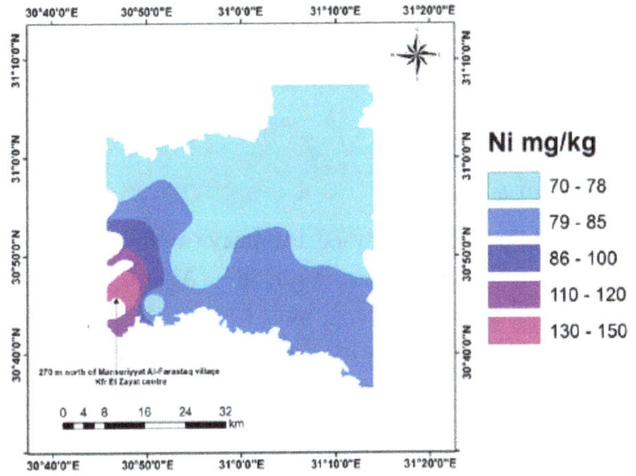

Fig. 6. Spatial interpolation of the weighted mean of nickel

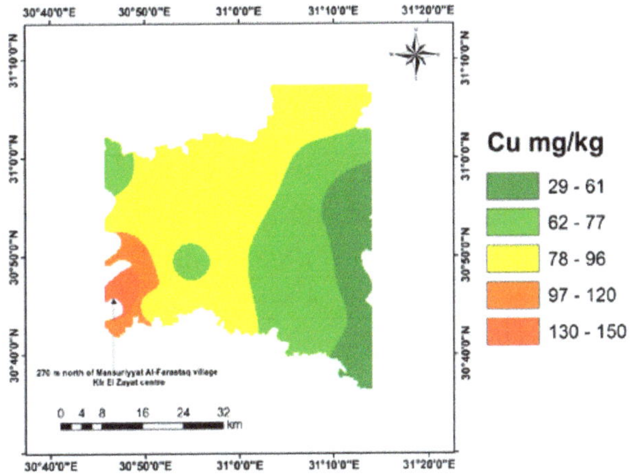

Fig. 7. Spatial interpolation of the weighted mean of copper

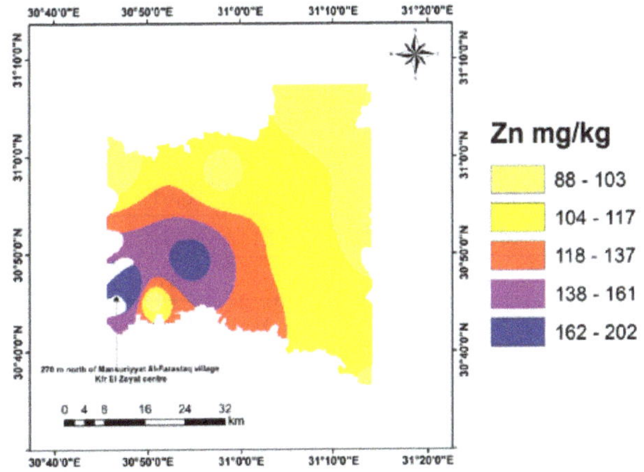

Fig. 8. Spatial interpolation of the weighted mean of zinc

village. This could be due to atmospheric deposition, originating from local industrial plants. The highest Zn concentrations were in the swales top-soil and moderately high terrace units.

2.3. Major soil oxides

Soil samples were analysed for heavy metals and major oxides (Table 7). Results for SiO_2, Al_2O_3, P_2O_5, K_2O, CaO, MgO, Na_2O and Fe_2O_3 were compared with average concentrations of major oxides in soil (Bohn *et al.* 2001) (Table 5). SiO_2 concentrations varied from 51.40–56.55% and were all less than the representative average value of 72.64% (Bohn *et al.* 2001). Al_2O_3 concentrations varied from 16.28–24.42% and the mean concentration of profiles ranged from 18.81–22.9%. Al_2O_3 concentrations in all samples exceeded the representative mean value of 13.22%. K_2O concentrations ranged from 1.05–1.62% with weighted mean values ranging from 1.14–1.35%, near the 1.2% representative mean. CaO concentrations ranged between 2.58–6.69% and the mean weighted value

Table 6. Igeo Index concentrations and associated contamination levels

Profile no	Depth, cm	V	C level	Cr	C Level	Ni	C Level	Cu	C level	Zn	C level
1	0–50	0.75	UN/M	0.77	UN/M	ND	UN	1.06	M	ND	UN
	50–85	0.97	UN/M	0.59	UN/M	0.05	UN/M	0.69	UN/M	ND	UN
	85–120	0.89	UN/M	0.67	UN/M	ND	UN	1.11	M	ND	UN
	120–150	0.71	UN/M	0.63	UN/M	ND	UN	0.68	UN/M	ND	UN
2	0–45	2.46	M/S	2.30	M/S	1.83	M	2.94	M/S	1.48	M
	45–85	0.85	UN/M	0.62	UN/M	ND	UN	0.44	UN/M	ND	UN
	85–110	0.89	UN/M	0.87	UN/M	ND	UN	0.92	UN/M	ND	UN
3	0–75	0.71	UN/M	0.70	UN/M	ND	UN	1.33	M	ND	UN
	75–100	0.83	UN/M	0.77	UN/M	ND	UN	0.96	UN/M	ND	UN
	100–150	0.82	UN/M	0.66	UN/M	0.12	UN/M	1.03	M	ND	UN
4	0–60	0.58	UN/M	0.60	UN/M	ND	UN	1.66	M	1.18	M
	60–100	0.93	UN/M	0.66	UN/M	ND	UN	ND	–	ND	UN
	100–120	0.72	UN/M	0.66	UN/M	ND	UN	1.00	M	ND	UN
	120–150	0.61	UN/M	0.77	UN/M	ND	UN	0.94	UN/M	ND	UN
5	0–45	0.52	UN/M	0.50	UN/M	ND	UN/M	1.34	M	ND	UN
	45–65	0.54	UN/M	0.60	UN/M	ND	UN	1.31	M	ND	UN
	65–110	0.73	UN/M	0.44	UN/M	ND	UN	1.34	M	ND	UN
	110–150	0.60	UN/M	0.51	UN/M	ND	UN	1.33	M	ND	UN
6	0–35	0.63	UN/M	0.41	UN/M	ND	UN	1.81	M	ND	UN
	35–65	0.67	UN/M	0.54	UN/M	0.03	UN/M	1.02	M	ND	UN
	65–100	0.71	UN/M	0.64	UN/M	0.17	UN/M	1.37	M	ND	UN
	100–150	0.70	UN/M	0.54	UN/M	0.18	UN/M	ND	–	ND	UN
7	0–55	0.75	UN/M	0.78	UN/M	0.02	UN/M	1.31	M	ND	UN
	55–110	0.77	UN/M	0.70	UN/M	ND	UN	1.24	M	ND	UN
	110–150	0.54	UN/M	0.68	UN/M	ND	UN	0.98	UN/M	ND	UN
8	0–30	0.69	UN/M	0.64	UN/M	ND	UN	ND	–	ND	UN
	30–60	0.83	UN/M	0.52	UN/M	ND	UN	ND	–	ND	UN
	60–100	0.87	UN/M	0.56	UN/M	0.13	UN/M	ND	–	ND	UN
	100–150	0.78	UN/M	0.56	UN/M	0.007	UN/M	1.19	M	ND	UN
9	0–45	0.89	UN/M	0.57	UN/M	0.06	UN/M	ND	–	ND	UN
	45–105	0.78	UN/M	0.59	UN/M	0.1	UN/M	0.98	UN/M	ND	UN
	105- 130	0.72	UN/M	0.68	UN/M	0.02	UN/M	0.99	UN/M	ND	UN
	130–150	0.77	UN/M	0.68	UN/M	ND	UN	0.98	UN/M	ND	UN

C Level = Contamination level; UN = Uncontaminated; UN/M = Uncontaminated/moderately contaminated; M = Moderately contaminated; M/S = Moderately/strongly contaminated; ND = Not detected

ranged between 3.70–5.95%. CaO and Na_2O concentrations exceeded the representative means of 1.44% and 0.99%, respectively. Fe_2O_3 concentration ranged between 9.73–12.23%, whereas the representative mean is 5.77%. P_2O_5 concentrations ranged from 0.15–0.49%. The weighted mean concentrations of P_2O_5 samples exceed the 0.18% representative mean (Table 8). Thus, these deltaic soils are predominantly siliceous, with slight enrichment of the alumina component.

2.4. Relationships between trace and major elements

V, Cr, Ni, Cu and Zn concentrations are significantly correlated (Table 9). There are no significant correlations between major elements and heavy metal concentrations, except for V, Ni and Zn. V and Ni have significant positive medium and strong correlations with Fe, respectively. This may indicate the sorption of these elements by Fe hydroxides. Ni has a strong positive association with Al, whereas a strong significant association was found between Zn and

Table 7. Summary of major oxide concentrations in soil samples of the study area

Profile No.	Mapping unit	Depth, cm	Major oxide concentrations (%)							
			SiO_2	Al_2O_3	P_2O_5	K_2O	CaO	MgO	Na_2O	Fe_2O_3
1	T3	0–50	52.67	20.10	0.31	1.29	3.66	4.18	1.44	10.93
		50–85	56.46	23.92	0.25	1.29	3.18	4.53	1.44	12.03
		85–120	55.12	21.48	0.23	1.16	3.88	4.43	1.77	11.54
		120–150	54.42	19.28	0.21	1.19	4.17	3.98	2.02	10.38
2	S	0–45	ND	ND	ND	ND	ND	ND	ND	ND
		45–85	52.26	18.57	0.24	1.21	4.98	4.66	1.55	11.16
		85–110	55.35	21.50	0.20	1.10	4.50	4.62	1.89	11.28
3	D1	0–75	54.33	20.74	0.39	1.45	4.53	4.83	1.48	10.51
		75–100	52.83	18.36	0.20	1.10	5.03	4.30	1.72	11.20
		100–150	54.03	20.96	0.18	1.08	4.63	4.59	1.66	11.03
4	T2	0–60	55.87	19.11	0.43	1.60	3.70	5.11	1.49	10.61
		60–100	56.55	21.37	0.20	1.31	4.26	5.26	1.47	10.77
		100–120	53.60	18.24	ND	1.13	6.69	4.96	1.70	10.78
		120–150	52.15	16.48	0.20	1.07	6.34	4.55	1.84	10.28
5	D2	0–45	52.70	18.80	0.46	1.30	5.36	5.16	1.42	10.54
		45–65	54.20	19.19	0.20	1.24	4.38	5.43	1.43	10.96
		65–110	54.88	19.60	0.17	1.20	3.56	5.57	1.50	11.03
		110–150	53.93	19.20	0.28	1.25	4.43	5.39	1.45	10.84
6	OB1	0–35	54.64	19.19	0.49	1.62	5.22	4.99	1.48	10.82
		35–65	55.46	22.14	0.29	1.35	4.03	4.99	1.39	11.44
		65–100	55.71	23.59	0.20	1.25	3.47	5.00	1.36	11.74
		100–150	56.04	23.64	0.19	1.18	3.40	4.99	1.33	11.75
7	OB2	0–55	56.43	19.43	0.24	1.39	4.74	5.88	1.60	10.72
		55–110	55.29	19.4	0.15	1.22	3.06	4.99	2.44	10.59
		110–150	53.4	17.14	0.17	1.28	4.31	4.62	2.80	9.72
8	L	0–30	52.38	19.40	0.32	1.22	4.23	4.35	1.59	10.38
		30–60	54.42	19.47	0.28	1.21	4.35	4.90	1.62	11.11
		60–100	55.01	23.20	0.21	1.09	3.45	4.80	1.50	11.67
		100–150	56.06	24.42	0.17	1.09	2.58	4.79	1.40	12.23
9	T1	0–45	51.40	17.93	0.31	1.47	4.67	4.93	1.52	11.06
		45–105	52.32	19.01	0.20	1.14	4.32	4.78	1.64	11.43
		105–130	52.36	21.47	0.16	1.05	3.12	4.52	1.70	11.73
		130–150	52.09	21.08	0.15	1.05	3.17	4.44	1.71	11.68

* ND = Not detected.

K_2O and a medium correlation with P_2O_5. Such associations indicate strong affinity for these elements for Fe, Al and K oxides. Al_2O_3 and Fe_2O_3 display a strong positive significant correlation. In addition, P_2O_5 is significantly correlated with K_2O. There are strong significant negative correlations between Al_2O_3 and CaO, between CaO and Fe_2O_3 and a medium significant negative correlation between Na_2O and Fe_2O_3.

Conclusions

This study shows that concentrations of V, Ni and Cr exceeded recommended limits in the soils of the Middle Nile Delta. Cu concentrations were very variable. Zn concentrations slightly exceed the recommended limit. V, Cr, Ni, Cu and Zn concentrations are significantly correlated. There are no significant correlations between major elements and heavy metal concentrations, except for V, Ni and Zn. V and Ni have significant positive medium and strong correlations with Fe, respectively. The Igeo Index of V showed that all soil samples were in the uncontaminated/moderately contaminated categories, except for the first horizon of Profile 2, which is classified as moderately/strongly contaminated, while the Igeo Index for Cr showed most soil samples are in the uncontaminated/moderately contaminated category. The Igeo Index for Ni reveals that all samples are in the uncontaminated and moderately contaminated categories. For Cu, soil samples were in three contamination categories (uncontaminated/moderately contaminated, moderately contaminated and moderately/strongly contaminated). All Zn concentrations were in the uncontaminated category, except for the first horizons of Profiles 2 and 4, which were moderately contaminated. The highest heavy metal concentrations dominate the south-west of El-Gharbia Governorate and is mainly attributed to human activities, especially pollution

Table 8. Mean weight of major oxide concentrations in soil samples and representative average limits (Bohn *et al.* 2001)

Profile No.	Mapping unit	Mean weight of major element concentrations (%)							
		SiO_2	Al_2O_3	P_2O_5	K_2O	CaO	MgO	Na_2O	Fe_2O_3
1	T3	54.47	21.14	0.25	1.23	3. 70	4.27	1.63	11.21
2	S	ND	ND	ND	ND	ND	ND	ND	ND
3	D1	53.97	20.40	0.28	1.27	4.64	4.66	1.58	10.79
4	T2	55. 06	19.07	ND	1.35	4.77	5.01	1.57	10.60
5	D2	53.88	19.19	0.28	1.24	4.44	5.38	1.45	10.82
6	OB1	55.51	22.29	0.28	1.33	5.95	4.99	1.38	11.47
7	OB2	55.20	18.81	0.18	1.29	4.01	5.21	2.23	10.40
8	L	54.71	22.09	0.23	1.14	3.49	4.72	1.50	11.48
9	T1	52.01	19.36	0.22	1.21	4.07	4.73	1.62	11.40
Conc. limits (%)		70.29	>13.22	0.18	1.20	1.44	0.99	0.99	5.77

*ND = Not detected

Table 9. Correlation coefficients between trace and major elements in soils of the study area

	V	Cr	Ni	Cu	Zn	SiO_2	Al_2O_3	P_2O_5	K_2O	CaO	MgO	Na_2O
Cr	0.97***											
Ni	0.97***	0.96***										
Cu	0.64***	0.71***	0.68***									
Zn	0.73***	0.75***	0.77***	0.68***								
SiO_2	0.19	0.11	0.28	0.18	0.19							
Al_2O_3	0.4*	−0.13	0.56***	−0.17	−0.09	0.61***						
P_2O_5	−0.33	−0.35	−0.26	0.29	0.49**	−0.06	−0.22					
K_2O	−0.25	−0.32	−0.19	0.32	0.56***	0.23	−0.18	0.78***				
CaO	−0.23	−0.15	−0.42**	0.14	−0.09	−0.38	−0.69***	0.37	0.1			
MgO	−0.25	0.35	0.27	0.24	0.21	0.41*	−0.03	0.14	0.39*	0.08		
Na_2O	−0.1	0.40*	−0.33	−0.05	−0.20	−0.16	−0.43	−0.37	−0.21	0.06	−0.30	
Fe_2O_3	0.51**	−0.17	0.56***	−0.18	−0.11	0.26	0.79***	−0.31	−0.35	−0.55***	−0.07	−0.51***

Note: *P < 0.05, **P < 0.01, ***P < 0.001, n = 33 soil samples from the Kfr Elzayat urban area (population in 1995 was 448,965). In terms of the distribution of heavy metals in the different physiographic units, the swales unit contained the highest values, as this is in Kfr El Zayat, which has many factories. We recommend that heavy metal contamination be studied within entire soil profiles and not just topsoils, because these metals affect soil and crop quality and can cause ground-water pollution. Protection against this hazard is vital for sustainable land management. Precise measures and efficient methods to improve soil and water quality must be conducted, in order to prevent soil and water pollution and to avoid the need for costly remediation in the future.

References

Abd Elgawad, M.; Hamdi, A. A.; Mahmoud, M. S.; Samir, I. G. 2007. Status of some heavy metals in Fayoum district soils, in *3rd Conference for Sustainable Agricultural Development*, 12–14 November 2007, Fayoum, Egypt, 507–526.

Abdel-Moati, M. A. R.; El-Sammak, A. A. 1997. Man-made impact on the geochemistry of the Nile Delta Lakes. A study of metal concentrations in sediments, Water, *Air and Soil Pollution* 97(3): 413–429. http://dx.doi.org/10.1007/BF02407476

Abdel-Sabour, M. F.; Abdou, F. M.; Elwan, I. M.; Al-Salama, Y. J. 2002. Effect of soil contamination due to wastewater irrigation on Cr fractions in some soils of Egypt, in *Proceedings of 6th Radiation Physics Conference*, Assuit, Egypt, 27–30.

Abu Khatita, A. M. 2011. *Assessment of soil and sediment contamination in the Middle Nile Delta area (Egypt)- Geo-Environmental study using combined sedimentological, geophysical and geochemical methods.* Alexander-Universität Erlangen-Nürnberg zur Erlangung, Germany.

Abu, Z. M. 2011. *Egyptian policies for using low quality water for irrigation.* Water Resource Center, Egypt, Cairo.

Ali, A. A.; Moghanm, F. S. 2013. Variation of soil properties over the landforms around Idku Lake, Egypt, *The Egyptian Journal of Remote Sensing and Space Sciences* 16(1): 91–101. http://dx.doi.org/10.1016/j.ejrs.2013.04.001

Atwia, M. G.; Abou Heleika, M. M.; El-Shishtawy, A. M.; Sharp, J. M. 2006. Hydrostratigraphy of the Central Nile Delta, Egypt, using geoelectric measurements, *EGS-AGU-EUG Joint Assembly*, Nice, France 5: 6–11.

Baghdady, N. H.; Sippola, J. 1984. Extractability of polluting elements Cd, Cr, Ni and Pb of soil with three methods, *Acta Agriculturae Scandinavica* 34(3): 345–348. http://dx.doi.org/10.1080/00015128409435402

Baker, D. E.; Senft, J. P. 1995. Copper, in B. J. Alloway (Ed.). *Heavy metals in soils.* London: Blackie. http://dx.doi.org/10.1007/978-94-011-1344-1_8

Ball, D. F. 1964. Loss-on-ignition as estimate of organic matter and organic carbon in non-calcareous soils, *Journal of Soil Science* 15(1): 84–92. http://dx.doi.org/10.1111/j.1365-2389.1964.tb00247.x

Bohn, L. H.; McNeal, L. B.; O'Connor, A. G. 2001. *Soil chemistry.* 2nd ed. New York: John Wiley.

Bowen, H. J. M. 1979. *Environmental chemistry of the elements.* London: Academic Press.

Bradl, H.; Xenidis, A. 2005. Chemical remediation techniques, in H. B. Bradl (Ed.). *Heavy metals in the environment: origin, interaction and remediation.* Amsterdam: Elsevier. http://dx.doi.org/10.1016/S1573-4285(05)80022-5

Brough, P. A. 1986. *Principles of geographical information systems for land resources assessment.* Oxford: Oxford University Press.

Climatologically normal for Egypt. 2011. The normal for El-Gharbia Governorate Station (1960–2011), Ministry of Civil Aviation: Meteorological.

Dawoud, M. A. 2004. Design of national groundwater quality monitoring network in Egypt, *Environmental Monitoring and Assessment* 96: 99–118. http://dx.doi.org/10.1023/B:EMAS.0000031718.98107.eb

Dobos, E.; Micheli, E.; Baumgardner, M. F.; Biehl, L.; Helt, T. 2000. Use of combined digital elevation models and satellite radiometric data for regional soil mapping, *Geoderma* 97: 367–391. http://dx.doi.org/10.1016/S0016-7061(00)00046-X

Dobos, E.; Norman, B.; Bruce, W.; Luca, M.; Chris, J.; Erika, M. 2002. The use of DEM and satellite images for regional scale soil database, in *17th World Congress of Soil Science (WCSS)*, 14–21 August 2002, Bangkok, Thailand.

Dumont, H. J. 2009. A description of the Nile Basin, and a synopsis of its history, ecology, biogeography, hydrology, and natural resources, in H. J. Dumont (Ed.). *The Nile: origin, environments, limnology and human use.* New York: Springer. http://dx.doi.org/10.1007/978-1-4020-9726-3_1

Fanos, A. M. 2002. *Report on the modified Mega Delta Workshop.*

FAO. 2006. *Guidelines for soil description.* 4th ed. FAO, Rome. ISBN 92-5-105521-1.

Govil, P. K.; Reddy, G. L. N.; Krishne, A. K. 2001. Contamination of soil due to heavy metals in Patanchera Industrial Development Area: Andra Pradesh, India, *Environmental Geology* 41: 461–469. http://dx.doi.org/10.1007/s002540100415

Gowd, S. S.; Reddy, M. R.; Govil, P. K. 2010. Assessment of heavy metal contamination in soils at Jajmau (Kanpur) and Unnao industrial areas of the Ganga Plain, Uttar Pradesh, India, *Journal of Hazardous Materials* 174: 113–121. http://dx.doi.org/10.1016/j.jhazmat.2009.09.024

He, J. Y.; Jia, X. 2004. Arc GIS geostatistical analyst application in assessment of MTBE contamination, in *ESRI User Conference*, Fremont, USA.

ITT. 2009. ITT *Corporation ENVI 4.7 software.* New York: White Plains.

Jankaite, A.; Vasarevičius, S. 2005. Remediation technologies for soils contaminated with heavy metals, *Journal of Environmental Engineering and Landscape Management* 13(2): 109–113.

Lagerkvist, J-S. B.; Oskarsson, A. 2007. Vanadium, in G. Nordberg, B. Fowler, M. Nordberg and L. Friberg, L. (Eds.). *Handbook on the toxicology of metals.* 3rd ed. San Diego: Academic Press.

Lillesand, T. M.; Kiefer, R. W. 2003. *Remote sensing and image interpretation.* 2nd ed. New York: John Wiley.

Lillesand, T. M.; Kiefer, R. W. 2007. *Remote sensing and image interpretation.* 5th ed. New York: John Wiley.

Ling, W.; Shen, Q.; Gao, Y.; Gu, X.; Yang, Z. 2007. Use of bentonite to control the release of copper from contaminated soils, *Australian Journal of Soil Research* 45(8): 618–623. http://dx.doi.org/10.1071/SR07079

Lu, S. G.; Bai, S. Q. 2010. Contamination and potential mobility assessment of heavy metals in urban soils of Hangzhou, China, *Environmental Earth Sciences* 60(7): 1481–1490. http://dx.doi.org/10.1007/s12665-009-0283-2

McLaughlin, M. J.; Zarcinas, B. A.; Stevens, D. P.; Cook, N. 2000. Soil testing for heavy metals. *Communications in Soil Science and Plant Analysis* 31(11–14): 1661–1700. http://dx.doi.org/10.1080/00103620009370531

Ohnesorge, F. K.; Wilhelm, M. 1991. Zinc, in E. Merian (Ed.). *Metals and their compounds in the environment, occurrence, analysis and biological relevance.* Verlagsgesellschaft, Weinheim, Basel: VCH.

Rahman, S.; Khanam, D.; Adyel T.; Islam, M. Sh.; Mohammad Ahsan, A.; Akbor, M. A. 2012. Assessment of heavy metal contamination of agricultural soil around Dhaka Export Processing Zone (DEPZ), Bangladesh: Implications of seasonal variations and indices, *Applied Sciences* 2(3): 584–601. http://dx.doi.org/10.3390/app2030584

Raven, P. H.; Berg, L. R.; Johnson, G. B. 1998. *Environment.* 2nd ed. New York: Saunders College Publishing.

Reimann, C.; de Caritat, P. 1998. *Chemical elements in the environment.* Berlin: Springer-Verlag. http://dx.doi.org/10.1007/978-3-642-72016-1

Romic, M.; Romic, D. 2003. Heavy metal distribution in agricultural top-soils in urban areas, *Environmental Geology* 43: 795–805.

Said, R. 1993. *The River Nile: geology, hydrology and utilization.* Oxford: Pergamon Press.

Shepard, D. 1968. *A two-dimensional interpolation function for irregularly-spaced data.* New York: ACM. http://dx.doi.org/10.1145/800186.810616

Sundermann, F. W. (Jr.); Oskarsson, A. 1991. Nickel, in E. Merian (Ed.). *Metals and their compounds in the environment; occurrence, analysis and biological relevance.* Weinteim, Basel: VCH.

Tong, S. T. Y. 1998. Indoor and outdoor household dust contamination in Cincinnati, Ohio, USA, *Environmental Geochemistry and Health* 20(3): 123–133. http://dx.doi.org/10.1023/A:1006561832381

USDA. 2010. *Keys to soil taxonomy.* 11th ed. United States Department of Agriculture, Natural Resources Conservation Service (NRCS).

WHO. 1996. *Trace elements in human nutrition and health.* World Health Organization: Geneva.

Wong, M. H.; Wong, J. W. C. 1986. Effects of fly ash on soil microbial activity, *Environmental Pollution* A(40): 127–144. http://dx.doi.org/10.1016/0143-1471(86)90080-2

Woo, K. S.; Jo, J. H.; Basu, P. K.; Ahn, J. S. 2009. Stress intensity factor by p-adaptive refinement based on ordinary Kriging interpolation, *Finite Elements in Analysis and Design* 45(3): 227–234. http://dx.doi.org/10.1016/j.finel.2008.10.002

Zeydan, B. A. 2005. The Nile Delta in a global vision, in *9th International Water Technology Conference, IWTC9 2005,* Sharm El-Sheikh, Egypt, 31–40.

Zheng, C. 2006. Using multivariate analyses and GIS to identify pollutants and their spatial patterns in urban soils in Galway, Ireland, *Environmental Pollution* 142(3): 501–511. http://dx.doi.org/10.1016/j.envpol.2005.10.028

Mohamed S. SHOKR. Assistant Lecturer, received the degrees of B.Sc. and M.Sc from Tanta University, Egypt. Currently, he is Assistant Lecturer in the Soils and Water Department, Tanta University and has a scholarship from the Egyptian Government to complete part of his Ph.D. studies in the University of Wolverhampton, UK. His research activities are concerned with pedology, soil sustainability, remote sensing and GIS.

Ahmed A. EL BAROUDY. Received the degrees of B.Sc., M.Sc. and Ph.D. from The University of Tanta, Egypt. Currently, he is Assoc. Prof. of Soil in the Soil and Water Department of the University of Tanta. His research activities are mainly concerned with pedology, soil degradation, desertification, land evaluation, remote sensing and GIS modelling. He has published widely in Soil Science, having authored 21 refereed and conference papers.

Michael A. FULLEN. Professor Fullen received the degrees of B.Sc. and M.Sc. from The University of Hull (UK), a Ph.D. from the UK Council for National Academic Awards (CNAA) and a D.Sc. from The University of Wolverhampton. Currently, he is a Professor of Soil Technology at the University of Wolverhampton, UK. His research activities are mainly concerned with soil erosion, soil conservation, desertification and desert reclamation and his fieldwork is mainly based in Europe and Asia. He has published widely in Soil Science (as of March 2016, he has authored one book, 221 refereed papers, 217 conference papers and 26 consultancy reports). He is a referee for 48 journals and a member of the Editorial Board of 24 journals.

Talaat R. EL-BESHBESHY. Received the degrees of B.Sc., M.Sc. and Ph.D. from The University of Minia, Egypt. Currently, he is Professor of Soil Science and Plant Nutrition in the Soil and Water Department of the University of Tanta. His research activities are mainly concerned with plant nutrition, hydroponics, soil fertility and soil chemistry. He has published widely in Soil Science, having authored 25 refereed and conference papers.

Ramadan R. ALI. Professor of Soil Science at the National Research Centre, Cairo, since 1999. He was appointed as a Research Assistant in the Soils and Water Use Department. In 1993, he was awarded the degree of M.Sc. and was then appointed as Assistant Researcher. He obtained a PhD in 2013 and was then appointed as a Researcher. During 2003–2015 he published 56 scientific papers. He became Assistant Professor in 2008 and then Professor in 2014. He also participated in 24 research projects, published three book chapters and supervised several M.Sc. and Ph.D. degrees.

Abd ELHALIM. Dr, received the degrees of B.Sc. and M.Sc. from The University of Tanta and a PhD from The University of Tanta through a scientific association with the University of Rostock, Germany. Currently, he is Assoc. Prof. of Soil Physics and Water Relations at the University of Tanta. His research activities are mainly concerned with soil physics, soil erosion, soil conservation, irrigation water management, evapotranspiration models and desert reclamation. He has published widely in Soil Science (he has authored 12 refereed papers and three conference papers).

Antonio J. T. GUERRA. Professor Guerra received the degrees of B.Sc. and M.Sc. from the Federal University of Rio de Janeiro (Brazil) and a PhD from King's College London (UK). He undertook his first Post-doctoral Fellowship at Oxford University (UK) and his second at the University of Wolverhampton (UK). Currently, he is Professor of Physical Geography at the Department of Geography, Federal University of Rio de Janeiro, where he co-ordinates LAGESOLOS (the Laboratory of Environmental Geomorphology and Soil Degradation). His research activities are mainly concerned with soil erosion, land degradation, soil rehabilitation and his fieldwork is mainly based in Rio de Janeiro and São Paulo States (Brazil). He has published nearly 100 papers in refereed papers in both Brazilian and international journals. He has co-edited 17 books on geomorphology, soil erosion, environmental geomorphology, tourism, and soil rehabilitation in Brazil.

Maria C. O. JORGE. Received the degree of B.Sc. from the Federal University of Paraná (Brazil) and M.Sc. from the State University of Rio Claro (Brazil). Currently she is completing her PhD at the Federal University of Rio de Janeiro and she has just finished the sandwich component of her PhD at the University of Wolverhampton (2015). So far she has published 14 refereed papers in both Brazilian and international journals. She has also co-edited two books on land degradation and soil rehabilitation in Brazil. Her research activities are concerned with geotourism, geoconservation and geodiversity and her field work is conducted in Ubatuba Municipality, São Paulo State, Brazil.

7

INFLUENCE OF FARMYARD MANURE ON RETENTION AND AVAILABILITY OF NICKEL, ZINC AND LEAD IN METAL-CONTAMINATED CALCAREOUS LOAM SOILS

Muhammad Abbas AZIZ[a], Hamaad Raza AHMAD[b], Dennis L. CORWIN[c], Muhammad SABIR[b], Khalid Rehman HAKEEM[d], Münir ÖZTÜRK[e]

[a]*Fauji Fertilizer Company, Dera Ghazi Khan, Pakistan*

[b]*University of Agriculture Faisalabad, 38040 Faisalabad, Pakistan*

[c]*USDA Agricultural Research Service, 20250 Washington, United States*

[d]*Department of Biological Sciences, Faculty of Science, King Abdulaziz University, Jeddah-21589, Saudi Arabia*

[e]*Botany Department & Centre for Environmental Studies, Ege University, Izmir, 35100 Bornova, Turkey*

Abstract. Continuous irrigation of soils with untreated effluents can result in the accumulation and translocation of some metals in the soils and plants. Application of farmyard manure (FYM) to such soils may increase or decrease their availability and retention time. Calcareous soils contaminated with 100, 200, and 400 mg kg^{-1} Ni, Zn, and Pb as chloride salts were used, and farmyard manure added (40 g kg^{-1} for 90 days) with moisture contents at field capacity. Soil samples were drawn at 30 day intervals, and metals extracted with (AB-DTPA) $C_{14}H_{23}NO_3O_{10}$. With FYM application of 400 mg kg^{-1}, Ni availability increased from 179 (day 30) to 240 mg kg^{-1} (day 90); Zn from 163 (day 30) to 230 mg kg^{-1} (day 90), but, Pb decreased from 214 to 161 mg kg^{-1}. FYM forms multi-dentate complex which greatly enhances the Ni and Zn solubility, and organic matter immobilizes Pb in the soil.

Keywords: availability, nickel, zinc, lead, farmyard manure, contaminated soil.

Introduction

Irrigation with raw city effluents is becoming a major threat to environment sustainability in many of the developing countries. Raw city effluent is taken as a good source of nutrients, however, depending upon its sources; it can introduce many of the heavy metal elements in soils (Murtaza *et al.* 2010). From soil, heavy metals can enter into the food chain through the consumption of contaminated food. Prolonged consumption of such contaminated food could pose severe threat to human health. Therefore, management and remediation of metal contaminated soils are very important for producing safe food from such soils. Remediation or cleaning up of contaminated sites with engineering based technologies requires heavy cost and skilled manpower. Moreover, such technologies are disruptive to site and are not considered as environmentally friendly because of the harsh impact of amendments on site properties. Therefore, it is need of time to identify soil amendments for not only improving soil properties but also restraining the mobility and availability of heavy metals to plants and leaching to ground water.

Phytoavailabilty of metals depend upon physical and chemical characteristics of soil such as pH, redox potential, cation exchange capacity, composition and concentration of organic amendments (Pinto et al.2015). Organic amendments like farmyard manure (FYM), press mud, compost etc. not only improve physical, chemical, biological and fertility property of soils, but can also be employed in altering the availability of metal contaminants in soils. Soil-applied animal manures increased the solubility of metals (Ni, Zn, and Pb) shortly after adding, through the formation of soluble organo-metallic complexes (Almas *et al.* 1999). There is also the possibility that metals become more tightly bound to organic materials, either through diffusion into or through occlusion in organic molecules (Davis 1984). Metals like Zn, Pb, and Ni reacted with soil organic matter by ion exchange, complexation, and precipitation reactions, but complexation reactions are thought to dominate (Clemente *et al.* 2006; Ping *et al.*

Corresponding author: Khalid Rehman Hakeem
E-mail: kur.hakeem@gmail.com; khakim@kau.edu.sa

2008). It is well known that the availability of heavy metals decreases with increasing retention of a heavy metal either by silicate clays, calcite, and/or soil organic matter. Such retention of heavy metals is ascribed to reactions between metal ions and soils like adsorption, precipitation, and/or complexation (Alexander 2000; Joner, Leyval 2001; Park *et al.* 2011). According to Walker *et al.* (2003) the retention of metals (i.e. Cu, Zn, Mn, Fe, and Pb) was greater with manure than with compost. On the other hand, the addition of sewage sludge, increased the mobile fraction of Zn, Cd, Cu, and Ni but decreased that of Pb compared to the control treatment (Usman *et al.* 2004). Immobilization of metals by interaction with humic substances could occur through the formation of insoluble complexes or through solid-phase complexation to humates present as coatings on clay surfaces (Stevenson, Cole 1999).

In order to understand the effect of FYM on the availability and retention of heavy metals, detailed information is needed. A large number of studies have been carried out to explore the role of incubation time, type and rate of organic amendments on metal chemical forms and/or fractions in acid and neutral soils (Bataillard *et al.* 2003; Davies *et al.* 2003; Lu *et al.* 2005), however, very little work has been conducted to study the effect of FYM on the availability and retention of heavy metals in calcareous metal contaminated soils. The objective of this study was to understand the effect of FYM on the availability and retention of Ni, Pb, and Zn in metal-contaminated calcareous loam soils at different incubation times.

1. Materials and methods

1.1. Soil sampling and analysis

Soil for the experiment was collected from the plough layer (0–15 cm) in bulk from a field located in the village 199 RB (Rakh Branch) Gattwala, Faisalabad.Soil belongs to Lyallpur soil series (loamy, mixed, hyperthermic Typic Calci-argids) having geographical coordinate viz. longitude 73°11′06″ and latitude 31°27′08″.

Table 1. Physical and chemical characteristics of soil used in the study

Soil characteristics	Unit	Value
pH_s	–	7.72
EC_e	dS m^{-1}	2.4
Organic matter	g kg^{-1}	5.8
$CaCO_3$	g kg^{-1}	54.7
Clay	g kg^{-1}	240.0
Saturation percentage	%	34.4
Textural class	–	Loam
AB-DTPA extractable metals		
Ni	mg kg^{-1}	0.33
Zn	mg kg^{-1}	3.0
Pb	mg kg^{-1}	2.9

Soil was air-dried, thoroughly mixed, and passed through a 2-mm sieve before analysis. These soil samples were analyzed (Table 1) for pH_s (Jenco pH meter 6718, Taipei Hsien, Taiwan, R.O.C.), electrical conductivity (EC_e) by a conductivity meter (HI 8033, Hanna Instruments, Woonsocket, USA), particle size distribution by the hydrometer method (Bouyoucos 1962), organic carbon by the Walkley-Black wet combustion method (Nelson, Sommers 1982), and lime ($CaCO_3$) by the calcimeter method (Moodie *et al.* 1959).

Metals were extracted with AB-DTPA. The AB-DTPA extracting solution was prepared by dissolving 79.10 g NH_4HCO_3 and 1.97g of DTPA in a liter of solution adjusted to pH 7.6 (Soltanpour 1985). Soil (10 g) was placed in a 250-mL Erlenmeyer flask and 20 mL of freshly prepared extracting solution was added. The mixture was shaken on a reciprocating shaker at 180 cycles per minute by keeping the flasks uncovered at room temperature. According to Soltanpour (1985) the extractable AB-DTPA solution allows the DTPA to chelate and extract metals efficiently from calcareous soils. The concentration of Ni, Zn, and Pb in the filtrate was determined by atomic absorption spectrophotometer (AAS; Thermo S-series, Hitachi, Tokyo, Japan). For artificial spiking of soils, solutions containing 100, 200, and 400 mg kg^{-1} Ni, Zn, and Pb as chloride salts were sprayed onto the soil followed by thorough mixing.

After thoroughly mixing the metal solutions with soil, it was placed in plastic pots (line with polyethylene sheet to check leakage of solution) at 1 kg soil per pot. Soil was equilibrated in a greenhouse at field capacity (moisture content approximately (1/3 bar) equivalent to 50% of the saturation percentage of soil for 35 days). After the 35-day of equilibration period, the soil was air-dried and passed through a 2-mm sieve again. Air-dried FYM (Table 2) collected from a nearby livestock farm was passed through a 2-mm sieve and mixed uniformly with metal contaminated soil in the amount of 40 g kg^{-1} soil. Plastic pots of 2-L capacity containing experimental soil were arranged in triplicate using a completely randomized design

Table 2. Properties of farmyard manure

Property	Unit	Value
Moisture content	g g^{-1}	0.13
pH		7.75
EC (1:10 FYM-water slurry)	dS m^{-1}	7.9
Organic carbon	g kg^{-1}	349
Total N	g kg^{-1}	16.8
Avail. Ni	mg kg^{-1}	0.10
Avail. Zn	mg kg^{-1}	0.19
Avail. Pb	mg kg^{-1}	0.22

and the lids of pots were kept loose to allow gaseous exchange and avoid the build up of CO_2 in pots. A similar set of treatments, but without amendments (control), was kept for comparison.

1.2. Analysis of farmyard manure (FYM)

The water soluble contents of Ni, Pb, and Zn in FYM were determined by preparing slurry with distilled water, i.e., 1:10 (w/v) ratio of FYM to distilled water. The pH and EC were also measured from the FYM-water slurry. The organic matter of FYM was determined by dry ashing while N was determined by the Kjehldal method (Jackson 1962).

1.3. Soil incubation and heavy metals determination

The pots were incubated at room temperature (25±2 °C) and moisture content (0.13 g g^{-1}) was maintained at field capacity throughout the experimentation and adjusted gravimetrically through the addition of distilled water. The soil was incubated for a 90- d period. Soil samples (30 g each) were taken at 30, 60, and 90 days and were extracted with AB-DTPA (Soltanpour 1985). The concentrations of Ni, Zn, and Pb were determined using atomic absorption spectrophotometer (AAS; Thermo S-series, Hitachi, Tokyo, Japan).

1.4. Statistical procedures

The significant differences between concentrations of Ni, Zn, and Pb in soils were assessed by ANOVA (Analysis of Variance) and DMR (Duncan's Multiple Range test) using the M STATC version 1.10 packages (MSTAT; East Lansing, MI.) (Russel, Eisensmith 1983) and statistical differences among treatments were defined at $P < 0.05$.

2. Results and discussion

2.1. Soil pH

The importance of soil pH on metal availability is well established as it influences the metal adsorption, retention, and transport. Soil pH was significantly affected by FYM and time during the study (Fig. 1). Soil pH decreased by

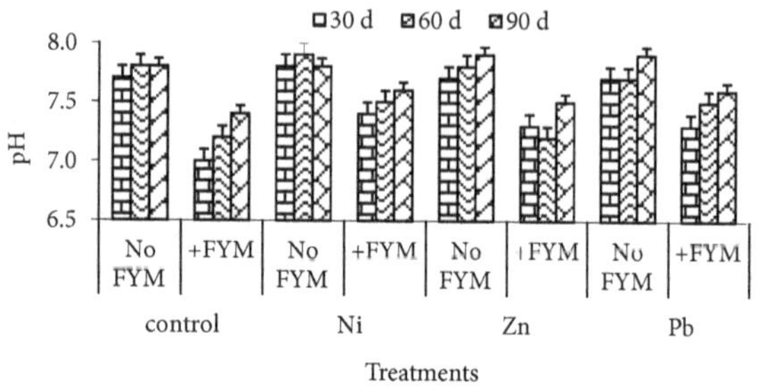

Fig. 1. Effect of FYM addition on soil pH at different incubation times. [FYM – farmyard manure]

the application of FYM. In the control treatment where no FYM was added it was 7.70 at day 30, which decreased to 7.00 by the application of FYM. Soil pH decreased from 7.80 to 7.40 by the application of FYM at day 90 in the control. Its values showed a similar trend for Ni, in the absence of FYM it was 7.80, which decreased to 7.60 with the addition of FYM at day 90. For Zn and Pb the soil pH decreased from 7.90 to 7.50 and 7.90 to 7.60, respectively, with the application of FYM at day 90.

Ostensibly, the decrease in soil pH with FYM application was due to the decomposition of organic matter into humic and fluvic acids, which lower the soil pH (Stevenson, Cole 1999). As soil pH decreases, metals like Ni and Zn must compete with the extra H^+ and Al^{3+} for positions on the exchange sites, solubility of these metals increases in the soil solution and a greater proportion is present as highly available free metal ions in the soil solution (McBride 1982; Sauve *et al.* 1997).

2.2. Availability of nickel

Availability of nickel was not significantly affected throughout the study period (Fig. 2). For the control, native Ni was <1 mg kg^{-1} and slight mobilization was observed with FYM at a first sampling time, i.e., at day 30 (1.07 mg kg^{-1}), which became 1.65 mg kg^{-1} at day 90. At 100 mg kg^{-1} Ni application rate, no major change in available Ni was observed without FYM, decreasing from 29.5 mg kg^{-1} on day 30 to 26.8 mg kg^{-1} at day 90. On the other hand, with manure application, the concentration

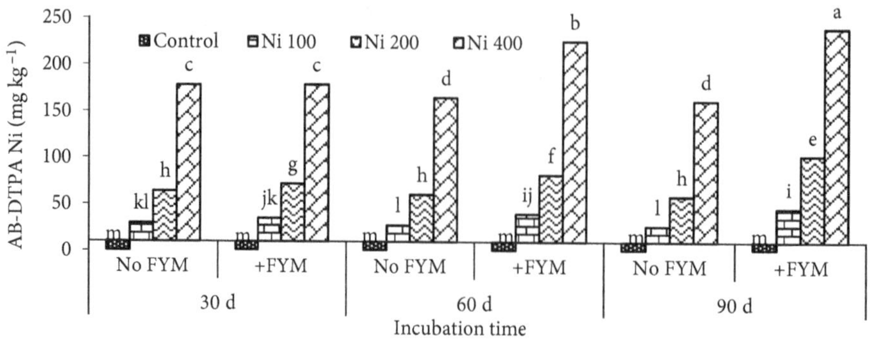

Fig. 2. Effect of incubation time on availability of Ni. FYM – farmyard manure; d – day. Lower case letters indicate the significance of treatments at $P < 0.05$

of Ni increased and remained higher than that in the unamended soil. With FYM addition, 34.8 mg kg^{-1} Ni was in available form at day 30, which increased to 46.2 mg kg^{-1} by the end of the experiment. At 200 and 400 mg kg^{-1} Ni application rates, Ni decreased in unamended soil, whereas manure application increased the Ni concentration. Martinez *et al.* (2003) observed a similar behavior between dissolved organic carbon and soluble Ni concentration in OM-rich soil, i.e., increasing the solubility with incubation time up to 100 days and no further change was observed until the conclusion of the experiment after 600 days. However, the experimental conditions were different as they used natural metal contaminated (Pb 120 mg kg^{-1}, Ni, 800 mg kg^{-1}, and Zn 1480 mg kg^{-1}) silt loam to sandy clay loam acidic grassland soils.

Ostensibly, the increase in available Ni with FYM was due to the formation of soluble Ni-organic complexes from decomposed organic matter (Fig. 2). The addition of FYM to soil results in the increased dissolved organic matter, which facilitated the mobility of metal in soil by acting as a carrier through formation of soluble metal-organic complexes (McCarthy, Zachara 1989; Mahdavi, Jafari 2010). As organic matter decomposes, more soluble organic compounds are released (Merritt, Erich 2003). Increase or decrease in Ni availability with organic matter depended upon its soluble (fulvic acid) or insoluble constituents like humic acid (McBride 1995). According to Stevenson and Cole (1999), humic and fulvic acids form both soluble and insoluble complexes with Ni. Due to their lower molecular weights and higher contents of acidic functional groups, Ni complexes of fulvic acids are more soluble than those of humic acids. Nickel extracted with DTPA was generally larger with increasing addition of exogenous humic acid solutions than that without humic acid application, which was attributed to the formation of Ni–humic complexes that ensured a temporary bioavailability of Ni and prevented their rapid transformation into insoluble species (Halim *et al.* 2003).

The decrease in soil pH from 7.80 to 7.60 with the addition of FYM at day 90 might be associated with organic matter (OM) decomposition responsible for increased concentration of Ni in soil as was also found by Heckman *et al.* (1987), Campbell and Beckett (1988). Similarly, Usman *et al.* (2004) also attributed the increase in concentration of Ni in incubated sludge-amended soil to the formation of soluble Ni-organic ligands associations and to a resultant decrease in soil pH (from 8.14 to 7.54) but the soil used was sandy with low organic carbon and not artificially metal incubated. The fast decomposition rate of organic matter could consume oxygen, thus possibly affect the redox conditions of soils, one of the main factors for increasing Ni solubility along with a change in soil pH and CO_2 concentration (Ross 1994).

2.3. Availability of zinc

Availability of Zn showed a similar pattern as was observed in the case of Ni. In the absence of added organic matter, Zn in the control soil was 3.7 mg kg^{-1} at day 30 and only 50% of applied Zn (1.78 mg kg^{-1}) was recovered at the end of the experiment (Fig. 3). The FYM application decreased extractability of the native Zn from 3.5 mg kg^{-1} to 2.4 mg kg^{-1}. In Zn spiked soil without FYM, Zn decreased at all the sampling times, whereas in manure amended soil, the availability of Zn gradually increased until the end of the experiment from 163at day 30 to 230 mg kg^{-1}at day 90 with400 mg kg^{-1} of Zn application, respectively.

The AB-DTPA extractable Zn in response to FYM added soil increased with time (Fig. 3). Previously Almas *et al.* (2000) observed the parallel behavior of Zn but under a different environment as they used acidic clay loam and marine loam soil texture for their experiment; they reported that the addition of organic matter (pig manure) increased the solubility of Zn through the formation of soluble organo-metallic complexes. Catlett *et al.* (2002) concluded that binding of Zn^{2+} with OM has a significant positive impact on its solubility in fine silty neutral to alkaline calcareous soils. The decomposition of organic matter resulted in the production of low molecular organic acid like acetic, formic, oxalic, lactic, propionic, malic,

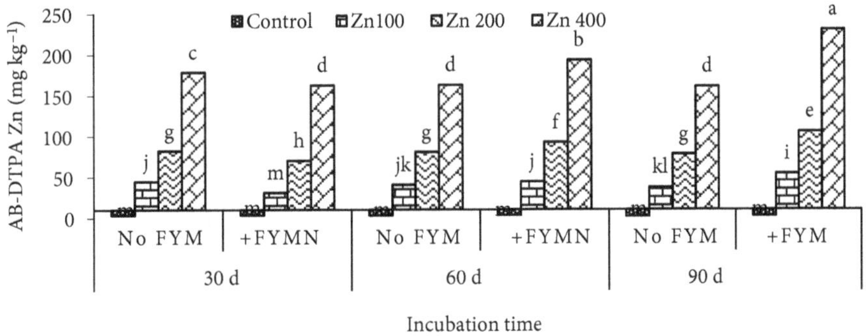

Fig. 3. Effect of incubation time on availability of Zn. FYM – farmyard manure; d – day. Lower case letters indicate the significance of treatments at $P < 0.05$

citric, and aconitic acids (Baziramakenga, Simrad 1998) to affect the pH perturbations as happened in our case the soil pH decreased 7.90 to 7.50 with the addition of organic matter at day 90 resulting increase in Zn availability.

At the end of our experiment, soluble organic components might have persisted resulting in higher availability of Zn. In addition, $CaCO_3$ in calcareous soils can lead to Zn precipitation as $ZnCO_3$ (Usman *et al.* 2004), but $ZnCO_3$ is soluble, hence more available. According to Lindsay (1979), minerals like ZnO (zincite) and $ZnCO_3$ (smithsonite) are too soluble to persist in soils. Antoniadis and Alloway (2002) reported increased extractability of Zn with an increase in dissolved organic carbon in sludge amended soils. In another study, Ashworth and Alloway (2004) found that Zn did not leach from columns filled with sewage sludge-amended sandy loam soil indicating its retention by soil. Pascual *et al.* (2004) reported lower Ni and Zn concentrations in unamended soil at the end of an incubation period of 64 days than those in the present study.

2.4. Availability of lead

Behavior of applied Pb was different from that exhibited by Ni and Zn (Fig. 4). With an incubation of FYM, the Pb availability decreased in metal–contaminated soil. In the unamended control, native Pb remained unchanged and its concentration was almost the same after 30, 60, and 90 days of incubation. The availability of Pb from the soil contaminated with 100, 200, and 400 mg kg^{-1} of Pb was 52, 98, and 224 mg kg^{-1}, respectively, after 30 days of incubation in the treatments with no FYM, which decreased by 40, 86, and 172 mg kg^{-1}, respectively, after 90 days of incubation. While the availability of Pb from the soil contaminated with Pb 100, 200 and 400 mg kg^{-1} on day 30 was 49, 104, and 214 mg kg^{-1} of the added Pb in the treatments with FYM, respectively. With FYM amended soil, available Pb was less than that in unamended soil and was 27, 79, and 161 mg kg^{-1} of the applied Pb at the end of the experiment, respectively. At the 100 mg kg^{-1} application rate, a decrease in availability was more than that at higher levels of Pb application.

In the present experiment, organic matter has an immobilization effect on Pb in soil. The AB-DTPA extractable Pb decreased at all levels of Pb addition but the decrease was more at the lowest rate of 100 mg Pb kg^{-1} of soil (Fig. 3). Similar findings where Clemente *et al.* (2006) reported EDTA-extractable Pb decreased from 42.7% of the total Pb in control samples to 37.3% in compost-treated samples and to 32.3% in manure treated samples. However, the experiment was conducted with different environment as they used sandy loam soil and followed McGrath and Cegarra (1992) procedure for metal extraction. Precipitation of Pb with $CaCO_3$ may also be responsible for the decreased availability of Pb (McBride 1994; Kumpiene *et al.* 2008). Over a 90-day incubation period, mobile Pb decreased to 0.62% of the total Pb contents in soil treated with compost (Usman *et al.* 2004). Precipitation of Pb, Zn, Cd, and Cu as insoluble compounds appeared to become important at approximately pH 6–7. At pH values of around 6 or 7, very little Pb is likely to be found in solution (Ram, Verloo 1985). This was attributed to the products like humic and fulvic acids resulting from decomposition of organic matter with which Pb formed stable organo-metallic complexes alone and in association with soil clays. As pointed out by Stevenson and Cole (1999), both fulvic and humic acids form soluble as well as insoluble complexes depending upon the degree of saturation. The importance of organic matter in limiting Pb availability has also been demonstrated by Strawn and Sparks (2000) and Saini and Gupta (2001).

Conclusions

From a soil and groundwater contamination point of view the retention of heavy metal in soil and their availability to plants is important. A calcareous loam soil was tested to investigate the effect of FYM on the availability of applied Ni, Zn, and Pb over a period of time. The availability of Ni and Zn increased with incubation time, but Pb availability decreased over time. Ostensibly, added FYM formed soluble organo-metallic complexes with Ni, and Zn increasing their availability with time. Whereas a decrease

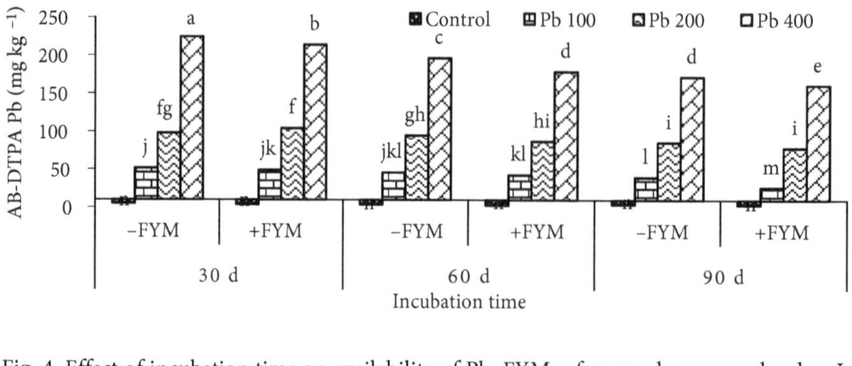

Fig. 4. Effect of incubation time on availability of Pb. FYM – farmyard manure; d – day. Lower case letters indicate the significance of treatments at $P < 0.05$

in Pb availability might be due to the presence of $CaCO_3$ resulting in its precipitation. The general order of affinity of heavy metal retention is as follows Cu > Zn > Pb.

Increased availability of Ni, and Zn with the application of farmyard manure could result in an increased risk to soil and groundwater resources. Continuous application of FYM with the passage of time not only significantly reduced metal sorption, but also increased their mobility through the formation of soluble metal complexes in alkaline soils and contaminates the groundwater through leaching of metals from these soils. During the application of FYM the nature, composition, physico-chemical properties of soil and microbial activities should be considered. More studies are needed to understand the phenomena of the increased availability of heavy metal by the addition of FYM and to prevent plants from their phytotoxic effect. However, the increased bioavailability of heavy metals by the application of farmyard manure can be used while phytoremediating the contaminated sites.

Disclosure statement

Authors declare no conflict of interest among them.

References

Alexander, M. 2000. Aging, bioavailability, and overestimation of risk from environmental pollutants, *Environmental Science & Technology* 34: 4259–4264. https://doi.org/10.1021/es001069+

Almas, A. R.; McBride, M. B.; Singh, B. R. 2000. Solubility and liability of cadmium and zinc in two soils treated with organic matter, *Soil Science* 165: 250–259. https://doi.org/10.1097/00010694-200003000-00007

Almas, A. R.; Singh, B. R.; Salbu, B. 1999. Mobility of cadmium and zinc in soil influenced by equilibration, time, temperature and organic matter, *Journal of Environmental Quality* 28: 1742–1750. https://doi.org/10.2134/jeq1999.00472425002800060008x

Antoniadis, V.; Alloway, B. J. 2002. The role of dissolved organic carbon in the mobility of Cd, Ni and Zn in sewage sludge-amended soils, *Environmental Pollution* 117: 515–521. https://doi.org/10.1016/S0269-7491(01)00172-5

Ashworth, D. J.; Alloway, B. J. 2004. Soil mobility of sewage sludge-derived organic matter, copper, nickel and zinc, *Environmental Pollution* 127: 137–144. https://doi.org/10.1016/S0269-7491(03)00237-9

Bataillard, P.; Cambier, P.; Picot, C. 2003. Short-term transformation of lead and cadmium compounds in soil after contamination, *European Journal of Soil Science* 54: 365–376. https://doi.org/10.1046/j.1365-2389.2003.00527.x

Baziramakenga, R.; Simrad, R. R. 1998. Low molecular weight aliphatic acid contents of composted manure, *Journal of Environmental Quality* 27: 557–561. https://doi.org/10.2134/jeq1998.00472425002700030012x

Bouyoucos, G. J. 1962. Hydrometer method improved for making particle-size analyses of soils, *Agronomy Journal* 54: 464–465. https://doi.org/10.2134/agronj1962.00021962005400050028x

Campbell, D. J.; Beckett, P. H. T. 1988. The soil solution in a soil treated with digested sewage sludge, *European Journal of Soil Science* 39: 283–298. https://doi.org/10.1111/j.1365-2389.1988.tb01215.x

Catlett, K. M.; Heil, D. M.; Lindsay, W. L. 2002. Ebinger MH. Soil chemical properties controlling zinc^{2+} activity in 18 Colorado soils, *Soil Science Society of America Journal* 66: 1182–1189. https://doi.org/10.2136/sssaj2002.1182

Clemente, R.; Escolar, A.; Bernal, M. P. 2006. Heavy metals fractionation and organic matter mineralization in contaminated calcareous soil amended with organic materials, *Bioresource Technology* 97: 1894–1901. https://doi.org/10.1016/j.biortech.2005.08.018

Davies, N. A.; Hodson, M. E.; Black, S. 2003. The influence of time on lead toxicity and bioaccumulation determined by OECD earthworm toxicity test, *Environmental Pollution* 121: 55–61. https://doi.org/10.1016/S0269-7491(02)00207-5

Davis, J. A. 1984. Complexation of trace metals by adsorbed natural organic matter, *Geochimica et Cosmochimica Acta* 48: 679–691. https://doi.org/10.1016/0016-7037(84)90095-4

Halim, M.; Conte, P.; Piccolo, A. 2003. Potential availability of heavy metals to phytoextraction from contaminated soils induced by exogenous humic substances, *Chemosphere* 52: 265–275. https://doi.org/10.1016/S0045-6535(03)00185-1

Heckman, J. R.; Angle, J. S.; Chaney, R. L. 1987. Residual effects of sewage sludge on soybean: I. accumulation of heavy metals, *Journal of Environmental Quality*16: 113–117. https://doi.org/10.2134/jeq1987.00472425001600020004x

Jackson, M. L. 1962. Chemical composition of soil, in F. E. Bean (Ed.). *Chemistry of soil.* New York: Van Nostrand Rchinheld Co., 71–144.

Joner, E. J.; Leyval, L. C. 2001. Time-course of heavy metal uptake in maize and clover as affected by root density and different mychorrhizal inoculation regimes, *Biology and Fertility of Soils* 33: 351–357. https://doi.org/10.1007/s003740000331

Kumpiene, J.; Lagerkvist, A.; Maurice, C. 2008. Stabilization of As, Cr, Cu, Pb and Zn in soil using amendments – a review, *Waste Manage* 28: 215–225. https://doi.org/10.1016/j.wasman.2006.12.012

Lindsay, W. 1979. *Chemical equilibrium in soils.* New York, NY, USA: John Wiley and Sons.

Lu, A.; Zhang, S.; Shan, X. 2005. Time effect on the fraction of heavy metals in soils, *Geoderma* 125: 225–234. https://doi.org/10.1016/j.geoderma.2004.08.002

Mahdavi, M.; Jafari, J. 2010. Environmental risks due to application of sewage sludge in farmlands, *Ozean Journal of Applied Sciences* 3: 303–311.

Martinez, C. E.; Jacobson, A. R.; McBride, M. B. 2003. Ageing and temperature effect on DOC and elemental release from a metal contaminated soil, *Environmental Pollution* 122: 135–143. https://doi.org/10.1016/S0269-7491(02)00276-2

McBride, M. B. 1982. Electron spins resonance investigation of Mn^{2+} complexation in natural and synthetic organics, *Soil Science Society of America Journal* 46: 1137–1143. https://doi.org/10.2136/sssaj1982.03615995004600060004x

McBride, M. B. 1994. *Environmental chemistry of soils.* New York, NY, USA: Oxford Univ. Press.

McBride, M. B. 1995. Toxic metal accumulation from the use of sludge: are USEPA regulations protective?, *Journal of Environmental Quality* 24: 5–18. https://doi.org/10.2134/jeq1995.00472425002400010002x

McCarthy, J. F.; Zachara, J. M. 1989. Subsurface transport of contaminants, *Environmental Science & Technology* 23: 496–502. https://doi.org/10.1021/es00063a001

McGrath, S. P.; Cegarra, J. 1992. Chemical extractability of heavy metals during and after long-term application of sewage sludge to soil, *European Journal of Soil Science* 43(2): 313–321. https://doi.org/10.1111/j.1365-2389.1992.tb00139.x

Merritt, K. A.; Erich, M. S. 2003. Influence of organic matter decomposition on soluble carbon and its copper-binding capacity, *Journal of Environmental Quality* 32(6): 2122–2131. https://doi.org/10.2134/jeq2003.2122

Moodie, C. D.; Smith, H. W.; McCreery, R. A. 1959. *Laboratory manual for soil fertility*. Washington State College, Mimeograph, USA, 31–39.

Murtaza, G.; Ghafoor, A.; Qadir, M.; Owens, G.; Aziz, M. A.; Zia, M. H.; SAIFULLAH. 2010. Disposal and use of sewage on agricultural lands in Pakistan: a review, *Pedosphere* 20: 23–34. https://doi.org/10.1016/S1002-0160(09)60279-4

Nelson, D. W.; Sommers, L. E. 1982. Total carbon, organic carbon, and organic matter, in A. L. Page, *et al.* (Eds.). *Methods of soil analysis. Part 2. Chemical and microbiological properties*. 2nd ed. Agronomy Monograph No. 9. American Society of Agronomy, Soil Science Society of America, 539–577.

Park, J. H.; Lamb, D.; Paneerselvam, P.; Choppala, G.; Bolan, N.; Chung, J. W. 2011. Role of organic amendments on enhanced bioremediation of heavy metal (load) contaminated soils, *Journal of Hazardous Materials* 185(2–3): 549–574. https://doi.org/10.1016/j.jhazmat.2010.09.082

Pascual, I.; Antolin, M. C.; Garcia, C.; Polo, A.; Sanchez-Diaz, M. 2004. Plant availability of heavy metals in a soil amended with a high dose of sewage sludge under drought conditions, *Biology and Fertility of Soils* 40(5): 291–299. https://doi.org/10.1007/s00374-004-0763-1

Ping, Li.; Xingxiang, W.; Taolin, Z.; Dongmei, Z.; Yuanqiu, H. E. 2008. Effects of several amendments on rice growth and uptake of copper and cadmium from a contaminated soil, *Journal of Environmental Sciences* 20(4): 449–455. https://doi.org/10.1016/S1001-0742(08)62078-1

Pinto, E.; Almeida, A. A.; Ferreira, I. M. P. L. V. O. 2015. Assessment of metal(loid)s phytoavailability in intensive agricultural soils by the application of single extractions to rhizosphere soil, *Ecotoxicology and Environmental Safety* 113: 418–424. https://doi.org/10.1016/j.ecoenv.2014.12.026

Ram, N.; Verloo, M. 1985. Effect of various organic materials on the mobility of heavy metals in soil, *Environmental Pollution Series B, Chemical and Physical* 10(4): 241–248. https://doi.org/10.1016/0143-148X(85)90017-5

Ross, S. M. 1994. Retention, transformation and mobility of toxic metals in soils, in S. M. Ross (Ed.). *Toxic metals in soil-plant system*. Chi Chester, UK: Jhon Wiley and Sons. Ltd., 63–152.

Russel, D. F.; Eisensmith, S. P. 1983. *MSTAT-C*. Crop Soil Science Department, Michigan State University, East Lansing, Mich. USA.

Saini, S. P.; Gupta, V. K. 2001. Lead and FYM influence on yield and Pb concentration of wheat sown on sandy and clay loam soils, *Journal of the Indian Society of Soil Science* 49: 508–510.

Sauve, S.; McBride, M. B.; Norvell, W. A.; Hendershot, W. H. 1997. Copper solubility and speciation of in situ contaminated soils: effects of copper level, pH and organic matter, *Water, Air, and Soil Pollution*100: 133–149. https://doi.org/10.1023/A:1018312109677

Soltanpour, P. N. 1985. Use of AB-DTPA to evaluate elements availability and toxicity, *Communications in Soil Science and Plant Analysis* 16: 323–338.

Stevenson, F. J.; Cole, M. A. 1999.*Cycles of soils: carbon, nitrogen, phosphorus, sulfur, micronutrients*. 2nd ed. New York, NY, USA: John Wiley and Sons.

Strawn, D, G.; Sparks, D. L. 2000. Effects of soil organic matter on the kinetics and mechanisms of Pb(II) sorption and desorption in soil, *Soil Science Society of America Journal* 64: 144–156. https://doi.org/10.2136/sssaj2000.641144x

Usman, A. R. A; Kuzyakov, Y.; Stahr, K. 2004. Dynamics of organic C mineralization and the mobile fraction of heavy metals in calcareous soils incubated with organic wastes, *Water, Air, and Soil Pollution* 158(1): 401–418. https://doi.org/10.1023/B:WATE.0000044864.07418.8f

Walker, D. J.; Clemente, R.; Rig, A; Bernal, M. P. 2003. The effects of soil amendments on heavy metal bioavailability in two contaminated Mediterranean soils, *Environmental Pollution* 122(2): 303–312. https://doi.org/10.1016/S0269-7491(02)00287-7

Muhammad Abbas AZIZ. Dr, PhD. Presently working for Fauji fertilizers, Pakistan. He has published ten papers in well-reputed refereed national and international journals. He also got William Roberts scholarship and HEC indigenous merit scholarship for PhD sponsored by Pakistan Government.

Hamaad Raza AHMAD. Dr, PhD. Working as Assistant professor at Institute of Soil and Environmental Sciences, University of Agriculture, Faisalabad, Pakistan. He has published more than fifty research articles in well-reputed refereed national and international journals. He has guided 21 MSc (Hons.) thesis and acted as committee member of 16 students. He also got Postdoc fellowship from HEC, Pakistan.

Dennis CORWIN. PhD. Working as Research Soil and Environmental Scientist at USDA-ARS United States Salinity Laboratory, Riverside.CA., YSA. He has more than 150 publications in well-reputed international journals. He got seven awards including American Society of Agronomy Fellow (2014) and Soil Science Society of America Fellow (2014). He is the Editor of many world reputed journals.

Muhammad SABIR. Dr, PhD. Working as Assistant professor at Institute of Soil and Environmental Sciences, University of Agriculture, Faisalabad, Pakistan. He has won Postdoc fellowships from HEC, Pakistan and Malaysian Governments. He has published more than fifty research articles in well-reputed refereed national and international journals. He has guided 23 MSc (Hons.) thesis and acted as committee member of 14 students.

Khalid Rehman HAKEEM. PhD. Currently, working as Associate Professor at King Abdulaziz University, Jeddah, Saudi Arabia, has worked at Universiti Putra Malaysia from (2012–2016) as Post Doc fellow and Fellow Researcher (Ass. Prof.) subsequently. He has more than nine (09) years of teaching/research experience in Biological sciences. He has a proven track record of over 60 research publications in highly reputed professional journals, in the fields of Plant stress physiology, soil science, ecology and environmental sciences, in addition to 50 chapters in international books and guest editorship of "The Malaysian Forester-Malaysia" and "Sains Malaysiana". Dr. Hakeem has published 21 books with International publishers and is a recipient of more than 5 honours and awards. Presently, Dr. Hakeem in studying the plant processes at Eco-physiological, as well as Proteomic levels.

Münir ÖZTÜRK. Dr, PhD. He is retired Professor of Ecology and Environmental Sciences. He has a proven track record of over 125 research publications in highly reputed professional journals, in the fields of ecology, environmental sciences and ethnobotany, in addition to 50 chapters in international books and guest editorship of "The Malaysian Forester-Malaysia", "Environmental Geology-Springer", "Journal of Environmental Biology-India" "Social and Behavioral Sciences Procedia-Elsevier". Has published more than 30 books (Daya Publishers, Cambridge Scholars, Birkhauser Verlag, Springer and Elsevier), and is a recipient of more than 15 honours and awards.

8

RUNOFF TREND AND POTENTIALITY IN MELAKA TENGAH CATCHMENT OF MALAYSIA USING SCS-CN AND STATISTICAL TECHNIQUE

Sharif Moniruzzaman SHIRAZI[a], MD Ibrahim ADHAM[b], Faridah OTHMAN[b], Noorul Hasan ZARDARI[c], Zubaidah ISMAIL[b]

[a]*Department of Civil Engineering, World University of Bangladesh, Dhaka, Bangladesh*
[b]*Department of Civil Engineering, Faculty of Engineering, University of Malaya, 50603 Kuala Lumpur, Malaysia*
[c]*Centre for Environmental Sustainability and Water Security (IPASA), Department of Hydraulics and hydrology, Faculty of Civil Engineering, Universiti Teknologi Malaysia, Skudai 81310, Johor Bahru, Malaysia*

Abstract. This study is focused to identify the surface runoff trends and potentiality of the five watersheds transforming the discrete runoff pattern to smooth patterns. Runoff potentiality was analyzed by Soil Conservation Service Curve Number (SCS-CN) technique. Considering Hydrologic Soil Group (HSG) and percentage of particular land use pattern, weighted CNs of five watersheds were found between 82 and 85. Monthly surface runoff trends were investigated by statistical autocorrelation, Mann-Kendall, Sen slope and lowess methods. According to the Mann-Kendall method, no statistical significant monotonic trends were found for all the watersheds. Smoothing curve analysis reveals that the monthly mean runoff is 30 mm, 34 mm, 39 mm, 28 mm and 37 mm and the percentage of runoff is 23%, 25%, 31%, 25% and 26% for the watersheds 1, 2, 3, 4 and 5, respectively. Degree of effect of several land use pattern with corresponding soil type was analyzed to assess the total runoff volume for contributing to the surface water resources. Result shows that 26% of the rainwater contributes to the surface runoff of Melaka Tengah catchment and provides the information for planning of surface water management and potentiality of groundwater recharge.

Keywords: curve number, SCS, surface runoff, Mann-Kendall test, Melaka Tengah catchment.

Introduction

The soil conservation service curve number (SCS-CN) technique is widely used in watershed management and environmental sector for calculating the surface runoff from particular rainfall amount. Climatic factors (e.g. precipitations, rainfall intensity, duration and distribution, relative humidity) and physiographic factors (e.g. topographic slope, watershed size, shape of a watershed, land use, soil group and moisture) influence the surface runoff in a watershed. Similarly, watershed characteristics have had a direct influence on the peak flow and runoff volume in any area (Komatsu *et al.* 2011; Zhang *et al.* 2010). High rainfall intensity on less permeable surface resulted in rapid runoff and ultimately causes flash flood in a watershed. Rainfall duration is another factor that increases runoff in a watershed. During the higher rainfall intensity and longer duration, maximum runoff pick is generated at the time of temporally variable of rainfall. Kinetic, spatial, temporal, overland flow of storm rainfall and climate change impact on storm events contribute their effect on surface runoff (Stephenson 1984; Singh 1997; Pourtouiserkani, Rakhshandehroo 2014; Okoli 2014). Increasing rate of urbanization and its adverse effect cause the surface runoff, and deteriorate groundwater and river water quality (Dimitriou, Moussoulis 2011; Khan *et al.* 2011; Shirazi *et al.* 2012). Land use patterns and soil type are also important factors for surface runoff (Wu *et al.* 2011; Bagdziunaite *et al.* 2011). The steep slope surface can produce higher velocity surface runoff due to short period of time for water infiltration. Hillel (1980) defines the infiltration rate into the soil on the basis of infiltration capacity. This infiltration capacity for a dry soil is higher at the initial period but decreases at a rapid rate due to conductivity of the soil layer or soil surfaces. However, textural characteristic of soil controls the infiltration rate as well as surface runoff.

Corresponding authors: Sharif Moniruzzaman Shirazi, Faridah Othman
E-mails: smshirazi@gmail.com; faridahothman@um.edu.my

The SCS-CN method is preferred to calculate the direct runoff due to its flexibility, simplicity and versatility (Melesse, Shih 2002; Gaudin *et al.* 2010; Hawkins 1978; Ragan, Jakson 1980; Slack, Welch 1980; Hawkins 1993; Lewis *et al.* 2000). SCS-CN (SCS 1972) also estimates the runoff volume for a particular rainfall depth of agricultural watershed area (USDA SCS 1985). This method is analyzed to clarify its theoretical and experimental basis (Ponce, Hawkins 1996) and to predict the land use changing effect on surface runoff in urban hydrology (Chen *et al.* 2009). This method along with functional data analysis technique predict the runoff potentiality of a watershed (Adham *et al.* 2014). Remote sensing data are used to derive runoff pattern and watershed evaluation (Zade *et al.* 2005; Kumar *et al.* 2014). The hydrologic soil group (HSG) defines the curve number (CN) which is input of the SCS-CN method. Grove *et al.* (1998) differentiated the effect of storm runoff depth using the composite and distributed curve numbers. Two-CN approach was also used to identify the variation of runoff curve number with rainfall and identification of spatial distribution SCS-CN parameter was analyzed in heterogeneous watersheds (Soulis, Valiantzas 2012, 2013). Individual land use under particular hydrologic soil group exhibits particular curve number and this number depends on several factors representing runoff potentiality of a watershed (Warren, Gray 2003; Carlesso *et al.* 2011; Sarauskiene *et al.* 2015). Due to the watershed complexity, weighted CN is easier to use with many complexes for large watershed area. This is occurred when same land use pattern belongs to different hydrologic soil groups. The CN for C and D soil groups contributes the greater runoff due to soil's textural pattern. However, area under C and D groups is the first indicator for rapid runoff of the watershed area rather than A and B soil groups. The main goal of this study was to define the weighted curve number of Melaka Tengah watershed in SCS-CN method for calculating the surface runoff. This runoff was analyzed by the Mann-Kendall, Sen slope, lowess statistical methods for detecting the trends of the runoff data set of the watersheds. The Mann-Kendall is the most widely used non-parametric method to detect monotonic trends in hydrological parameters (Antonopoulos *et al.* 2001; Caloiero *et al.* 2011; Hirsch *et al.* 1982; Mohsin, Gough 2010; Río *et al.* 2011). This method is less affected by the existence of outliers in the data series. This non-parametric method was used to detect the rainfall trend in Australia (Chowdhury, Beecham 2010, 2012). Jonsdottir *et al.* (2006) applied the Mann-Kendall method to detect long-term variability of rainfall, temperature and discharge of Iceland Rivers. Annual and seasonal rainfall trends of time series data were investigated by using Mann-Kendall method and showed a significant decreasing trend for the time series in Greece (Feidas *et al.* 2007). Henceforth, this study introduced lowess linear fit for smoothing the discrete runoff data for predicting the surface runoff pattern of this region. Finally, monthly mean runoff was calculated from the smooth curve pattern. Therefore, an attempt has been made to find out runoff trend and potentiality of Melaka Tengah catchment for the best water resources management practice in future.

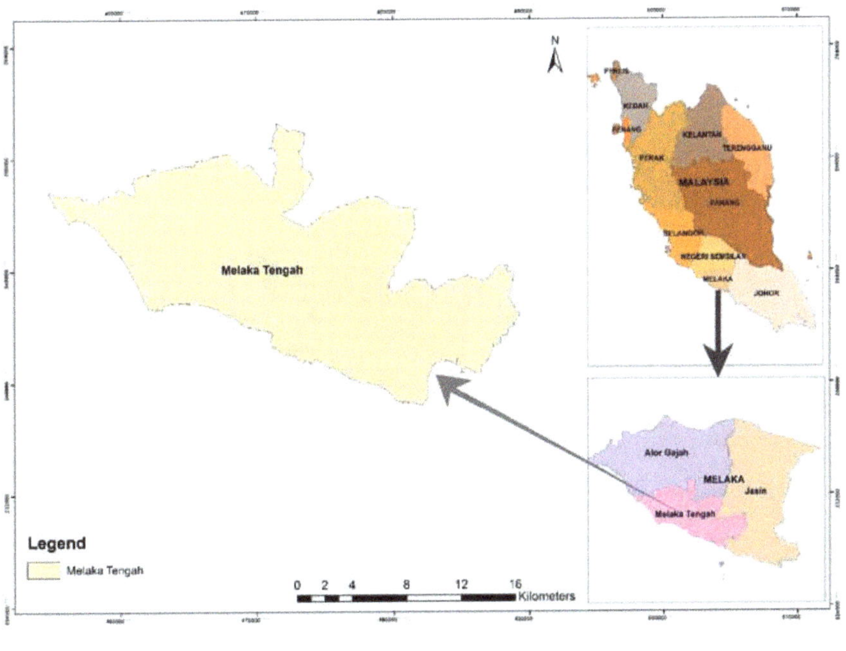

Fig. 1. Location map of the study area

1. Methodology

1.1. Study area

Melaka Tengah catchment with an area of 314 sq. km consists of five watersheds and located in Melaka state of Peninsular Malaysia (Fig. 1). The catchment contains different soils and land use patterns. Elevation of this catchment varied from 20–140 m above sea level. Characteristics of five watersheds are mainly fern shaped in nature (Fig. 2). It also shows the location of rainfall stations (RS) in the Melaka Tengah catchment. Runoff of Melaka Tengah catchment is influenced by practicing ten land use patterns and their land management criteria. Hydrologic soil group was classified into four types A, B, C and D (Mays 2005). These types of soils were grouped on the basis of water infiltration rate through the soil. In general, sandy soil is under the category A having the highest rate of infiltration and less runoff potential. On the contrary, clay soils that have the lowest infiltration rate and high runoff potential were placed in hydrologic soil group D.

The land use and soil group map of Melaka Tengah catchment were collected from the Drainage and Irrigation Department (DID), Malaysia and prepared by using geographical information system (GIS) technique. This soil group with land use pattern exhibits the specific CN values of each category. CN of various land use pattern varies from cultivated land to industrial and residential areas. Surface runoff curve number (CN) depends on soil type, land cover, hydrologic soil group and condition. Based on soil characteristics, soil group is classified for analysis of runoff (Table 1). CN values are used for various land uses on this soil type. Weighted CN can be calculated for a watershed made up of several soil types and land uses.

Table 1. Different hydrologic soil group classification (Mays 2005)

Soil characteristics	Soil Group
Deep sand, deep loess and aggregated silt	A
Shallow loess, sandy loam	B
Clay loams, shallow sandy loam, soils low in organic content and soils usually high in clay	C
Soils that swell significantly when wet, heavy plastic clay and certain saline soils	D

1.2. SCS-CN method

The SCS-CN method is that surface runoff relates to soil-land cover complexes and storm rainfall through a curve number. This method defines that the ratio of actual soil retention after runoff begins to potential maximum retention is equal to the ratio of direct runoff to available rainfall. This method is suitable for the watersheds with area less than 250 square kilometer (Ponce 1989). In this study, five watersheds of Melaka are valid to apply this method to estimate surface runoff due to having area less than 250 square kilometer for each watershed. The Soil Conservation Service Curve Number method (SCS-CN) was used to calculate the runoff of different watersheds in Melaka (Eq. (1)):

$$R_{wi} = \frac{\{P-(I_a)_{wi}\}^2}{\{P-(I_a)_{wi}\}+S_{wi}}, \tag{1}$$

where wi indicates different watershed number, R_{wi} – Runoff (mm), P – Precipitation (mm), S_{wi} – Potential

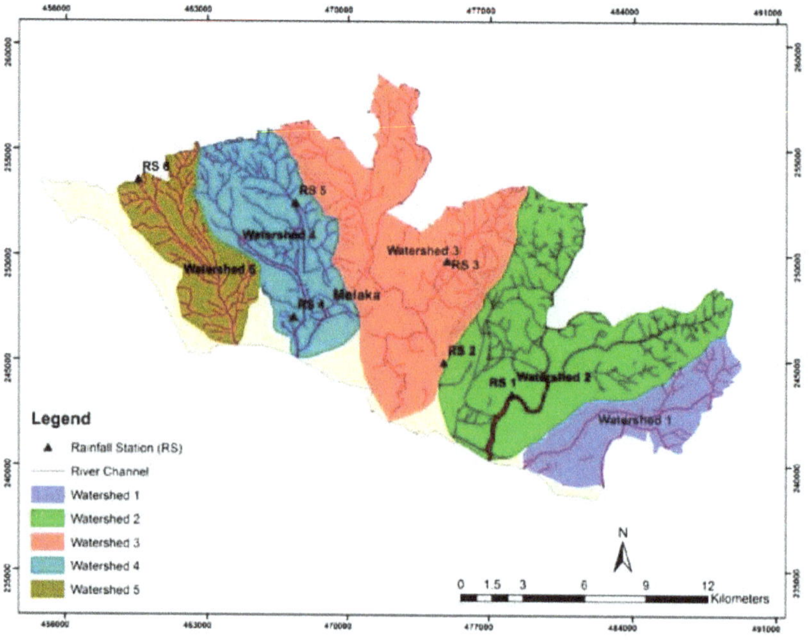

Fig. 2. Watersheds and river channel network at Melaka Tengah catchment

maximum retention after runoff begins (mm) and $(I_a)_{wi}$ – Initial abstraction which is the water losses before surface runoff begins. $(I_a)_{wi}$ is to be approximated in watershed studies by Equation 2:

$$(I_a)_{wi} = 0.2S_{wi}. \quad (2)$$

Therefore, S_{wi} and P are to be allowed to yield the runoff amount and expressed as (Eq. (3)):

$$R_{wi} = \frac{(P - 0.2S_{wi})^2}{P + 0.8S_{wi}}. \quad (3)$$

For suitability of Equation 3, S_{wi} is defined in the form of a dimensionless runoff curve number representing the runoff potential of the watershed. It is related to the land cover-soil complex characteristics governed by the soil moisture condition, soil type, and land use and treatment (USDA SCS 1985). In Melaka Tengah watersheds, S_{wi} exhibits this relation through the weighted curve number (CN_{wi}). SCS-CN method estimates the potential retention through empirical studies and the equation is expressed as (Eq. (4)):

$$S_{wi} = \frac{25400}{CN_{wi}} - 254, \quad (4)$$

where CN_{wi} is a weighted runoff curve number. It is a dimensionless number. The higher value of weighted runoff curve number indicates higher runoff factor or runoff potential of a watershed. S_{wi} is an intrinsic model parameter being independent of initial moisture condition, and found that $(I_a)_{wi}$ and S_{wi} are independent of each other for $(I_a)_{wi}$ is not an intrinsic (Michel *et al.* 2005).

1.3. Mann-Kendall method

Mann-Kendall is a non-parametric trend analysis method which is mostly used to detect the trends in climatological and hydrological time series (Mann 1945; Kendall 1975). The Mann-Kendall test statistic S can be given as:

$$S = \sum_{i=1}^{n-1} \sum_{j=i+1}^{n} sign(x_j - x_i); \quad (5)$$

$$Sign(x_j - x_i) = \begin{Bmatrix} +1 if (x_j - x_i) > 0 \\ 0 if (x_j - x_i) = 0 \\ -1 if (x_j - x_i) < 0 \end{Bmatrix}, \quad (6)$$

where x_i and x_j are the ranked values of the data and n is the length of data series. The variance of the statistic is provided for independently and identically distributed data, and is given as:

$$Var(S) = \frac{n(n-1)(2n+5) - \sum_{i=1}^{m} t_i (t_i - 1)(2t_i + 5)}{18}, \quad (7)$$

where m is the number of ties values and t_i is the number of ties of extent i. The Mann-Kendall test statistic, Z_s, can be estimated as:

$$Z_s = \begin{Bmatrix} \frac{S-1}{\sqrt{Var(s)}}, \; if\, S > 0 \\ 0, \; if\, S = 0 \\ \frac{S+1}{\sqrt{Var(s)}}, \; if\, S < 0 \end{Bmatrix}, \quad (8)$$

positive Z value indicates a positive trend whereas a negative value represents a down word trend in the data series. This test is conducted at the specific significance level (α = 0.05). In order to test the significance of trends at significance level, Z_s is compared with $Z_{1-\alpha/2}$ and if $s > Z_{1-\alpha/2}$, then the trend is significance at that level. Here $Z_{1-\alpha/2}$ is obtained from the standard normal cumulative distribution.

1.4. Lowess method

Lowess method is known as locally weighted polynomial regression (Cleveland 1979). A low degree polynomial is fitted to a sub-set of the data at each point in the data set with explanatory variables values near the point whose response is being estimated. Lowess linear fit method was appliefor smoothing the runoff data set. This method computes the residuals indicating the fitting criterion of the curve and also determines the regression weight function for the data points within the span. This regression uses a first degree polynomial for lowess. The regression weights are shown by the function (Eq. (9)):

$$W_i = \left(1 - \left|\frac{X - X_i}{d(X)}\right|^3\right)^3, \quad (9)$$

where X – predictor value which is associated with the response value to be smoothed, X_i – nearest neighbors of X as defined by the span, and $d(X)$ – distance along the abscissa from X to the most distance predictor value within the span. In this smoothing procedure, neighboring outlier values reveal the bulk of the data. This provides the goodness of fit resulting in fitting accuracy of the smooth curve.

2. Results and discussion

Five watersheds of Melaka Tengah catchment are associated with eleven soil series. Each soil series exhibits different soil textures and is classified on the basis of soil characteristics and hydrological soil group classification (Table 2). Analyzing the soil characteristics of Melaka Tengah catchment, soil type falls under only C and D hydrological soil groups (Fig. 3). According to the soil characteristics, ten land use patterns are found in the study area. The land use pattern controls the runoff potentiality of catchment area where as tree-palm-permanent crops and urban-settlement occupy 36% and 26% of the total watershed area and contributes the runoff of this region.

Table 2. Hydrological soil group classification of Melaka Tengah catchment

Soil Mapping Unit	Soil Texture	Hydrologic Soil Group (HSG)
Melaka	Clay	D
Kranji	Clay	D
Melaka Prang Association	Clay	D
Rengam	Sandy clay loam	C
Linau-Telok-Local Alluvium Complex	Sandy clay loam	C
Munchong-Melaka-Serdang Association	Clay	D
Melaka-Munchong-Tavy Association	Clay	D
Melaka-Munchong Association	Clay	D
Local Alluvium Complex	Loam to clay	C

Runoff curve numbers were defined on the basis of land cover and hydrological soil condition. CN was computed for various land cover on different hydrologic soil groups in Melaka Tengah catchment (Table 3). This area exhibited only hydrological soil groups C and D, and CN values of these two groups represented less differences among them. Normally runoff estimation should be made with distributed CN values (Grove *et al.* 1998; Soulis, Valiantzas 2012, 2013). However, CN values variation in the land cover and hydrologic soil group complexes present in the studied watershed is generally low. Accordingly, weighted curve number was considered valid for the particular watershed. This weighted runoff curve number can be calculated by weighting the CN's of the different subareas in proportion to the land cover associated with each CN value for a catchment (Wong *et al.* 2001; Gumbo *et al.* 2002). The actual weighted CN_{wi} number of watershed area is the summation of two weighted curve numbers and expressed as (Eq. (10)):

$$CN_{wi} = \sum_{i=1}^{n} \left\{ \sum (CN_{Ci} \times A_{Ci}) + \sum (CN_{Di} \times A_{Di}) \right\}, \quad (10)$$

where, CN_{wi} - Area weighted runoff curve number for particular watershed area; CN_{Ci} - CN for a particular land use for soil group C; CN_{Di} - CN for a particular land use for soil group D; and A_{Ci} and A_{Di} - land use pattern percentage for soil groups C and D.

S_{wi} was calculated by using the equation 4 after getting the weighted CN_{wi} of each watershed. The daily event rainfall data of 2006 to 2012 were considered to analyze the runoff of Melaka Tengah catchment. The runoff depth (R_{wi}) was calculated for each watershed after putting rainfall data and S_{wi} values in Equation 3. The monthly mean runoff, weighted CN_{wi} value and S_{wi} of each watershed of Melaka Tengah catchment are shown in Table 4.

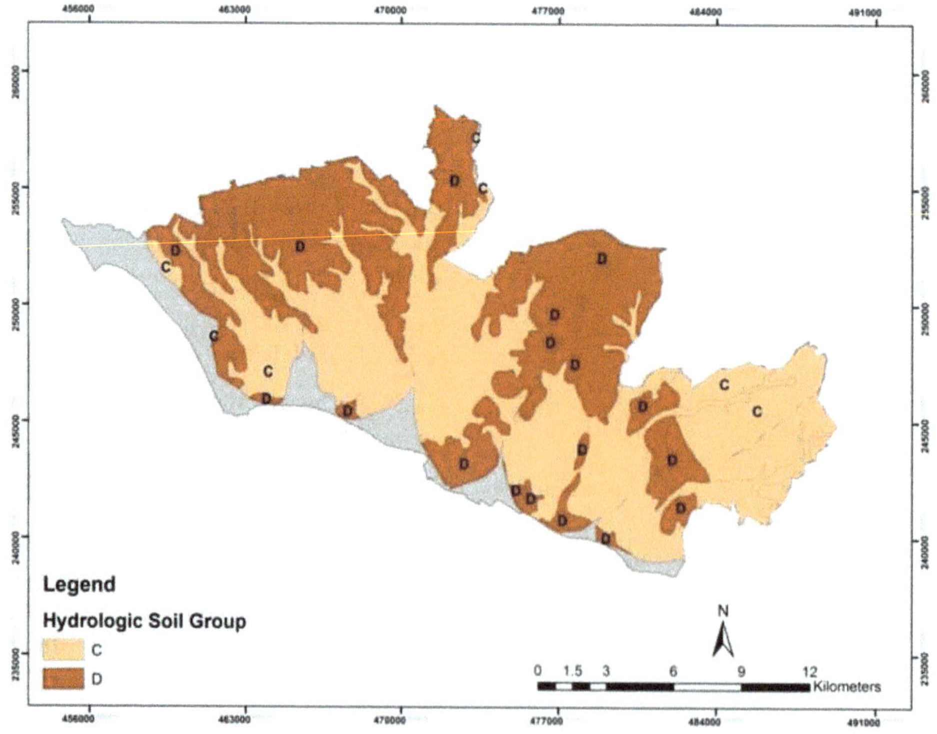

Fig. 3. Hydrologic soil group map of Melaka Tengah catchment

Table 3. CN for particular soil category and percentage of land cover of watershed area

Watershed	Land use Category	Land covered by soil group C (%)	Land covered by soil group D (%)	Total Area (km^2)	CN for Hydrologic Soil Group C	CN for Hydrologic Soil Group D
1	Animal husbandry areas	0.46	0.00	28.29	82	86
	Forest land	1.91	0.00		70	77
	Horticultural lands	11.63	2.55		78	81
	Ideal grassland	3.46	0.81		74	80
	Short-term crops	1.20	0.99		82	86
	Swamps, marshland and wetland forest	1.24	0.57		83	86
	Tree, palm and other permanent crops	48.21	16.19		82	86
	Urban, settlement and associated non-agricultural area	8.40	1.29		90	92
	Others	1.10	0.00		–	–
2	Animal husbandry areas	2.99	0.06	79.73	82	86
	Forest land	0.37	2.06		70	77
	Horticultural lands	11.44	2.87		78	81
	Ideal grassland	2.71	0.83		74	80
	Short-term crops	3.93	0.00		82	86
	Swamps, marshland and wetland forest	5.32	3.43		83	86
	Tree, palm and other permanent crops	32.11	9.98		82	86
	Urban, settlement and associated non-agricultural area	6.61	13.37		90	92
	Water body	0.00	0.60		98	98
	Others	1.30	0.00		–	–
3	Animal husbandry areas	0.08	0.00	85.57	82	86
	Forest land	0.00	8.32		70	77
	Horticultural lands	6.73	4.02		78	81
	Ideal grassland	2.31	1.15		74	80
	Short-term crops	3.82	2.49		82	86
	Swamps, marshland and wetland forest	8.61	0.25		83	86
	Tree, palm and other permanent crops	0.00	18.22		82	86
	Urban, settlement and associated non-agricultural area	26.22	16.27		90	92
	Water body	0.00	0.40		98	98
	Others	0.00	1.11		–	–

End of Table 3

Watershed	Land use Category	Land covered by soil group C (%)	Land covered by soil group D (%)	Total Area (km^2)	CN for Hydrologic Soil Group C	CN for Hydrologic Soil Group D
4	Animal husbandry areas	0.00	0.23	48.06	82	86
	Forest land	0.00	0.29		70	77
	Horticultural lands	13.84	2.04		78	81
	Ideal grassland	5.20	0.12		74	80
	Short-term crops	9.42	0.65		82	86
	Swamps, marshland and wetland forest	1.92	0.00		83	86
	Tree, palm and other permanent crops	6.67	34.98		82	86
	Urban, settlement and associated non-agricultural area	11.51	12.77		90	92
	Others	0.37	0.00		–	–
5	Animal husbandry areas	0.12	0.00	29.98	82	86
	Forest land	0.00	1.42		70	77
	Horticultural lands	4.87	13.74		78	81
	Ideal grassland	7.47	2.99		74	80
	Short-term crops	17.68	0.47		82	86
	Swamps, marshland and wetland forest	1.08	0.00		83	86
	Tree, palm and other permanent crops	7.97	25.47		82	86
	Urban, settlement and associated non-agricultural area	5.30	10.88		90	92
	Others	0.54	0.00		–	–

Table 4. Weighted CN_{wi}, S_{wi} and monthly mean runoff for each watershed

Watershed	Weighted CN_{wi}	Value of S_{wi} (mm)	Mean runoff (mm)
1	82	55.76	30
2	83	52.02	34
3	85	44.82	39
4	84	48.38	28
5	83	52.02	37

Daily runoff was calculated using SCS-CN method from the daily rainfall event data for period 2006 to 2012. After calculating the daily runoff, monthly runoff was estimated by summing the daily runoff data (Fig. 4). Eighty four data sets were prepared for monthly runoff analysis. These data sets presented the monthly rainfall-runoff pattern of watersheds in the Melaka Tengah catchment. In watersheds 1, 2, 3, 4 and 5 the rainfall varies 0–333 mm, 0–365 mm, 0–467 mm, 0–322 mm and 0–374 mm, respectively. The corresponding runoff varies 0–97 mm, 0–120 mm, 0–199 mm, 0–125 mm and 0–127 mm, respectively for watersheds 1, 2, 3, 4 and 5. This analysis shows that the changing pattern of monthly runoff varies with the particular monthly rainfall pattern for the watersheds.

2.1. Autocorrelation

It is important to define the autocorrelation properties of the runoff data series before selecting the trend detection test. Figure 5 represents the autocorrelation of monthly surface runoff of five watersheds. From the analysis, it shows that almost all of the autocorrelations fall within the 95% confidence limit. A few lags just fall outside the 95% confidence limit. This analysis indicates that all watersheds

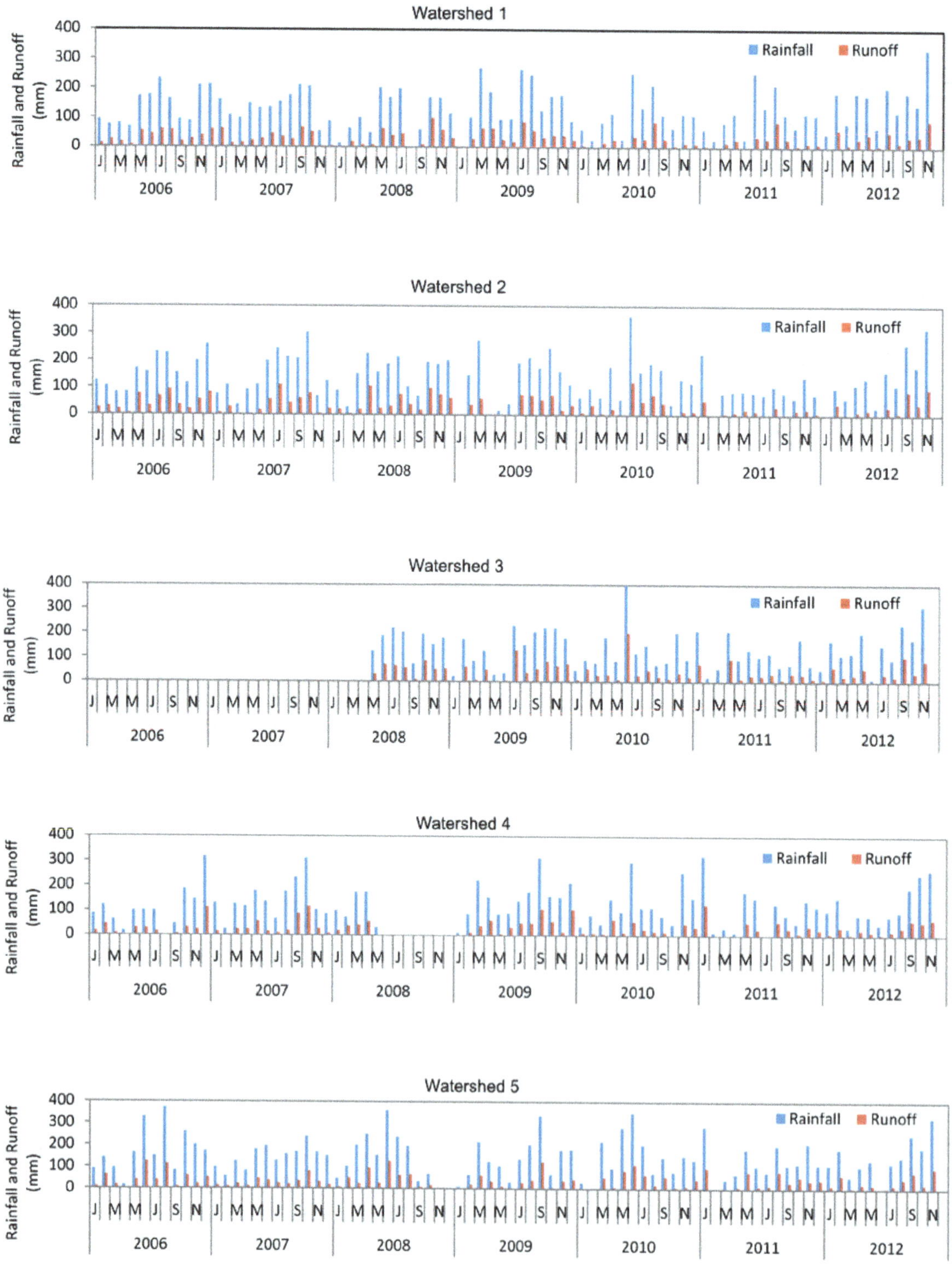

Fig. 4. Monthly rainfall and runoff of watershed area

show non-significant autocorrelation at various lags. Both significant and non-significant lag 1 autocorrelations were conducted for daily, monthly, seasonal and annual temporal resolution rainfall (Rashid *et al.* 2013).

2.2. Monthly runoff trend

Monthly surface runoff trend analysis from 2006 to 2012 was conducted for the watersheds of Melaka Tengah catchment. Monthly values for all watersheds revealed that runoff varies with seasonal effect and most of the values amounted to about high during from November to February and May to August. Runoff peaks in these months and declines thereafter. Therefore, no statistically significant changes in trend line for the monthly runoff during the observed for all the watersheds. The negative trend of monthly runoff was observed for the watersheds 1, 2 and 5 with a Sen slope of about - 0.116, - 0.243 and - 0.036, respectively, while upward runoff trend line was observed for watersheds 3 and 4 with the Sen slope of about 0.312 and 0.009, respectively (Fig. 6).

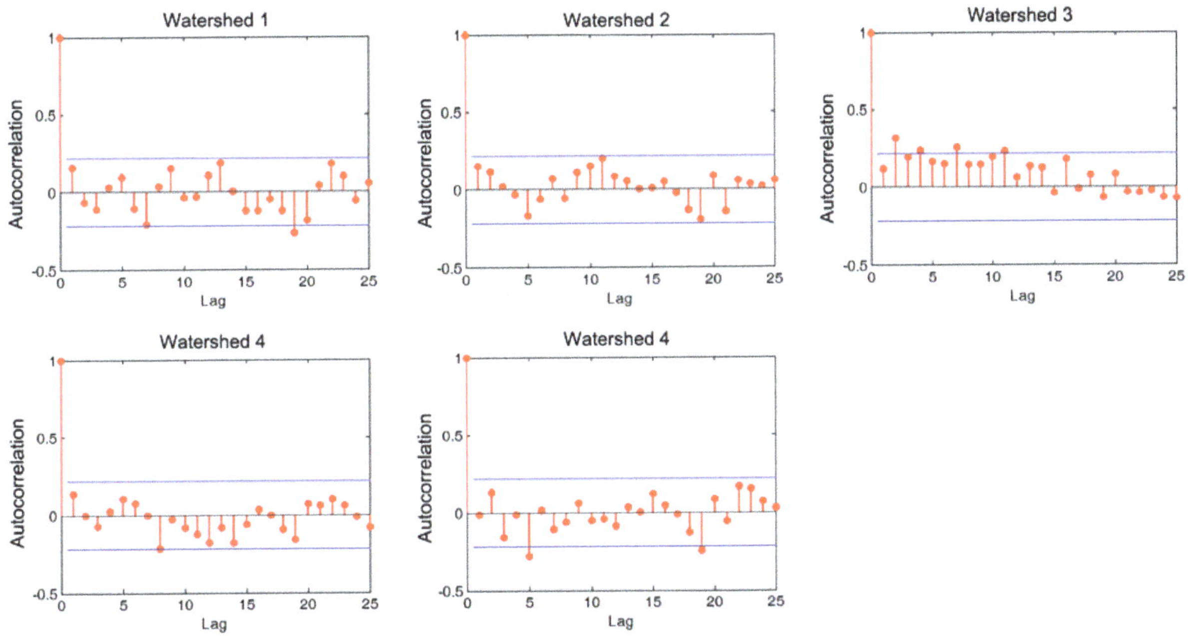

Fig. 5. Autocorrelation function for different lags of monthly runoff of watersheds

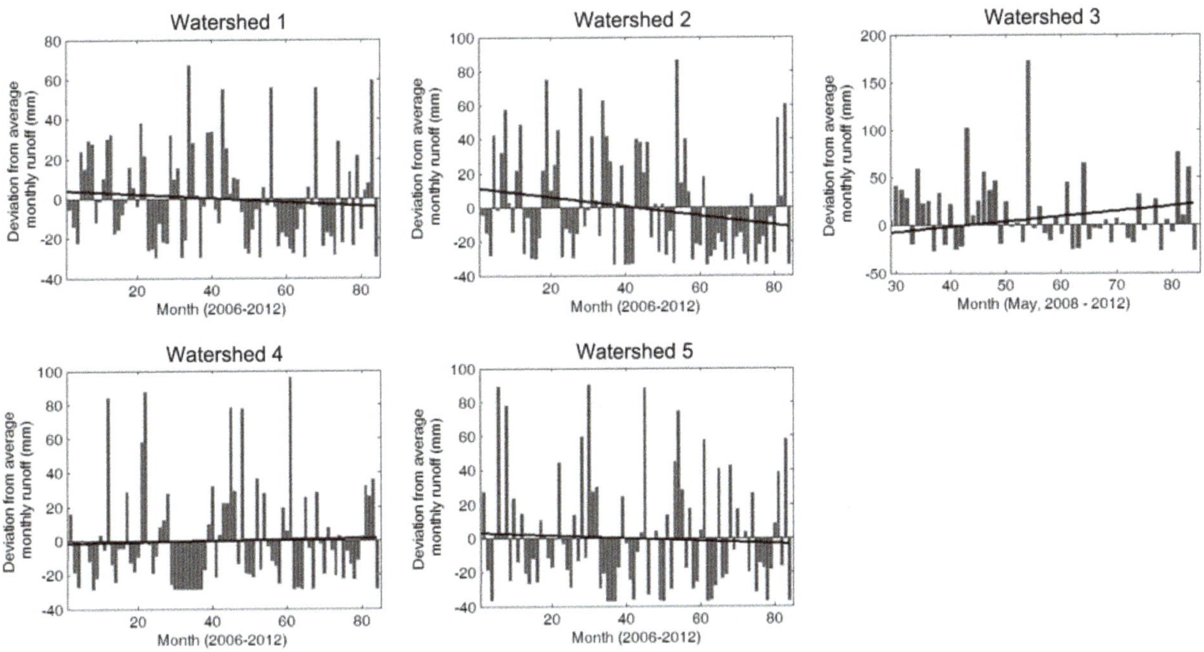

Fig. 6. Monthly deviation from average surface runoff amounts in watershed

2.3. Mann-Kendall trend analysis

The Mann-Kendall test provides the information of existence of monotonic trends either positive or negative. This method was applied to detect any monotonic trend was existing in the runoff data series. Table 5 represents the Mann-Kendall test and Sen slope analysis of the monthly runoff trends. The two-tailed p-value indicates the null (H_0) or alternative (H_1) hypothesis of the test. If the null hypothesis is rejected, alternative hypothesis indicates that there is an increasing or decreasing trend in the runoff data series.

The p-value was greater than the significance level (α = 0.05) for the watersheds 1, 4 and 5. It accepted the null (H_0) hypothesis representing no trends in this time series. However, Sen slope of these trends was -0.116, 0.009 and -0.036 for the watersheds 1, 4 and 5, respectively. On the other hand, the null hypothesis was rejected for watersheds 2 and 3, and it accepted the alternative hypothesis because of the p-value was less than the significance level. This indicated that there was a decreasing and increasing trends which was statistically significance for the watersheds 2 and 3, respectively. Sen slope value of these two watersheds also were -0.243 and 0.312 indicating upward and down ward direction of the runoff trend.

Table 5. Mann-Kendall trend and Sen slope analysis of watersheds

Watershed	Kendall's tau	*p*-value (two-tailed)	H_0 hypothesis (α = 5%)	Mann-Kendall trend (α = 5%)	Sen slope
1	-0.089	0.230	Accept	No	-0.116
2	-0.178	0.017	Reject	Yes (-)	-0.243
3	0.414	<0.0001	Reject	Yes (+)	0.312
4	0.031	0.673	Accept	No	0.009
5	-0.037	0.624	Accept	No	-0.036

2.4. Smooth curve and runoff potentiality

Figure 7 shows the smooth curve runoff for five watersheds. It reveals that the runoff values varied from one peak to another peak but all peaks occurred during the months of November to February and May to August when most of the rainfall occurred. From this analysis it implies that this runoff exhibits the watershed characteristics in the form of different degree of effect for the study area. The estimated monthly rainfall-runoff relation for the watershed area indicated a good correlation between both criteria (Fig. 8). Relation was found to be linear (R^2 = 0.73

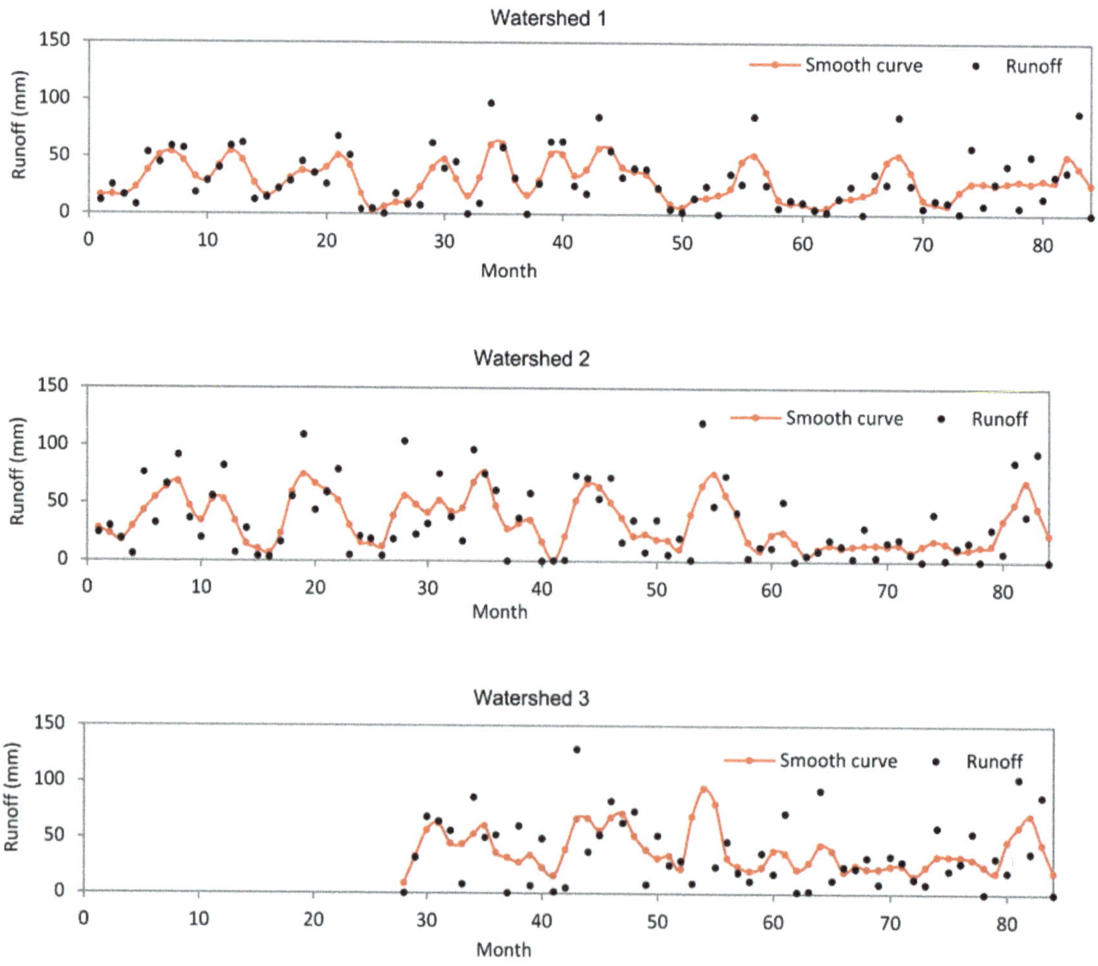

Fig. 7 →

→

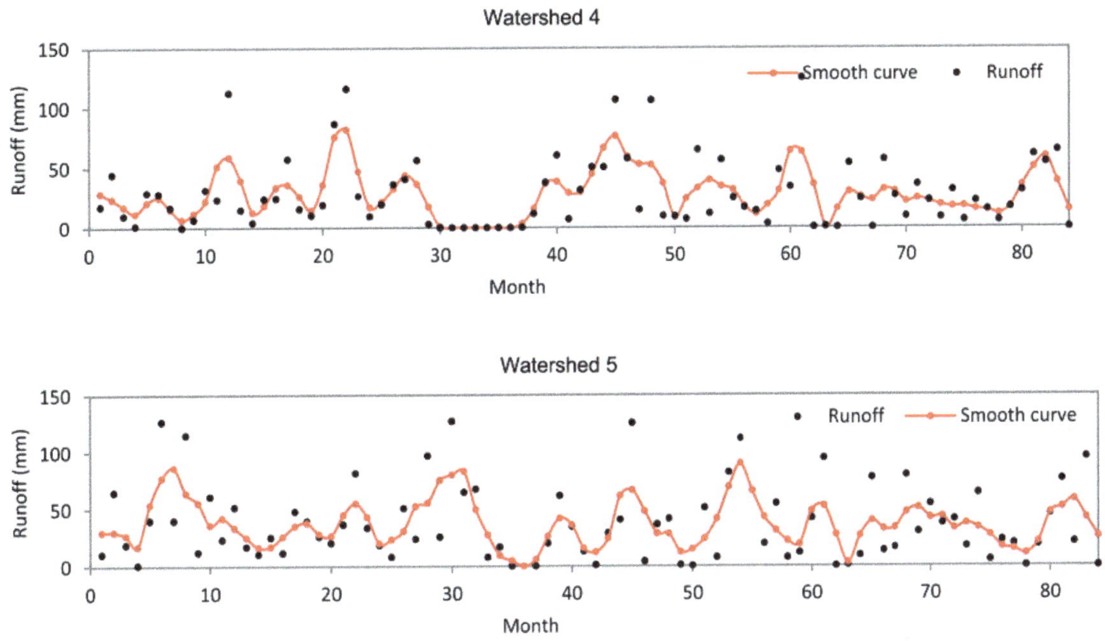

Fig. 7. Runoff and smoothing curve of five watersheds

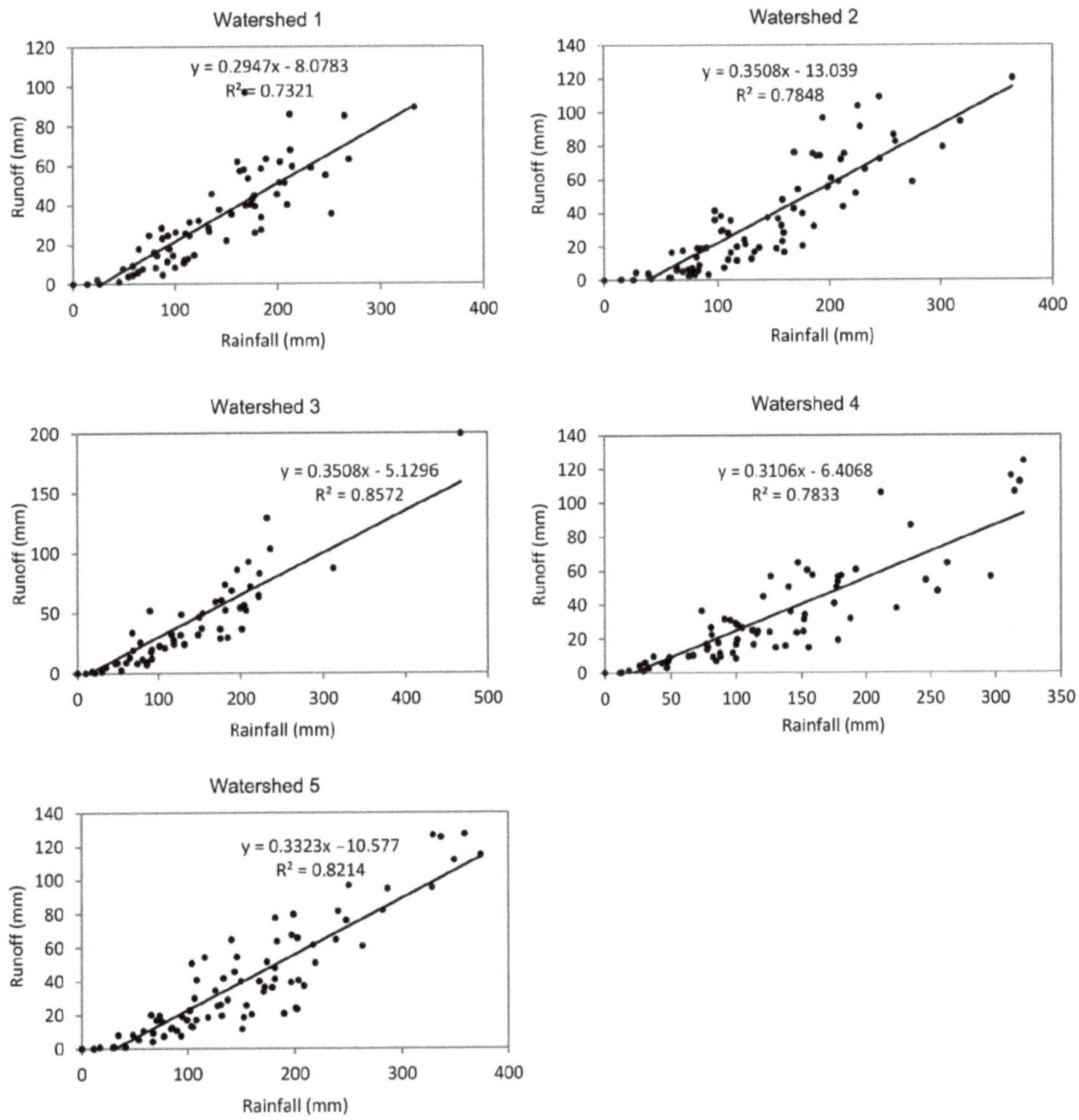

Fig. 8. Estimated monthly rainfall runoff relationship of the study area

to 0.85). This analysis indicated that the surface runoff due to rainfall in Melaka Tengah catchment may be predicted using SCS-CN method. The runoff varied with the different value of CN in Melaka Tengah catchment. Henceforth, runoff volume was calculated for a particular watershed, and total runoff volume in Melaka Tengah catchment was calculated (Table 6) and it revealed that watershed 3 contributes monthly 3.34 Mm^3 of surface runoff.

Table 6. Monthly runoff of particular watershed area of Melaka

Watershed	Runoff R_{wi} (m)	Rainfall P (m)	Watershed Area A (m^2)	Rainfall Volume $P \times A$ (Mm^3)	Runoff Volume $R_{wi} \times A$ (Mm^3)	Runoff (%)
1	0.030	0.13	2829×10^4	3.65	0.85	23
2	0.034	0.13	7973×10^4	10.68	2.71	25
3	0.039	0.13	8557×10^4	10.87	3.34	31
4	0.028	0.11	4806×10^4	5.38	1.35	25
5	0.037	0.14	2998×10^4	4.29	1.11	26
			Average			26

Conclusions

The SCS-CN method was considered to assess the surface runoff of five watersheds in Melaka Tengah catchment. Soil of this region falls mostly under C and D soil groups on the subject of soil texture and hydrological condition. Considering of several sub-catchments with different CNs, the area averaged weighted curve number was computed for the entire watershed and varied between 82 and 85 for different watersheds. It is recommended that weighted curve number may apply for the hydrological soil groups C and D, and less differences of CN values among them. Whereas, distributed CN value is the consideration of hydrological soil groups for accurate surface runoff. Monthly runoff trends over the time period were investigated by using autocorrelation, Mann-Kendall and Sen slope estimation methods. According to the Mann-Kendall test, watersheds 1, 4 and 5 showed no trend at the significant level ($\alpha = 0.05$), while watersheds 2 and 3 showed a trend with Sen slope - 0.243 and 0.312, respectively. The trend analysis revealed that there was no significantly increasing and decreasing trend for these runoff data series. Moreover, the runoff value varied with seasonal and indicated the discrete pattern of the data set. Lowess linear fit method was applied for discrete runoff data and to provide the smoothing of curves. Five smooth curves one for each watershed indicated the surface runoff pattern and these patterns were compared among the watersheds. These curves fluctuated due to the seasonal variation of monsoonal rainfall. As a result, most of the highest peaks of the curve were observed during November to February and May to August, when maximum rainfall occurred. Based on the smoothing curves, the monthly mean runoff was identified and the value is 30 mm, 34 mm, 39 mm, 28 mm and 37 mm for the watersheds 1, 2, 3, 4 and 5, respectively. Watershed 3 had most of the surface runoff of this region to contribute runoff water to the river. About 26% volume of water from rainfall directly goes to the river through surface runoff. This trend and volume of water provides the firsthand information for rainwater distribution and contribution. It may be helpful for planning of surface water management and for contribution and potentiality of groundwater recharge.

Acknowledgements

The study has been supported by project vote Q.J130000.2509.08H38, Vot. 08H43 (FRGS Grant) and Vot. 4L827 (TRGS Grant) of Universiti Teknologi Malaysia (UTM). The study was partially supported by JSPS-ACP (Japan Society for the Promotion of Science – Asian Core Program) Integrated Watershed Management (IWM).

References

Adham, M. I.; Shirazi, S. M.; Othman, F.; Yusop, Z.; Ismail, Z. 2014. Runoff potentiality of a watershed through SCS and Functional Data Analysis Technique, *Scientific World Journal*, Article ID 379763. http://dx.doi.org/10.1155/2014/379763

Antonopoulos, V. Z.; Papamichail, D. M.; Mitsiou, K. A. 2001. Statistical and trend analysis of water quality and quantity data for the Strymon River in Greece, *Hydrology and Earth System Sciences Discussions* 5(4): 679–692. http://dx.doi.org/10.5194/hess-5-679-2001

Bagdziunaite, L. L.; Litvinaitis, A.; Saulys, V. 2011. Patterns of river runoff change considering the size of the basin, *Journal of Environmental Engineering and Landscape Management* 10(4): 326–334. http://dx.doi.org/10.3846/16486897.2011.634057

Caloiero, T.; Coscarelli, R.; Ferrari, E.; Mancini, M. 2011. Trend detection of annual and seasonal rainfall in Calabria (Southern Italy), *International Journal of Climatology* 31(1): 44–56. http://dx.doi.org/10.1002/joc.2055

Carlesso, R.; Spohr, R.B.; Eltz, F.L.F. 2011. Runoff estimation in southern Brazil based on Smith's modified model and the Curve Number method, *Agricultural Water Management* 98: 1020–1026. http://dx.doi.org/10.1016/j.agwat.2011.01.012

Chen, Y.; Xu, Y.; Yin, Y. 2009. Impact of land use change scenarios on storm-runoff generation in Xitiaoxi basin, China, *Quaternário Internacional* 1: 1–8. http://dx.doi.org/10.1016/j.quaint.2008.12.014

Chowdhury, R.; Beecham, S. 2010. Australian rainfall trends and their relation to the southern oscillation index, *Hydrological Processes* 24(4): 504–514.

Chowdhury, R.; Beecham, S. 2012. South Australian rainfall-trends and climate drivers, in *Water and Climate: Policy Implementation Challenges; Proceedings of the 2nd Practical Responses to Climate Change Conference*, 1–3 May 2012, Canberra.

Cleveland, W. S. 1979. Robust locally weighted regression and smoothing scatterplots, *Journal of the American Statistical Association* 74(368): 829–836.

Dimitriou, E.; Moussoulis, E. 2011. Land use change scenarios and associated groundwater impacts in a protected peri-urban area, *Environmental Earth Science* 64: 471–482. http://dx.doi.org/10.1007/s12665-010-0871-1

Feidas, H.; Noulopoulou, C.; Makrogiannis, T.; Bora-Senta, E. 2007. Trend analysis of precipitation time series in Greece and their relationship with circulation using surface and satellite data: 1955–2001, *Theoretical and Applied Climatology* 87(1–4): 155–177. http://dx.doi.org/10.1007/s00704-006-0200-5

Gaudin, R.; Celette, F; Gary, C. 2010. Contribution of runoff to incomplete off season soil water refilling in a Mediterranean vineyard, *Agricultural Water Management* 97: 1534–1540. http://dx.doi.org/10.1016/j.agwat.2010.05.007

Grove, M.; Harbor, J.; Engel, B. 1998. Composite vs. Distributed curve numbers: effects on estimates of storm runoff depths, *Journal of the American Water Resources Association* 34: 1015–1023. http://dx.doi.org/10.1111/j.1752-1688.1998.tb04150.x

Gumbo, B.; Nelson, M.; George, S.; Hubart, H. G. S. 2002. Coupling of digital elevation model and rainfall-runoff model in storm drainage network design, *Physics and Chemistry of the Earth* 27(11–22): 755–764.

Hawkins, R. H. 1978. Runoff curve numbers with varying site moisture, *Journal of Irrigation and Drainage Division* ASCE 104(IR4): 389–398.

Hawkins, R. H. 1993. Asymptotic determinations of runoff curve numbers from data, *Journal of Irrigation and Drainage Engineering* 119: 334–345. http://dx.doi.org/10.1061/(ASCE)0733-9437(1993)119:2(334)

Hillel, D. 1980. *Applications of soil physics*, vol. 7. New York, NY: Academic Press.

Hirsch, R. M.; Slack, J. R.; Smith, R. A. 1982. Techniques of trend analysis for monthly water quality data, *Water Resources Research* 18(1): 107–121. http://dx.doi.org/10.1029/WR018i001p00107

Jonsdottir, J.; Jónsson, P.; Uvo, C. 2006. Trend analysis of Icelandic discharge, precipitation and temperature series, *Nordic Hydrology* 37(4–5): 365–376. http://dx.doi.org/10.2166/nh.2006.020

Kendall, M. 1975. *Rank correlation methods*. London, UK: Charles Griffin.

Khan, H. H.; Khan, A.; Ahmed, S.; Perrin, J. 2011. GIS-based impact assessment of land-use changes on groundwater quality: study from a rapidly urbanizing region of South India, *Environmental Earth Science* 63: 1289–1302. http://dx.doi.org/10.1007/s12665-010-0801-2

Komatsu, H.; Shinohara, Y.; Kume, T.; Otsuki, K. 2011. Changes in peak flow with decreased forestry practices: analysis using watershed runoff data, *Journal of Environmental Management* 92(6): 1528–1536. http://dx.doi.org/10.1016/j.jenvman.2011.01.010

Kumar, G.; Sena, D. R.; Kurothe, R. S. 2014. Watershed impact evaluation using remote sensing, *Current Science* 106: 1370–1378.

Lewis, M. J.; Singer, M. J.; Tate, K. W. 2000. Applicability of SCS curve number method for a California oak woodlands watershed, *Journal of Soil and Water Conservation* 53(2): 226–230.

Mann, H. B. 1945. Nonparametric tests against trend, *Econometrica* 13(3): 245–259. http://dx.doi.org/10.2307/1907187

Mays, L. W. 2005. *Water recourses engineering*. John Wiley & Sons, Inc., 262–268.

Melesse, A. M.; Shih, S. F. 2002. Spatially distributed storm runoff depth estimation using Landsat images and GIS, *Computers and Electronics in Agriculture* 37: 173–183. http://dx.doi.org/10.1016/S0168-1699(02)00111-4

Michel, C.; Andreassian, V.; Perrin, C. 2005. Soil Conservation Service Curve Number method: how to mend a wrong soil moisture accounting procedure, *Water Resources Research*, 41. Article ID W02011. http://dx.doi.org/10.1029/2004WR003191

Mohsin, T.; Gough, W. A. 2010. Trend analysis of long-term temperature time series in the Greater Toronto Area (GTA), *Theoretical and Applied Climatology* 101(3–4): 311–327. http://dx.doi.org/10.1007/s00704-009-0214-x

Okoli, C. S. 2014. Hydraulics of slope erosion by overland flow. A case study: Nnobi-Onitsha hillslope site, Anambra State, S. E. Nigeria, *Scientia Iranica A.* 21(2): 251–262.

Ponce, V. M. 1989. *Engineering hydrology – principles and practices*. San Diego State University, San Diego.

Ponce, V. M.; Hawkins, R. H. 1996. Runoff curve number: Has it reached maturity?, *Journal of Hydrologic Engineering*, ASCE 1(1): 11–19. http://dx.doi.org/10.1061/(ASCE)1084-0699(1996)1:1(11)

Pourtouiserkani, A.; Rakhshandehroo, G. H. 2014. Investigating climate change impact on extreme rainfall events. Case study: Chenar-Rahdar Basin, Fars, Iran, *Scientia Iranica A.* 21(3): 525–533.

Ragan, R. M.; Jackson, T. J. 1980. Runoff synthesis using Landsat and SCS model, *Journal of Hydraulics Division*, ASCE 106(HY5): 667–678.

Rashid, M.; Beecham, S.; Chowdhury, R. 2013. Assessment of statistical characteristics of point rainfall in the Onkaparinga catchment in South Australia, *Hydrology and Earth System Sciences Discussions* 10(5): 5975–6017. http://dx.doi.org/10.5194/hessd-10-5975-2013

Río, S. D.; Herrero, L.; Fraile, R.; Penas, A. 2011. Spatial distribution of recent rainfall trends in Spain (1961–2006), *International Journal of Climatology* 31(5): 656–667. http://dx.doi.org/10.1002/joc.2111

Sarauskiene, D.; Kriauciuniene; Reihan, A.; Klavins, M. 2015. Flood pattern changes in the rivers of the Baltic countries, *Journal of Environmental Engineering and Landscape Management* 23(1): 28–38. http://dx.doi.org/10.3846/16486897.2014.937438

Shirazi, S. M.; Imran, H. M.; Akib, S. 2012. GIS-based DRASTIC method for groundwater vulnerability assessment: a review, *Journal of Risk Research* 15(8): 991–1011. http://dx.doi.org/10.1080/13669877.2012.686053

Singh, V. P. 1997. Effect of spatial and temporal variability in rainfall and watershed characteristics on stream flow hydrograph, *Hydrological Processes* 11(12): 1649–1669. http://dx.doi.org/10.1002/(SICI)1099-1085(19971015)11:12<1649::AID-HYP495>3.0.CO;2-1

Slack, R. B.; Welch, R. 1980. Soil conservation service runoff curve number estimates from Landsat data, *Water Resources Bulletin* 16(5): 887- 893. http://dx.doi.org/10.1111/j.1752-1688.1980.tb02504.x

Soil Conservation Service (SCS). 1972. *National engineering handbook. Section 4. Hydrology*. U.S. Government Printing Office.

Soulis, K. X.; Valiantzas, J. D. 2012. Variation of runoff curve number with rainfall in heterogeneous watersheds: the Two-CN system approach, *Hydrology and Earth System Sciences* 16: 1001–1015. http://dx.doi.org/10.5194/hess-16-1001-2012

Soulis, K. X.; Valiantzas, J. D. 2013. Identification of the SCS-CN parameter spatial distribution using rainfall-runoff data in heterogeneous watersheds, *Water Resources Management* 27: 1737–1749. http://dx.doi.org/10.1007/s11269-012-0082-5

Stephenson, D. 1984. Kinematic study of effects of storm dynamics on runoff hydrographs, *Water SA* 10(4): 189–196.

USDA-Soil Conservation Service (USDA SCS). 1985. *National engineering handbook. Section 4. Hydrology*. USDA-SCS, Washington, D.C.

Warren, V.; Gary, L. L. 2003. *Introduction to hydrology*. 15th ed. Prentice Hall.

Wong, T. H.; Mansor, S. B.; Mispan, M. R.; Sulaiman, W. N. A.; Ahmad, N. 2001. An application of remote sensing in hydrology – a case study in Malaysia, in *Geoscience and Remote Sensing Symposium IGARSS '01*, 9–13 July 2001, IEEE. http://dx.doi.org/10.1109/IGARSS.2001.977933

Wu, Y. S.; Hasi, E.; Wugetemole, X. W. 2011. Characteristics of surface runoff in a sandy area in southern Mu Us sandy land, *Chinese Science Bulletin* 57: 270–275. http://dx.doi.org/10.1007/s11434-011-4728-0

Zade, M.; Ray, S. S.; Dutta, S. 2005. Analysis of runoff pattern for all major basins of India derived using remote sensing data, *Current Science* 88: 1301–1305.

Zhang, X.; Wenhong, C.; Qingchao, G.; Sihong, W. 2010. Effects of landuse change on surface runoff and sediment yield at different watershed scales on the Loess Plateau, *International Journal of Sediment Research* 25(3): 283–293. http://dx.doi.org/10.1016/S1001-6279(10)60045-5

Sharif Moniruzzaman SHIRAZI, Dr, Professor, Department of Civil Engineering, World University of Bangladesh, Dhaka, Bangladesh. Dr. Shirazi obtained his Ph.D. in the department of Civil and Environmental Engineering of Saitama University, Japan. He obtained B.Sc. and M.Sc. in Irrigation Engineering from Bangladesh Agricultural University. He worked as a JSPS Postdoctoral Fellow at Geosphere Research Institute of Saitama University. He also worked as an Associate Professor, Centre for Environmental Sustainability and Water Security (IPASA), Research Institute for Sustainable Environment, Faculty of Civil Engineering, Universiti Teknologi Malaysia and as a Senior Lecturer in the department of Civil Engineering, Faculty of Engineering, University of Malaya (UM). Dr. Shirazi has 25 years research experience in the area of physico-chemical properties of soil, irrigation and water management, runoff modelling, groundwater vulnerability and impact of climatic parameters on crop production. He has about 65 technical paper published in international journals.

MD Ibrahim ADHAM, PhD Candidate, Faculty of Civil Engineering, University of Malaya. Master of Science in Geology at Rajshahi University of Bangladesh. Research publication 7 and research interest in Soil erosion, runoff modelling, and environmental pollution.

Faridah OTHMAN, Dr, Associate Professor, Faculty of Civil Engineering, University of Malaya. Publications: over 60 scientific publications. Research interest in Water quality modelling, environmental pollution, water balance, water resources estimation and management.

Noorul Hasan ZARDARI, Dr, Senior Lecturer, Department of hydraulics and hydrology, Faculty of Civil Engineering, Universiti Teknologi Malaysia. PhD in water resources. Publication: over 40 scientific articles. His areas of expertise include water resources management, irrigation, groundwater hydrology, decision-making and policy in water resources allocations.

Zubaidah ISMAIL, Dr, Professor, Faculty of Civil Engineering, University of Malaya. Publications: over 70 scientific publications. Research interest in mathematical modeling in hydrology, flash flood management and water quality.

9

THE DIFFERENT FACES OF BIOCHAR: CONTAMINATION RISK VERSUS REMEDIATION TOOL

Isabel HILBER[a], Ana Catarina BASTOS[b], Susana LOUREIRO[b], Gerhard SOJA[c], Aleksandra MARSZ[c], Gerard CORNELISSEN[d, e, f], Thomas D. BUCHELI[a]

[a]*Environmental Analytics, Natural Resources and Agriculture, Agroscope ISS, Reckenholzstrasse 191, 8046 Zurich, Switzerland*

[b]*Department of Biology & Centre for Environmental and Marine Studies (CESAM), University of Aveiro, 3810-193 Aveiro, Portugal*

[c]*Environmental Resources and Technologies, AIT Austrian Institute of Technology GmbH, 3430 Tulln, Austria*

[d]*Department of Environmental Engineering, Norwegian Geotechnical Institute NGI, P.O. Box 3930 Ullevål Stadion, 0806, Oslo, Norway*

[e]*Department of Environmental Sciences (IMV), Norwegian University of Life Sciences (NMBU), P.O. Box 5003, 1432 Ås, Norway*

[f]*Department of Applied Environmental Sciences (ITM), Stockholm University, 10691 Sweden*

Abstract. This article reviews the different aspects of biochar as source and sink of organic and inorganic contaminants. Biochar can contain organic contaminants such as polycyclic aromatic hydrocarbons or heavy metals. As the distribution coefficients of the biochar especially for contaminants are high, the freely dissolved concentrations are low and with that also the bioavailability. The link between biochar's inherent contaminants and toxicity to soil meso- and macro-fauna remains unclear, with data being often contradictory and influenced by feedstock and pyrolysis conditions. The biochar's potential to remediate contaminated soils has mainly been addressed in lab studies, but rarely in the field. This far, results have been contradicting. Many studies reported successful immobilization of contaminants but some not. In summary, the ambivalent face of the biochar with regard to contaminants prevails. In future, long term field studies are needed to properly address the sustainability of biochar in this respect.

Keywords: biochar inherent organic and inorganic contaminants, bioavailability, bioaccessibility, ecotoxicity, environmental risks.

Introduction

Besides its potential agricultural benefits (Atkinson *et al.* 2010; DeLuca *et al.* 2015; Ippolito *et al.* 2015; Jeffery *et al.* 2015; Thies *et al.* 2015; Van Zwieten *et al.* 2015; Whitman *et al.* 2015), biochar can contribute to remediation and recovery of contaminated soils through sorption of heavy metals and/or organic contaminants (Beesley *et al.* 2011; Ahmad *et al.* 2016; Kupryianchyk *et al.* 2016; Li *et al.* 2016; Shen *et al.* 2016). This can be envisioned as the positive face of biochar. Nonetheless, biochar may contain contaminants itself, either introduced by its feedstock (e.g., heavy metals), or co-produced during (improper) pyrolysis (e.g., polycyclic aromatic hydrocarbons; PAHs). The over-time release of contaminants inherent to biochar and the potential consequences to soil and the wider environment represent the negative face. Due to the manifold agricultural and environmental management procedures in which biochar is involved, it is important to rigorously assess its potential hazards and benefits over the relevant timescales. This paper reviews the different aspects of

Corresponding author: Thomas D. Bucheli
E-mail: thomas.bucheli@agroscope.admin.ch

biochar and discusses and contextualizes hazards and benefits of biochar in light of the latest scientific research. *Pro et contra* of contaminants in biochar and its suitability for remediation purposes will be summarized and denoted with a "smiley", a "frowny", and – where not clear – with an ambivalent emoticon. Furthermore, current knowledge gaps will be listed and future research needs identified.

1. Biochar – associated contaminants – total concentrations

1.1. Organic compounds

During pyrolysis, a multitude of (natural) organic compounds are involved either as precursor, intermediate or terminal products in any of the three physical states: syngas, bio-oil and biochar (Bucheli *et al.* 2015). These authors concluded that formation of polychlorinated aromatic compounds under fully pyrolytic conditions is not likely, whereas polychlorinated dibenzo-p-dioxins, -furans (PCDD/Fs) may be formed if suitable precursors, such as salt in food waste, are present in the feedstock. Hale *et al.* (2012) reported of up to 92 ng kg^{-1} PCDD/Fs (0.15 ng kg^{-1} toxic equivalents (TEQ)) from biochars whose feedstock was, amongst many, food waste. These concentrations are very low as the European Biochar Certificate (EBC 2012) set the TEQ of PCDD/Fs <20 ng kg^{-1}. However, during pyrolysis of the feedstock, PAHs are formed (Bucheli *et al.* 2015). The authors reviewed 21 studies that reported PAHs in biochars and contextualised the concentrations as determined by pyrolysis type and temperature, feedstock, and residence time of the feedstock in the reactor. These concentrations ranged from a low 100 μg kg^{-1} to 100 mg kg^{-1}, with outliers in the thousands of mg kg^{-1} for the sum (Σ) of the 16 US EPA PAHs. Generally, the evaluation was difficult, since different extraction methods were used to quantify the PAHs at that time. These are now standardized by the EBC and the International Biochar Initiative (IBI 2015b). Nevertheless, trends could be identified. For instance, elevated PAH concentrations were found at high (>800 °C) temperatures that are typical of gasification (Bucheli *et al.* 2015), which is due to the re-condensation of the PAHs on relatively small fractions of solid residues (Hale *et al.* 2012; Schimmelpfennig, Glaser 2012). In contrast, Yargicoglu *et al.* (2015) could not detect PAHs after the gasification of wood pellets. They hypothesize that this might be due to some oxygen during the gasification leading to a more complete combustion of the wood and less PAH formation. Additionally, the pelleting prior to pyrolysis might have removed the small particles where PAH preferably re-condensate on. Most recently, Buss *et al.* (2016) convincingly confirmed the hypothesis expressed in Bucheli *et al.* (2015) that recondensation of pyrosynthesized PAHs onto biochar is responsible for elevated PAH concentrations in biochar, and that effective separation of syngases from solid feedstock residues (biochar) during pyrolysis leads to biochar with PAHs in the low mg kg^{-1} concentration range. Irrespective of the pyrolysis type, recondensates also contain a large variety of (semi-)volatile organic compounds (VOC, (Spokas *et al.* 2011; Buss *et al.* 2015)). The most encountered VOC were acetic and formic acid, phenols, o-, m- and p-cresol, 2,4-dimethylphenol (Buss *et al.* 2015) and acetone, methyl ethyl ketone, methyl acetate, benzene, and trichloroethene (Spokas *et al.* 2011). Polycyclic aromatic hydrocarbons in biochar peaked at temperatures of 400–500 °C, without a clear influence of the feedstock type (Bucheli *et al.* 2015). The impact of residence time was also minor. Some individual studies reported a tendency of decreasing PAH concentrations with increasing residence time (Bucheli *et al.* 2015) but was not observed for slow pyrolysis in the study of Buss *et al.* (2016).

1.2. Heavy metals

Heavy metals (elements with a specific density >5 g mL^{-1}; (Morris 1992)) and metalloids that occur in the input material for pyrolysis become concentrated during carbonization, because of volatilization of gaseous carbon products (syngases), and a corresponding mass loss of the residual solid material (Brown *et al.* 2015). This concentration process has to be considered especially in the case of sewage sludge as input material (Agrafioti *et al.* 2013). Waste water originating from industrial areas may produce sewage sludge with elevated heavy metal concentrations that will further concentrate during pyrolysis. In such cases, a subsequent gasification step of the char product involving the addition of chlorides results in enhanced volatilization of metal chloride complexes (Luan *et al.* 2013) that can decrease the heavy metal content in the char residues (Dong *et al.* 2015; Li *et al.* 2015).

Besides sewage sludge also vegetation–derived feedstock may contain heavy metals that will not be lost during pyrolysis. Depending on the geological conditions and the resulting soil concentrations, plant root uptake may lead to elevated concentrations in the above-ground tissues. Atmospheric deposition may cause metal enrichment in plant tissues, too, but it will mainly remain on leaf surfaces. Green waste residues usually are separated into fine fractions, including leaf materials that are rather composted than carbonized and woody fractions that are preferably used for pyrolysis or gasification. Therefore vegetation-derived biochar is more likely to contain heavy metals from plant uptake if the plants are grown on soils with high geogenic metal concentrations or on anthropogenically contaminated land, or if the plant is a species that is able to translocate the contaminants into the shoots (Greger *et al.* 2007; Enell *et al.* 2016).

1.3. Guide/threshold values of contaminants in biochar

The two main guidelines, EBC and IBI, denote the organic and inorganic contaminant content threshold values, which are summarized in Table 1 (for further information on the IBI values we point to the references therein). The minimum threshold values of the IBI and the values for premium grade biochars of the EBC for inorganic and organic contaminants are alike. The maximum threshold values of the IBI are factors higher (2x for Pb, 60x for Cu, 25x for the Σ16 US EPA PAHs) than the concentrations set by the EBC for all contaminants. This difference reflects the management of the contaminants in different countries as the IBI based their maximum values mainly on Australian and US regulations, while the EBC based theirs on those of Germany and Switzerland. After all, national standards need to be met also. For instance Austria regulates the heavy metal contents of plant materials for quality control of pyrolysis feedstock (ÖNORM S 2211–1 2016). According to this standard, the biochar must not exceed certain threshold concentrations if agricultural applications are envisaged (in mg kg^{-1}_{dw}): Pb 100, Cd 3, Cu 150, Ni 100, Hg 1, Zn 500, Cr_{tot} 100. For use in organic agriculture there are lower thresholds than in conventional agriculture (in mg kg^{-1}_{dw}): Pb 45, Cd 0.7, Cu 70, Ni 25, Hg 0.4, Zn 200, Cr_{tot} 70. The EBC as well as the IBI also give guidelines for producers concerning biochar properties, quality assurance and control (QA/QC), analytical methods, among others. Also here, we refer to the recommendations and guidelines listed in the EBC and IBI for the specific analytical procedures.

The QA/QC in Europe is coordinated by the q.inspecta agency which awards certificates to producers according to the EBC guidelines. Q.inspecta is a governmental accredited company with inspections of production plants in individual countries carried out by independent national inspection agencies. Producers with a capacity <20 t y^{-1} are exempt from the yearly inspections of the production process. These small scale producers are to assure the quality of their biochar via self-declaration and a detailed description of the complete production process. The IBI Certification Programme (IBI 2015a) provides biochar manufacturers the opportunity to certify their biochar(s) as having met the minimum criteria established in the IBI Biochar Standards. The biochar producer shall apply online for the certification procedure, the sample will be tested by a laboratory according to the IBI Standards and the certificate shall be renewed, if obtained, after a year.

Table 1. List of inorganic and organic thresholds of contaminants recommended in biochars of the European Biochar Certificate (EBC 2012) and the International Biochar Initiative (IBI 2015b) guidelines. For details, see text

Contaminant	EBC	EBC	IBI	IBI
Heavy metals	Premium grade g	Basic grade	Minimum	Maximum
		gt^{-1}_{dw}		
Pb	<120	<150	121	300
Cd	<1.0	<1.5	1.4	39
Cu	<100	<100	143	6000
Ni	<30	<50	47	420
Hg	<1	<1	0.8	17
Zn	<400	<400	416	7400
Cr	<80	<90	93	1200
Co	–	–	34	100
Mo	–	–	5	75
Se	–	–	2	200
Organic compounds				
Sum of the 16 US EPA PAHs	<4	<12	6	300
Benzo[*a*]pyrene	–		3 g t^{-1} TEQ_{dw}[a]	
PCB	<0.2		0.2	1.0
Dioxins & furans	<20 ng kg^{-1} TEQ		17 ng kg^{-1} TEQ	

Note: [a] Toxicity equivalents TEQ.

1.4. Analytical aspects

The biochar community endeavours to develop and apply recommended methods for the quantification of organic and inorganic contaminants in biochar. Prior to the quantification, it is crucial to sample the biochar lot appropriately to obtain a representative sample that can be sent to the laboratory. In essence, a (biochar) sample is representative if its distributional heterogeneity is reduced. This can be achieved by a thorough mixing and splitting of the biochar lot for several times (Bucheli *et al.* 2014; Hilber *et al.* 2017). After that, the biochar lot shall be sampled several times and the subsamples joined to one bulk sample and further prepared (EBC) prior to be sent to the laboratory. The representative mass of the sample depends on its particle size in the way the bigger the particle the more of the sample (mass) has to be taken. The ISO norm (ISO 10381-8 2006 2006) provides, as a rule of thumb, a sample particle size to sample mass relation for the appropriate mass to be taken to the laboratory (Hilber *et al.* 2017).

To estimate the reliability of analytical results, 22 laboratories in 12 countries participated in an interlaboratory comparison within the COST Action TD1107 (Bachmann *et al.* 2016). Three biochars produced by slow pyrolysis from wood chips, paper sludge and wheat husks mixture, and from sewage sludge were analysed on 38 physical-chemical parameters (macro- and micro-elements, heavy metals, PAHs, pH, electrical conductivity, and specific surface area) by the participants with their preferential methods. The data were evaluated with the PROLab Plus™ (Version 2015.11.3.0) software that compares Zu scores, and the performance was valued according to EBC. A Zu-score is a point-based system that identifies a result of a parameter measured by a laboratory as inconspicuous, conspicuous or as an outlier (Bachmann *et al.* 2016). In summary, while the intra-laboratory reproducibility was generally good, the inter-laboratory reproducibility was mostly not (Bachmann *et al.* 2016). Only the carbon content (C) and pH showed a mean reproducibility standard deviation (SD) <10%. Those of hydrogen (H), ash and phosphorous (P) were between 10 and 20%. The variability of all other parameters was too high to be reliable. This result seems to be the consequence of participant laboratories using their own methods, because the EBC guidelines are quite recent (first version published in 2012). The study showed an urgent need to improve and standardize methods for biochar characterization. Future QA/QC measures require the production of a set of representative biochar reference materials and professional round robin tests with laboratories that use biochar reference methods (Bachmann *et al.* 2016). A recent review of benefits and risk of biochar clearly identified the need to develop biochar quality standards (Kuppusamy *et al.* 2016).

The QA/QC in the above section covers the technical aspects of certification. However, it can extend far beyond a pure labelling for a high quality biochar product. The sustainability inherent to a certification scheme should also include the application component, for instance, in the form of an optimal biochar dose (OBD), a contamination control, a life cycle assessment, among other aspects (Verheijen *et al.* 2015a). The authors of this new and sustainable certification concept identified future challenges such as e.g., (i) trade-offs between soil functions, (ii) the longer mean residence time in soil compared to traditional soil amendments, and (iii) limited mechanistic understanding of the nature and extent of the effects of biochars in a soil-crop system. Although a considerable effort is required to achieve such an comprehensive and practical certification system, sustainable biochar production and application should be considered as two sides of the same coin (Verheijen *et al.* 2015a).

2. Biochar-associated contaminants – bioavailable concentrations

Previous studies suggested that biochar total concentrations of organic (Hale *et al.* 2012) and inorganic (Freddo *et al.* 2012) contaminants are often comparable to those in other soil additives or natural soils, including in natural reference substrates such as Lufa 2.2 soil (Bastos *et al.* 2014b). Despite the specific mechanisms having not yet been identified, it is likely that the potential to induce toxicity to exposed organisms can be attributed to water–extractable organic and inorganic contaminants in certain biochars, individually or interacting synergistically and/or antagonistically (e.g., Smith *et al.* 2013; Bastos *et al.* 2014a). According to the equilibrium partitioning theory, toxicity to soil and sediment dwelling organisms upon exposure to toxicants occur via the aqueous phase, this being the main uptake route for most representative soil organisms (Belfroid *et al.* 1995; Janssen *et al.* 1997).

2.1. The strong sorption capacity of biochar

While the definition of bioavailability in the section above refers to the effects of contaminants in biochar for the ecosystem, this chapter sheds light on the exposure assessment by means of traditional phase distribution models, assessed by chemical proxies that are used to evaluate the bioavailability and -accessibility of contaminants in biochar.

2.1.1. Sorption and desorption of organic compounds

As discussed in Section 1.1., PAHs are among the priority contaminants in biochar. Therefore, the following section focuses on PAHs in biochars. Nevertheless, the sorption behavior of a variety of organic contaminants and biochar has been tested, including personal care products, pesticides, and pharmaceuticals (Yao *et al.* 2012; Jung *et al.* 2013; Zheng *et al.* 2013a; Lian *et al.* 2014; Xie *et al.* 2014; Ahmed *et al.* 2015; Calisto *et al.* 2015). Similar to the aqueous phase that determines the uptake route for organisms in ecotoxicity, it is the freely dissolved (aqueous) concentration (also C_{free} or C_W), thus molecules that are not bound to particles or dissolved organic carbon (Reichenberg, Mayer 2006), that determines the bioavailability of an organic contaminant. The C_{free} over the total solid phase concentration (C_S) determines the distribution coefficient (K_D) of compounds in a matrix in relation to water. In a recent ring-trial, the binding of PAHs in 21 biochars was found to be strong (Cornelissen *et al.* 2017). Hence, the biochar-to-water distribution coefficient, $K_{biochar}$, of PAHs was compared with that of organic carbon (organic carbon-to-water partition coefficient, K_{OC}) or an examplary soil which contained 5% OC (Fig. 1). The K_{OC} example is taken to represent the average PAH sorption characteristics for amorphous OC not containing any black carbon (Cornelissen *et al.* 2005). The example soil is simply taken to represent a soil containing 5% of this "average" amorphous OC. It can be observed from Figure 1 that the 21 biochars sorbed PAHs on average 400±200 times more strongly than for the example "average" soil, and 20±10 times more strongly than "average" OC.

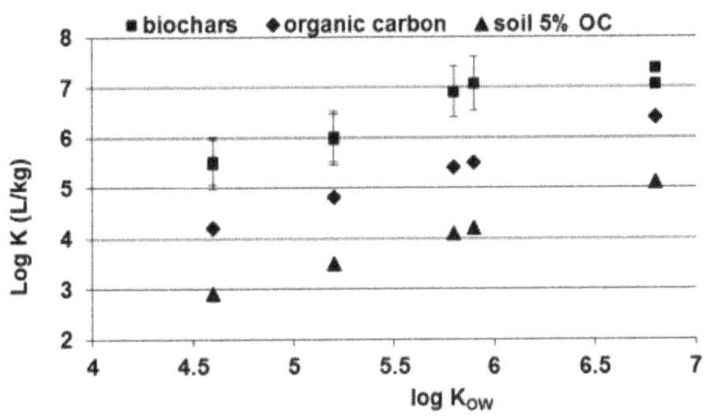

Fig. 1. Sorption (expressed as logarithmic solid-water distribution coefficients, K) to 21 biochars (average and standard deviation), organic carbon (without black carbon) and an examplary soil containing 5% of this organic carbon, as a function of PAH hydrophobicity expressed as octanol–water partition constant (log K_{OW}) (source: Cornelissen *et al.* 2017)

Table 2. PAH octanol-water (K_{OW}) partition constants (Jonker, Koelmans 2002) (numbers without footnote), organic carbon-water ($K_{OC\ amorphous}$) partition coefficients without black carbon calculated ($K_{OC} = 0.35K_{OW}$) from Seth *et al.* (1999), experimentally measured charcoal–water distribution coefficients ($K_{charcoal}$) of Jonker and Koelmans (2002), and biochar-water distribution coefficients ($K_{biochar}$) of medians of 59 biochars (Hale *et al.* 2012)

Compound	Log K_{OW}	Log $K_{OC\ amorphous}$	Log $K_{charcoal}$	Log $K_{biochar}$
Naphthalene	3.3[a]	2.8	n.d.	5.0
Acenaphthylene	4.0[a]	3.5	n.d.	n.d.
Acenaphthene	4.2[a]	3.7	n.d.	n.d.
Fluorene	4.3[a]	3.8	n.d.	5.0
Phenanthrene	4.6	4.1	6.2	5.4
Anthracene	4.6	4.1	7.1	5.5
Fluoranthene	5.2	4.7	6.3	5.7
Pyrene	5.2	4.7	6.4	5.7
Benzo[*a*] anthracene	5.9	5.4	7.5	6.1
Chrysene	5.8	5.3	7.2	5.9
Benzo[*e*]pyrene	6.4	5.9	7.9	n.d.
Benzo[*b*] fluoranthene	5.8	5.3	7.9	6.7
Benzo[*k*] fluoranthene	6.2	5.7	8.1	6.6
Benzo[*a*]pyrene	6.0	5.5	9.1	6.2
Indeno [1,2,3-*cd*] pyrene	n.d.	n.d.	n.d.	7.6
dibenzo[*a,h*] anthracene	n.d.	n.d.	n.d.	n.d.
Benzo[*ghi*] perylene	6.9	6.4	8.2	n.d.

Note: [a] values from Schwarzenbach *et al.* (2003), [b] n.d. = not detected.

Early evidence for limited bioavailability came from Jonker and Koelmans (2001). As shown in Table 2, the charcoal-to-water distribution coefficients ($K_{charcoal}$) from Jonker and Koelmans (2001) for PAHs were comparable to that for $K_{biochar}$. These were about 100–1000 times higher than octanol-water partition constants (K_{OW}), which were proportional to $K_{OC\ amorphous}$ ($= 0.35K_{OW}$) from Seth *et al.* (1999) as indicated in Figure 1. By defining organic matter-bound PAHs as bioavailable indicates that only 0.1–1% of charcoal-bound PAHs are bioavailable.

As a consequence from the definition of the partition coefficient ($K_D = C_S/C_W$), either C_S is high and/or C_W is low to obtain such high $K_{biochar}$. Indeed, the C_{free} in a study of Hale *et al.* (2012) ranged from 0.2–10 ng/L, with the highest concentration of 162 ng/L listed for a biochar produced via fast pyrolysis (gasification). These C_{free} are lower than environmental background levels reported by Lang *et al.* (2015). Thus, the bioavailability of biochar-bound PAHs is probably low in the short-term after soil application.

From the chemical sorption point-of-view, a compound is bioaccessible when it has slow desorption rates due to physical obstruction and strong binding of the contaminant to the matrix (Reichenberg, Mayer 2006). It is technically challenging to quantify bioaccessibility of organic contaminants from highly sorptive materials as for instance biochar (Mayer *et al.* 2016). Several papers state the so called infinite sink conditions via depletive methods as a prerequisite to determine the bioaccessibility of hydrophobic organic compounds (i) in matrices that have high $K_{i,D}$s (>3 log units) (Cornelissen *et al.* 2001; Hilber *et al.* 2009; Mayer *et al.* 2011; Gouliarmou, Mayer 2012; Collins *et al.* 2013; Mayer *et al.* 2016).

Here, we define the sorption capacity ratio (SCR) for a given contaminant i as the ratio of the sink material m_{sink} (e.g., a polymer) times its $K_{i,D}$ to the sample material m_{sample} times its $K_{i,D}$ (eq. 1). The product of $K_{i,D}$ and m_{sink} must be bigger than the one of the sample so that the sink can act as such for a contaminant (i.e., $SCR > 1$). In practice, a method can only be depletive if the SCR for a contaminant used in a desorption experiment is factors or magnitudes greater than 1 (i.e., $SCR >> 1$).

$$\text{with } SCR = \frac{m_{sink}}{m_{sample}} \cdot \frac{K_{i,sink}}{K_{i,\,sample}}. \qquad (1)$$

Mayer *et al.* (2016) showed that evaluating the bioaccessibility in matrices with such high K_Ds as biochar (Table 1) is very challenging because the usual sink materials such as Tenax® or silicone rods do not provide a high enough SCR. The only matrix fulfilling the infinite sink condition (SCR > 1) in presence of biochar is charcoal (Table 1) or activated carbon (AC, similar K_Ds as charcoal), if used in excess (i.e., $m_{sink} > m_{sample}$). To this end, Mayer *et al.* (2016) used so called "contaminant traps", i.e., glass jars with a silicone/AC layer, as sinks for PAHs from biochars.

The mass ratio was 500 originating from 50 g AC in the trap layer and 0.1 g biochar. Although this method is appropriate because the infinite sink condition is given (SCR > 1), the bioaccessibility cannot be assessed directly because the trapped PAHs in the silicone/AC layer cannot be extracted. Instead, it is assessed indirectly by extraction of the desorption resistant PAHs after the incubation of biochar to the contaminant traps (Mayer *et al.* 2011), and in comparison with PAHs quantified in biochar controls in absence of the sink. Results by Mayer *et al.* (2016) revealed that the contaminants in two investigated biochars were highly desorption resistant. All PAHs scattered around the 1:1 line indicating equal concentrations in the control (–trap) and the container with the sink layer (+trap, Fig. 2). Thus, the bioaccessibility of PAHs in biochars was very limited (Mayer *et al.* 2016). Nonetheless, a range of biotic (e.g., biological activity) and abiotic (e.g. time scale, climate, natural organic matter) factors may act on biochar's physical and chemical properties over time, resulting in increased biochar mobility (Jaffé *et al.* 2013; Verheijen *et al.* 2014) and/or bioaccessibility/bioavailability of a contaminant fraction.

2.1.2. *Sorption of heavy metals*

Before any beneficial or adverse effect of biochar and/or contaminated soils appear, the elements have to become plant available for root uptake (Semple *et al.* 2004). Thus, the elements need to be dissolved in the pore water. Here the sorption ability of biochar plays an important role. The organic matrix of biochar may immobilize and thereby reduce the bioavailability of heavy metals either existing in biochar or in the surrounding soil solution (Park *et al.* 2011). The ability of immobilizing heavy metals is achieved by different physical and chemical binding mechanisms of the biochar and are direct sorption mechanisms. The high sorption capacity of biochar to metal elements is made possible by high specific surface area, high cation exchange capacity (CEC) and long residence time in soil (Beesley *et al.* 2015). Biochar surfaces may be modified during pyrolysis or by post-treatments with reagents inducing artificial aging to increase the binding capacity deliberately and for specific purposes (Chen *et al.* 2011; Fristak *et al.* 2014).

Biochar that has been freshly produced at high temperatures (>500 °C) contains relatively few functional groups (e.g., carbonyl, hydroxyl, carboxyl or phenolic groups). However, subsequently the reactions of biochar with oxidizing additives, atmospheric oxygen, microbial metabolism or enzymatic breakdown may create or modify functional groups. These provide a negative charge to the biochar surface and a high CEC. Biochar produced at lower temperature is characterized by higher CEC even as fresh material (Beesley, Marmiroli 2011). The sorption of metals frequently is accompanied by a release of protons (Uchimiya *et al.* 2010c).

Electrostatic interactions between positively charged metal cations and the π-electrons of the ligands of the aromatic structures of biochars may also contribute to immobilization of heavy metals (Gomez-Eyles *et al.* 2013). As the aromatic condensation of biochar increases with higher pyrolysis temperature and the oxygenated functional groups decrease (McBeath *et al.* 2015), the proportion of cations sorbed by electrostatic binding increases. Biochar produced at lower temperatures conserves more functional groups because of incomplete pyrolysis, but these are less resistant to modification or degradation when deployed as soil

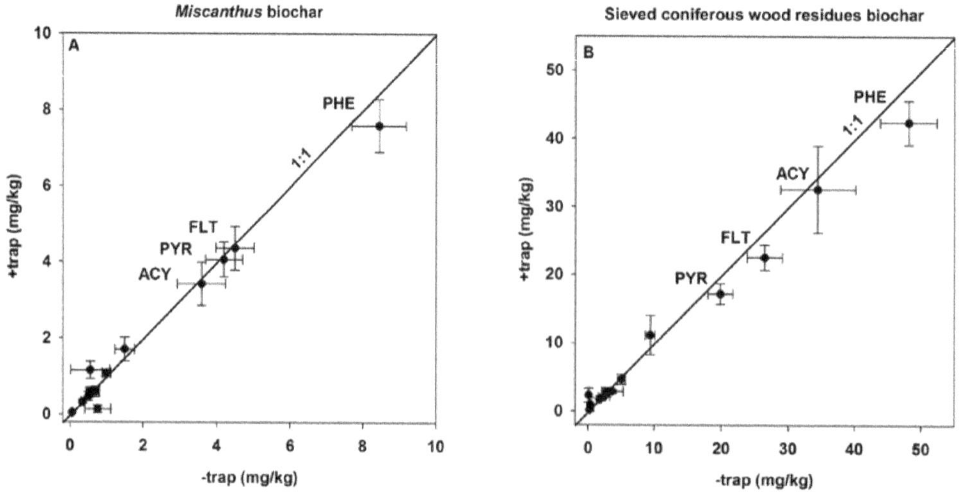

Fig. 2. Scatter plots of remaining PAH contents of two biochars (panel A and B) after their incubation in either contaminant traps (+trap) or similar controls without the infinite sink (–trap). Error bars indicate the standard deviation of duplicates and fall within the symbol if not visible. Individual PAH that showed the highest concentrations are phenanthrene (PHE), fluoranthene (FLT), acenaphthylene (ACY), and pyrene (PYR). (source: Mayer *et al.* (2016) modified. "This is an unofficial adaptation of an article that appeared in an ACS publication. ACS has not endorsed the content of this adaptation or the context of its use.")

additive, thereby releasing again previously sorbed metals (Mukherjee *et al.* 2011).

Indirect mechanisms rely on the effects of biochar on soil characteristics, which subsequently affect the release or binding of heavy metals or metalloids. The mixing of biochar with soil frequently results in pH-changes (Kloss *et al.* 2014c). Minerals in the feedstock become constituents of the ash fraction of biochar. They cause a high pH of biochar and facilitate the precipitation of Cu, Cd, Pb or Zn as phosphates or carbonates at the biochar surface (Beesley *et al.* 2015). Non-modified plant-based biochars usually have pH-values in the range of 8–9.5 and reduce the solubility of metals by the formation of metal hydroxides (e.g. $Zn(OH)_2$) (Brümmer 1986). The situation is quite different for metalloids as, e.g., As. Its solubility and availability increases at higher pH (Moreno-Jiménez *et al.* 2012). Antimony and Mo, too, occur as anionic species in the soil and behave similarly to As. Cationic metals that are bound to negatively charged surfaces in soil increase their solubility at low pH as CEC and high proton density are positively related. This example shows that biochar may act positively for cationic and negatively for anionic metal species in the soil. However, there exist modification options for biochar that may reverse the negative face of biochar for anionic forms (Pan *et al.* 2015).

Biochar in soil may (at least in the short term and if the pyrolysis temperature was not high) increase dissolved OC. Some metals and metalloids like Cu and As preferably form complexes with organic materials and thereby enhance their mobility (Kloss *et al.* 2014a). Biochar may also become a source for phosphorus that is bioavailable, depending on the feedstock and pyrolysis conditions. Enhanced P-supply of plants may decrease the uptake of As in the plant roots as both elements use the same uptake system in the roots (Moreno-Jiménez *et al.* 2012; Beesley *et al.* 2013). Increased P supply may also enhance the precipitation of Pb-phosphate if soil pore water is contaminated with Pb, e.g., by mining activities (Karami *et al.* 2011).

The oxygenated functional groups at biochar surfaces may modify the oxidation status of redox-sensitive metals and thus also their speciation. As an example, the ecotoxicologically undesired Cr(VI) may be modified to the less harmful Cr(III) by sorption to biochar surfaces (Bolan *et al.* 2012).

2.2. Contaminant bioavailability and potential ecotoxicity of biochar

Microbial biomass and activity have shown to increase upon biochar amendment (Van Zwieten *et al.* 2010; Domene *et al.* 2014; Domene *et al.* 2015b), as it provides refuge (Jaafar *et al.* 2014, 2015) and nutrients in the case of fresh biochars from nutrient-rich feedstocks (Jeffery *et al.* 2011) or reduces bioavailability of toxic compounds in their environment (e.g., Elad *et al.* 2011; Graber *et al.* 2014). Short-term effects have also been reported on avoidance behaviour and survival of earthworms (*Eisenia fetida, Aporrectodea caliginosa*), collembolans (*Folsomia candida*) and enchytraeids (*Enchytraeus crypticus*) as well as on earthworm weight loss and abundance of ant communities, at biochar application rates of 0–14% (w/w) (Liesch *et al.* 2010; Li *et al.* 2011; Busch *et al.* 2012; Hale *et al.* 2013; Tammeorg *et al.* 2014; Castracani *et al.* 2015; Domene *et al.* 2015a, 2015b). Chronic toxicity studies have been performed to a much lesser extent, focusing primarily on reproduction of collembolans and earthworms, for up to 28 and 56 days, respectively (Liesch *et al.* 2010; Li *et al.* 2011; Marks *et al.* 2014; Domene *et al.* 2015a). However, the link between biochar's inherent contaminants and toxicity to soil meso- and macro-fauna remains unclear, with data being often contradictory and influenced by feedstock, pyrolysis conditions, application rates, soil type, pH and the exposed organism (Lehmann *et al.* 2011; Ameloot *et al.* 2013). Yet, further to the aforementioned direct effects, biochar amendment can also impact soil biological activity indirectly. For instance, biochar (10% v:v) applied to contaminated mine soil increased soil toxicity (based on *Vibrio fischeri* luminescence inhibition and *Lolium perenne* germination), due to solubilisation of As in pore water (>2500 μg/L) linked to changes in pH and soluble phosphate (Beesley *et al.* 2014). However, by combining biochar with compost, the amended soil toxicity was effectively reduced, due to decreased metal extractability and increased soluble nutrients (e.g. P; Beesley *et al.* 2014).

Indirect effects from biochar amendment also include the potential risk to aquatic communities by runoff or leaching from amended soils, due to the inherent mobility of soil (Verheijen *et al.* 2015b) and black carbon (Jaffé *et al.* 2013) particles, as well as evidence of biochar dust subjected to wind erosion (Silva *et al.* 2015) and emitted as aerosols (Genesio *et al.* 2016). Recently, Bastos *et al.* (2014a) assessed the effects on aquatic organisms upon exposure to water-extracts of biochar-amended soil at 80 t/ha (mid-point of the biochar concentration range (Jeffery *et al.* 2011)). Although aqueous extracts contained metals (Σmetals 96.3 μg/L) and PAHs (Σ16 EPA PAHs 106 ng/L) within EU surface water quality targets (2008/105/EC), there was reduced bioluminescence by the marine bacterium *V. fischeri* and 20–25% acute mobility impairment of the invertebrate *Daphnia magna* (Bastos *et al.* 2014a). In contrast, no effects were observed on the growth of the microalgae *Raphidocellis subcapitata* (Bastos *et al.* 2014a).

Interestingly, toxicity to *V. fischeri* exposed to biochar leachates was reduced, by decreasing biochar's inherent PAHs concentrations (and possibly other dissolvable organic contaminants) after pyrolysis, by thermal post-treatment (Kołtowski, Oleszczuk 2015). In contrast, biochar post-treatment did not have a substantial effect on phytotoxicity or *D. magna's* survival (Kołtowski, Oleszczuk 2015). A recent study by Buss *et al.* (2015) revealed that the presence of VOCs in biochar, such as organic acids and phenols, caused germination inhibition of cress, due to phytotoxic leachates in amended soil, as they are more water soluble and therefore, more mobile than PAHs. Ecotoxicological responses to water or solvent-based extracts in the above studies were also dependent on biochar characteristics and concentration of the extract (Oleszczuk *et al.* 2013; Smith *et al.* 2013). Potential impacts associated to biochar ageing in pristine (i.e. non-contaminated) soil on biota-mediated ecosystem functions, at the relevant time scales, can only be speculated at present, where specific effects and mechanisms are expected to depend on a combination of biochar characteristics, soil type and on-site environmental factors. For instance, various reports suggest that ageing of pyrogenic organic matter may further improve some of its attributed benefits to soil, including the provision of microbial habitat (e.g. Glaser 2007; Mukherjee *et al.* 2014). Yet, it is also possible that physical, chemical and biological weathering of pyrogenic organic matter (e.g. Hamer *et al.* 2004; Mukome *et al.* 2004; Brodowski *et al.* 2005; Cheng *et al.* 2006; Brodowski *et al.* 2007; Hale *et al.* 2011; Sorrenti *et al.* 2016) overtime, may contribute to enhanced desorption and/or bioavailability of some of its inherent toxic elements in amended soils, in the long-term. According to the recently assembled thematic task force on ecotoxicology and functional biodiversity within COST Action 1107 (Verheijen *et al.*, this issue; Tammeorg *et al.*, this issue), achieving an acceptable level of scientific understanding of biochar effects on terrestrial and aquatic ecosystems involves further testing of biochar properties and combinations of biochar-soil type, climatic conditions, and a functionally diverse range of representative organisms (Tammeorg *et al.*, this issue). It is also necessary to consider the relevant spatial and temporal scales to biochar, as a soil amendment and an environmental management tool (Verheijen *et al.* 2012; Verheijen *et al.* 2014). Specifically, biochar ecotoxicological studies will need to include representative biochar ageing procedures and modeling approaches that account for the relevant biochar characteristics and environmental factors, to more easily derive and extrapolate long-term effects on biota, in natural systems.

3. Biochar for remediation of contaminated soils

Biochar can be used for the remediation of soils (or sediments) that are contaminated by inorganic and organic compounds. Its high sorption capacity in combination with its high surface area are appropriate to immobilize contaminants. Thus, the contaminants will not be removed from the matrix but sequestered. It is assumed that the contaminants will not be released into the matrix until the biochar is degraded. A detailed discussion about the stability of biochar over time is beyond the focus of this paper. However, mean residence times (MRT) of pyrogenic organic matter (PCM) studied under different field conditions varied widely and ranged from 6 to 5448 years (Lehmann *et al.* 2015). Part of this variability owes to the fact that the models to calculate the MRT differed, or that PCM properties or experimental approaches were different (Lehmann *et al.* 2015).

3.1. Biochar for remediation of soils contaminated with organic compounds

While a considerable number of studies and reviews have been published on the suitability of AC (Hale *et al.* 2015) and references therein) for the reclamation of organically contaminated soils, Hale *et al.* (2015) pointed out in their overview that relatively few laboratory studies were carried out with biochar and, at that time, no field study results were published. Another review by Xie *et al.* (2015) confirmed that the majority of the studies with biochar were conducted in the laboratory. Biochars used for remediation were better suited to immobilize organic than inorganic contaminants (Xie *et al.* 2015) and AC reduced organic contaminants more effectively than biochar (Hale *et al.* 2015). The modest effect was due to the relatively low K_D values of the biochars in comparison to *in situ* K_{OC}. In comparison to these, the K_{OC} values listed in Table 2 are not containing carbonaceous geosorbents such as black carbon, coal or kerogen that naturally occur in organic matter and exhibit a 10–100 times higher sorption capacity than amorphous OC (Cornelissen *et al.* 2005).

A recent field study over two consecutive years where biochar and granulated AC (GAC) were added at 2.5% each in different plots to a field soil that was contaminated with 39 mg kg^{-1} dichlorodiphenyltrichloroethane (DDT) showed that biochar could significantly reduce DDT accumulation in earthworms (49%), whereas the GAC did not and the invertebrates showed toxic effects (Denyes *et al.* 2016). None of the char amendments reduced DDT uptake in plants (*Cucurbita pepo spp. pepo*). The C_{free} was assessed by the polyoxymethylene (POM) passive sampler proxy and correlated well with the availability of the earthworms, but not with the phytoavailability (Denyes *et al.* 2016). So, different endpoints (earthworms, plants, proxy) under real world conditions indicated contradicting results and the form and type of char influences the outcome. In summary, the limited data of only a recent field study with biochar leads to inconclusive results.

Equilibrium times are, besides the weathering of biochars in soils or sediments, important and need to be taken into account before judging over success or failure of remediation of contaminated sites with biochar (or AC).

Positive or immobilization effects cannot be expected to be reached immediately, as equilibration times may be very long for biochar or AC amendment. A polychlorinated biphenyl-contaminated sediment that was amended with 2% AC reached equilibrium after about 2000 to 2500 days (Werner *et al.* 2006). When the batch experiment was mixed, the C_{free} was reduced to a tenth of its original concentration after only 50 days, but when it was left unmixed the same result was reached not until about 900 days (Werner *et al.* 2006).

3.2. Biochar for remediation of soil contaminated with heavy metals

The sorption characteristics of biochar may reduce the bioavailability of undesired elements and reduce ecotoxic effects (Karer *et al.* 2015). Nutrients from organic soil additives in combination with biochar may enhance the productivity of marginal lands with contaminated soils. As for other soil remediation technologies, it is important to survey the site-specific soil and contaminants conditions and to check the efficacy of the intended remediation technology at small scale. Different mixtures of contaminants might require bespoke biochars or combinations of biochars with different sorption characteristics. As shown in Figure 3, that contains results of mostly lab studies, the dominance of the positive or the negative face of biochar for remediation purposes depends on the element and its speciation. Whereas metals occurring predominantly in cationic form will be bound by biochar and therefore show a smiley, anionic metals and metalloids will not be immobilized by biochar and show a frowny. Elements like As, B and Mo that predominantly occur in anionic form are rather mobilized because of negative surface charges of biochar, pH effects and competition with DOC for binding sites (Beesley *et al.* 2014).

In the following we focus on recent field studies about remediation of heavy metal contaminated sites. Similar to Figure 3, the most investigated heavy metals are Cd and Pb in paddy rice soils (Cui et *al.* 2011; Chen *et al.* 2016) or Zn (and Ni) (Shen *et al.* 2016). The uptake of Cd into rice grain was reduced by 57–86% over two seasons (Chen *et al.* 2016) and by 17–62% over two years (Cui *et al.* 2011). The study of Shen *et al.* (2016) revealed that 0.5–2% biochar amendment showed reduced carbonic acid leachates of Ni and Zn by 83–98% over three years, which was comparable with the cement based treatment carried out at the same site in parallel. Despite the positive aspect that the contaminant uptake into plants was reduced in the above studies, plants react differently to the biochar amendment e.g., different rice cultivars reacted differently on the Cd uptake (Chen *et al.* 2016) or no grass could be planted on the field (Shen *et al.* 2016). These outcomes indicate that though biochar has excellent sorption capacities for heavy metals, other factors like the plant species, the contaminant itself, and the influence of the biochar on soil conditions such as pH, available nutrients, earthworm habitats, etc. (Beesley *et al.* 2015) are important for a successful reclamation of a site.

Finally, Shaheen and Rinklebe (2015) compared different emerging amendments as, e.g., biochar, chitosan, nano-hydroxyapatite and organo-clay and low cost alternative material such as limestone, zeolite, bentonite and others in their immobilizing effects of a Cd and Pb floodplain soil as expressed by rapeseed (*Brassica napus*)

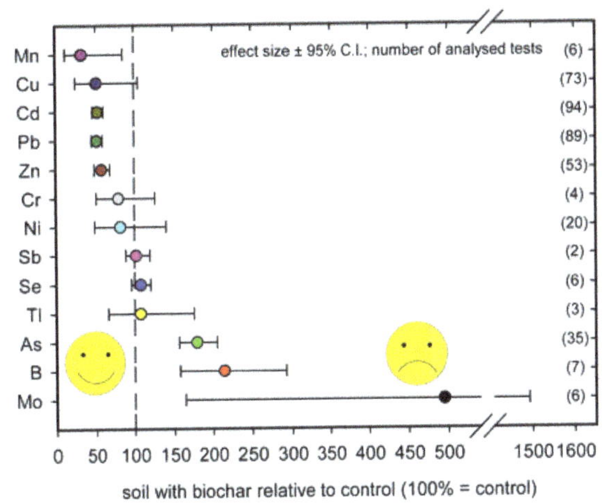

Fig. 3. Changes in the availability of elements from soil to soil solutions (studied with extractants simulating bioavailability for plant roots) in response to biochar additions in lab and field experiments. 100% means identical elemental availability in soils with and without biochar (= treatment and control). 50% means that in soils with biochar the elements were only half as available as in soils without biochar. 200% means that in soils with biochar the elements had double the availability compared to soil without biochar. The reduced availability (up to max. 100%) is a positive effect and therefore marked with a smiley. All studies that showed enhanced availabilities of heavy metals (>100%) after biochar amendments were marked with a frowny. Effect sizes (±95% confidence interval, C.I.) for different elements were calculated with MetaWin 2.1 software. Numbers in brackets give the numbers of control-treatment-comparisons in peer-reviewed literature that fulfilled the data quality requirement for meta-analyses. Data were taken from the following references (Hartley *et al.* 2009; Namgay *et al.* 2010; Uchimiya *et al.* 2010a; Uchimiya *et al.* 2010b; Beesley, Dickinson 2011; Beesley, Marmiroli 2011; Beesley *et al.* 2011; Cao *et al.* 2011; Fellet *et al.* 2011; Karami *et al.* 2011; Park *et al.* 2011; Sizmur *et al.* 2011; Trakal *et al.* 2011; Uchimiya *et al.* 2011a; Uchimiya *et al.* 2011c, 2011b, 2011d, 2012a, 2012b; Buss *et al.* 2012; Choppala *et al.* 2012; Cui *et al.* 2012; Debela *et al.* 2012; Jiang *et al.* 2012a, 2012b; Mendez *et al.* 2012; Zheng *et al.* 2012; Beesley *et al.* 2013; Bian *et al.* 2013; Bolan *et al.* 2013; Chang *et al.* 2013; Gartler *et al.* 2013; Houben *et al.* 2013a, 2013b; Jiang, Xu 2013; Khan *et al.* 2013; Melo *et al.* 2013; Moon *et al.* 2013; Park *et al.* 2013; Sneath *et al.* 2013; Uchimiya, Bannon 2013; Xu, Zhao 2013; Zheng *et al.* 2013b; Ahmad *et al.* 2014; Beesley *et al.* 2014; Bian *et al.* 2014; Brennan *et al.* 2014; Kelly *et al.* 2014; Kloss *et al.* 2014b; Liang *et al.* 2014; Lucchini *et al.* 2014a, 2014b; Rees *et al.* 2014; Wagner, Kaupenjohann 2014; Waqas *et al.* 2014)

uptake in a pot experiment. The biochar and chitosan amendments decreased the Cd concentrations in the plant, but so did zeolite and bentonite. In contrast, the amendment with limestone increased the Cd concentration in *B. napus*. The authors suggested that limestone increased the carbonate fraction ($CaCO_3$ content in limestone is 98.5%), which induced Cd precipitation and complexation with carbonates, and was then mobilized in the acid rhizosphere zone and taken up into the plant (Shaheen, Rinklebe 2015). The researchers proposed to profit from this enhanced phytoextraction for Cd for the use of rapeseed as bioenergy crop. Biochar would indirectly remediate Cd-contaminated soils by phytoremediation. The results of the emerging amendments, except organo-clay, showed also a decreased uptake of Pb into the plant, and so did all low cost amendments except zeolite.

It can be concluded that field– and long–term studies with biochar and contaminated soils are very scarce for both organic contaminants and heavy metals and represent future research needs (Zhang *et al.* 2013). Another aspect of biochar amendment might be the reduced mineralization versus the immobilization and low bioavailability and -accessibility of the contaminants. These effects of opposite directions, immobilization and microbial degradation of biochar or AC amendment need to be carefully and systematically evaluated. Furthermore, the oral exposure of biochar contaminants, in other research field called bioavailability, needs to be assessed in future, similar to works published with soot (Gouliarmou *et al.* 2013; Zhang *et al.* 2015).

Conclusions

A qualitative overall assessment of biochar (mainly) in relation to contaminants, is compiled in Table 3. The positive, negative, and unclear aspects of biochar in these respects are depicted with corresponding emoticons. If the assessment was positive a smiley is put, if negative a frowny and if unclear an indifferent emoticon is placed. The indifferent face prevails while the smiley and frowny are both two times encountered. The indifferent expression reflects outcomes of biochar that are not yet clear as for instance the consequences of the interplay of biochar with the environment over time or for remediation purposes. The behavior of soil, sediment dwelling, or aquatic organisms to biochar in their habitat is species dependent. Effects to biochar might be positive, adverse, or the organism might avoid a biochar amended area. The same holds for plants that root in biochar amended soils. Yield might increase or decrease according to different plant species and in some studies even plant subspecies reacted differently to the same amendment in the same soil.

According to the overall assessment of biochar as a sink or source of contaminants and its suitability for pollutant remediation purposes, the following questions reflecting knowledge gaps and future research directions were identified:

1. How can high quality biochar and a robust contaminant quantification in biochar be assured?
2. How can robust and effect-based assessments of contaminants and contaminant mixtures to biota (plants, soil/sediment dwelling organisms) be achieved?
3. How can the above two points be harmonized towards an international protocol?
4. Long term impacts: What is the influence of time (years, decades) in biochar amended soils/sediments to:
   a. Biota (habitat, biodiversity changes, etc.)?
   b. Biochar itself and its immobilization of organic and inorganic contaminants?
   c. The contaminant's bioavailability and –accessibility and to their ecotoxicological effects?

The first three points address the QA/QC where the analytical methods in the lab (Bachmann *et al.* 2016) including as well as sustainable certification procedures (Verheijen *et al.* 2015a) need to be improved and standardized or, in the latter two cases, even included. Here, we would like to open up the radar to other contaminants such as VOC as pointed out by Buss *et al.* (2015).

The long-term effects of biochar in ecosystems need to be addressed in future that require especially long-term field experiments as also already highlighted in the review of Kuppusamy *et al.* (2016). Ecological effects of biota in soils or sediments can be studied as well as the attenuation effect of the biochar's sorption capacity, the weathering and degradation of biochar and with it the fate of the contaminants inherent in the biochar and the matrix. Researchers from different disciplines would need to collaborate to address as completely as possible the whole topic. Thus, we propose more international research programs as the one of COST Action TD1107 (ref of this volume of JEELM) that help to decide over the positive or negative face of biochar, its inherent contaminants and appropriateness for remediation.

Funding

This work was supported by the Polish-Swiss Research Programme through the Swiss Contribution to the enlarged European Union [project number PSPB-135/2010]; the Swiss National Science Foundation through the r4d programme [project number IZ01Z0_147314/1]; the Portuguese Science Foundation (FCT) for the fellowship of A.C. Bastos [SFRH/BPD/98231/2013]; the financial support to CESAM [UID/AMB/50017/2013], FCT/MEC through national funds and the co-funding by FEDER, within the PT2020 Partnership Agreement and Compete 2020; and the

Table 3. Qualitative overall assessment of contaminant-related aspects of biochar, of biochar itself, and its suitability for remediation of contaminated sites. Emoticons indicate whether the outcome per topic is positive, negative or unclear

Topic	Organic compounds	Heavy metals
Total concentrations inherent in biochars	PAHs can easily exceed the EBC guideline but less the maximum value of IBI; mechanisms of PAH formation are about to be understood and recommendations to minimize PAHs are communicated	Not a problem if the feedstock is not polluted by heavy metals → increase of heavy metals due to concentration in the pyrolysed material
Bioavailability and -accessibility of inherent contaminants	Generally low	Low due to low total concentrations
Sorption capacities	Very high (1–100000 times stronger than organic carbon) but - 1–100 times weaker if pores are blocked by organic matter - 1–10 times weaker if the mass transfer of the contaminants to the biochar is incomplete - 10–100 times weaker due to strong sorption to carbonaceous geosorbents in the soil or sediment	High for cationic heavy metals but low for anionic metals and metalloids such as As, Sb, Mo.
Remediation	Can be an attractive alternative to activated carbon. However, negative sides can be: - Long equilibrium time - Reduced sorption due to fouling and pore blockage - pH increase → overliming effect - Anionic metals and metalloids are mobilized - Success contaminant and crop species/cultivar dependent - Ecological biodiversity altered and remediated areas avoided by some species The remediation with biochar must be tailored for every situation as there does not seem to exist an allin-one solution. What is therefore the… - Correct dose? - Appropriate particle size? - Appropriate soil incorporation technique? - Appropriate feedstock for the least ecotoxicological effects? - Appropriate pre- or post-pyrolysis modification of the feedstock or biochar? - Appropriate contaminant?	

COST Action TD1107 Biochar as option for sustainable resource management.

Disclosure statement

The authors declare that they don't have any competing financial, professional, or personal interests from any other parties.

References

ÖNORM S 2211-1. 2016. *Austrian Standards: Pflanzenkohle Teil 1: Ausgangsmaterialien, Qualitätsanforderungen und Untersuchungsmethoden'*. Austrian Standards, Vienna.

Agrafioti, E.; Bourkas, G.; Kalderis, D.; Diamadopoulos, E. 2013. Biochar production by sewage sludge pyrolysis, *Journal of Analytical and Applied Pyrolysis* 101: 72–78. https://doi.org/10.1016/j.jaap.2013.02.010

Ahmad, M.; Lee, S. S.; Lim, J. E.; Lee, S. E.; Cho, J. S.; Moon, D. H.; Hashimoto, Y.; Ok, Y. S. 2014. Speciation and phytoavailability of lead and antimony in a small arms range soil amended with mussel shell, cow bone and biochar: EXAFS spectroscopy and chemical extractions, *Chemosphere* 95: 433–441. https://doi.org/10.1016/j.chemosphere.2013.09.077

Ahmad, M.; Ok, Y. S.; Rajapaksha, A. U.; Lim, J. E.; Kim, B.-Y.; Ahn, J.-H.; Lee, Y. H.; Al-Wabel, M. I.; Lee, S.-E.; Lee, S. S. 2016. Lead and copper immobilization in a shooting range soil using soybean stover- and pine needle-derived biochars:

chemical, microbial and spectroscopic assessments, *Journal of Hazardous Materials* 301: 179–186. https://doi.org/10.1016/j.jhazmat.2015.08.029

Ahmed, M. B.; Zhou, J. L.; Ngo, H. H.; Guo, W. 2015. Adsorptive removal of antibiotics from water and wastewater: progress and challenges, *Science of the Total Environment* 532: 112–126. https://doi.org/10.1016/j.scitotenv.2015.05.130

Ameloot, N.; Graber, E. R.; Verheijen, F. G. A.; De Neve, S. 2013. Interactions between biochar stability and soil organisms: review and research needs, *European Journal of Soil Science* 64(4): 379–390. https://doi.org/10.1111/ejss.12064

Atkinson, C. J.; Fitzgerald, J. D.; Hipps, N. A. 2010. Potential mechanisms for achieving agricultural benefits from biochar application to temperate soils: a review, *Plant and Soil* 337(1–2): 1–18. https://doi.org/10.1007/s11104-010-0464-5

Bachmann, H. J.; Bucheli, T. D.; Dieguez-Alonso, A.; Fabbri, D.; Knicker, H.; Schmidt, H.-P.; Ulbricht, A.; Becker, R.; Buscaroli, A.; Buerge, D.; Cross, A.; Dickinson, D.; Enders, A.; Esteves, V. I.; Evangelou, M. W. H.; Fellet, G.; Friedrich, K.; Gasco Guerrero, G.; Glaser, B.; Hanke, U. M.; Hanley, K.; Hilber, I.; Kalderis, D.; Leifeld, J.; Masek, O.; Mumme, J.; Carmona, M. P.; Calvelo Pereira, R.; Rees, F.; Rombola, A. G.; de la Rosa, J. M.; Sakrabani, R.; Sohi, S.; Soja, G.; Valagussa, M.; Verheijen, F.; Zehetner, F. 2016. Toward the standardization of biochar analysis: the COST Action TD1107 interlaboratory comparison, *Journal of Agricultural and Food Chemistry* 64(2): 513–527. https://doi.org/10.1021/acs.jafc.5b05055

Bastos, A. C.; Prodana, M.; Abrantes, N.; Keizer, J. J.; Soares, A. M. V. M.; Loureiro, S. 2014a. Potential risk of biochar-amended soil to aquatic systems: an evaluation based on aquatic bioassays, *Ecotoxicology* 23(9): 1784–1793. https://doi.org/10.1007/s10646-014-1344-1

Bastos, A. C.; Prodana, M.; Oliveira, J. M. M.; Calhôa, C. F.; Santos, M. J. G.; Soares, A. M. V. M.; Loureiro, S. 2014b. Water-extractable priority contaminants in LUFA 2.2 soil: back to basics, contextualisation and implications for use as natural standard soil, *Ecotoxicology* 23(9): 1814–1822. https://doi.org/10.1007/s10646-014-1335-2

Beesley, L. Dickinson, N. 2011. Carbon and trace element fluxes in the pore water of an urban soil following greenwaste compost, woody and biochar amendments, inoculated with the earthworm *Lumbricus terrestris*, *Soil Biology and Biochemistry* 43(1): 188–196. https://doi.org/10.1016/j.soilbio.2010.09.035

Beesley, L.; Inneh, O. S.; Norton, G. J.; Moreno-Jimenez, E.; Pardo, T.; Clemente, R.; Dawson, J. J. C. 2014. Assessing the influence of compost and biochar amendments on the mobility and toxicity of metals and arsenic in a naturally contaminated mine soil, *Environmental Pollution* 186: 195–202. https://doi.org/10.1016/j.envpol.2013.11.026

Beesley, L. Marmiroli, M. 2011. The immobilisation and retention of soluble arsenic, cadmium and zinc by biochar, *Environmental Pollution* 159(2): 474–480. https://doi.org/10.1016/j.envpol.2010.10.016

Beesley, L.; Marmiroli, M.; Pagano, L.; Pigoni, V.; Fellet, G.; Fresno, T.; Vamerali, T.; Bandiera, M.; Marmiroli, N. 2013. Biochar addition to an arsenic contaminated soil increases arsenic concentrations in the pore water but reduces uptake to tomato plants (*Solanum lycopersicum* L.), *Science of the Total Environment* 454–455: 598–603. https://doi.org/10.1016/j.scitotenv.2013.02.047

Beesley, L.; Moreno-Jimenez, E.; Fellet, G.; Melo, L.; Sizmur, T. 2015. Biochar and heavy metals, in J. S. Lehmann, S. Joseph (Eds.). *Biochar for environmental management: science, technology and implementation*. London, UK: Routledge.

Beesley, L.; Morenoi-Jimenez, E.; Gomezi-Eyles, J. L.; Harris, E.; Robinson, B.; Sizmur, T. 2011. A review of biochars' potential role in the remediation, revegetation and restoration of contaminated soils, *Environmental Pollution* 159(12): 3269–3282. https://doi.org/10.1016/j.envpol.2011.07.023

Belfroid, A. C.; Seinen, W.; van Gestel, C. A. M.; Hermens, J. L. M.; van Leeuwen, K. J. 1995. Modeling the accumulation of hydrophobic organic chemicals in earthworms – application of the Equilibrium Partitioning Theory, *Environmental Science and Pollution Research* 2: 5–15. https://doi.org/10.1007/BF02987505

Bian, R.; Chen, D.; Liu, X.; Cui, L.; Li, L.; Pan, G.; Xie, D.; Zheng, J.; Zhang, X.; Zheng, J.; Chang, A. 2013. Biochar soil amendment as a solution to prevent Cd-tainted rice from China: results from a cross-site field experiment, *Ecological Engineering* 58: 378–383. https://doi.org/10.1016/j.ecoleng.2013.07.031

Bian, R.; Joseph, S.; Cui, L.; Pan, G.; Li, L.; Liu, X.; Zhang, A.; Rutlidge, H.; Wong, S.; Chia, C.; Marjo, C.; Gong, B.; Munroe, P.; Donne, S. 2014. A three-year experiment confirms continuous immobilization of cadmium and lead in contaminated paddy field with biochar amendment, *Journal of Hazardous Materials* 272: 121–128. https://doi.org/10.1016/j.jhazmat.2014.03.017

Bolan, N.; Mahimairaja, S.; Kunhikrishnan, A.; Choppala, G. 2013. Phosphorus-arsenic interactions in variable-charge soils in relation to arsenic mobility and bioavailability, *Science ot the Total Environment* 463–464: 1154–1162.

Bolan, N. S.; Kunhikrishnan, A.; Choppala, G. K.; Thangarajan, R.; Chung, J. W. 2012. Stabilization of carbon in composts and biochars in relation to carbon sequestration and soil fertility, *Science of the Total Environment* 424: 264–270. https://doi.org/10.1016/j.scitotenv.2012.02.061

Brennan, A.; Jimenez, E. M.; Puschenreiter, M.; Alburquerque, J. A.; Switzer, C. 2014. Effects of biochar amendment on root traits and contaminant availability of maize plants in a copper and arsenic impacted soil, *Plant and Soil* 379(1): 351–360. https://doi.org/10.1007/s11104-014-2074-0

Brodowski, S.; Amelung, W.; Haumaier, L.; Abetz, C.; Zech, W. 2005. Morphological and chemical properties of black carbon in physical soil fractions as revealed by scanning electron microscopy and energy-dispersive X-ray spectroscopy, *Geoderma* 128(1–2): 116–129. https://doi.org/10.1016/j.geoderma.2004.12.019

Brodowski, S.; Amelung, W.; Haumaier, L.; Zech, W. 2007. Black carbon contribution to stable humus in German arable soils, *Geoderma* 139(1–2): 220–228. https://doi.org/10.1016/j.geoderma.2007.02.004

Brown, R.; del Campo, B.; Boateng, A. A.; Garcia-Perez, M.; Masek, O. 2015. Fundamentals of biochar production, in J. Lehmann, S. Joseph (Eds). *Biochar for environmental management: science*, technology and implementation. London, UK: Routledge.

Brümmer, G. W. 1986. Heavy metal species, mobility and availability in soils, in M. Bernhard, F. E. Brinckman, P. J. Sadler (Eds). *The importance of chemical "speciation" in environmental processes. Volume 33 of of the Dahlem Workshop Reports*. Berlin, Heidelberg: Springer Berlin Heidelberg, 169–192. https://doi.org/10.1007/978-3-642-70441-3_11

Bucheli, T. D.; Bachmann, H. J.; Blum, F.; Bürge, D.; Giger, R.; Hilber, I.; Keita, J.; Leifeld, J.; Schmidt, H.-P. 2014. On the heterogeneity of biochar and consequences for its representative sampling, *Journal of Analytical and Applied Pyrolysis* 107: 25–30. https://doi.org/10.1016/j.jaap.2014.01.020

Bucheli, T. D.; Hilber, I.; Schmidt, H. 2015. Polycyclic aromatic hydrocarbons and polychlorinated aromatic compounds in biochar, in J. Lehmann, S. Joseph (Eds.). *Biochar for environmental management: science, technology and implementation.* London, UK.

Busch, D.; Kammann, C.; Grünhage, L.; Müller, C. 2012. Simple biotoxicity tests for evaluation of carbonaceous soil additives: establishment and reproducibility of four test procedures, *Journal of Environmental Quality* 41(4): 1023–1032. https://doi.org/10.2134/jeq2011.0122

Buss, W.; Graham, M. C.; MacKinnon, G.; Mašek, O. 2016. Strategies for producing biochars with minimum PAH contamination, *Journal of Analytical and Applied Pyrolysis* 119: 24–30. https://doi.org/10.1016/j.jaap.2016.04.001

Buss, W.; Kammann, C.; Koyro, H. W. 2012. Biochar reduces copper toxicity in *Chenopodium quinoa Willd.* in a sandy soil, *Journal of Environmental Quality* 41(4): 1157–1165. https://doi.org/10.2134/jeq2011.0022

Buss, W.; Masek, O.; Graham, M.; Wüst, D. 2015. Inherent organic compounds in biochar – their content, composition and potential toxic effects, *Journal of Environmental Management* 156: 150–157. https://doi.org/10.1016/j.jenvman.2015.03.035

Calisto, V.; Ferreira, C. I. A.; Oliveira, J. A. B. P.; Otero, M.; Esteves, V. I. 2015. Adsorptive removal of pharmaceuticals from water by commercial and waste-based carbons, *Journal of Environmental Management* 152: 83–90. https://doi.org/10.1016/j.jenvman.2015.01.019

Cao, X.; Ma, L.; Liang, Y.; Gao, B.; Harris, W. 2011. Simultaneous immobilization of lead and atrazine in contaminated soils using dairy-manure biochar, *Environmental Science & Technology* 45(11): 4884–4889. https://doi.org/10.1021/es103752u

Castracani, C.; Maienza, A.; Grasso, D. A.; Genesio, L.; Malcevschi, A.; Miglietta, F.; Vaccari, F. P.; Mori, A. 2015. Biochar-macrofauna interplay: searching for new bioindicators, *Science of the Total Environment* 536: 449–456. https://doi.org/10.1016/j.scitotenv.2015.07.019

Chang, Y. -T.; Hsi, H. -C.; Hseu, Z.-Y.; Jheng, S.-L. 2013. Chemical stabilization of cadmium in acidic soil using alkaline agronomic and industrial by-products, *Journal of Environmental Science and Health Part A – Toxic/Hazardous Substances & Environmetnal Engineering* 48: 1748–1756.

Chen, B.; Chen, Z.; Lv, S. 2011. A novel magnetic biochar efficiently sorbs organic pollutants and phosphate, *Bioresource Technology* 102(2): 716–723. https://doi.org/10.1016/j.biortech.2010.08.067

Chen, D.; Guo, H.; Li, R.; Li, L.; Pan, G.; Chang, A.; Joseph, S. 2016. Low uptake affinity cultivars with biochar to tackle Cd-tainted rice – a field study over four rice seasons in Hunan, China, *Science Of The Total Environment* 541: 1489–1498. https://doi.org/10.1016/j.scitotenv.2015.10.052

Cheng, C. H.; Lehmann, J.; Thies, J. E.; Burton, S. D.; Engelhard, M. H. 2006. Oxidation of black carbon by biotic and abiotic processes, *Organic Geochemistry* 37(11): 1477–1488. https://doi.org/10.1016/j.orggeochem.2006.06.022

Choppala, G. K.; Bolan, N. S.; Megharaj, M.; Chen, Z.; Naidu, R. 2012. The influence of biochar and black carbon on reduction and bioavailability of chromate in soils, *Journal of Environmental Quality* 41: 1175–1184. https://doi.org/10.2134/jeq2011.0145

Collins, C. D.; Mosquera-Vazquez, M.; Gomez-Eyles, J. L.; Mayer, P.; Gouliarmou, V.; Blum, F. 2013. Is there sufficient "sink" in current bioaccessibility determinations of organic pollutants in soils?, *Environmental Pollution* 181: 128–132. https://doi.org/10.1016/j.envpol.2013.05.053

Cornelissen, G.; Gustafsson, O.; Bucheli, T. D.; Jonker, M. T. O.; Koelmans, A. A.; Van Noort, P. C. M. 2005. Extensive sorption of organic compounds to black carbon, coal, and kerogen in sediments and soils: mechanisms and consequences for distribution, bioaccumulation, and biodegradation, *Environmental Science & Technology* 39(18): 6881–6895. https://doi.org/10.1021/es050191b

Cornelissen, G.; Hale, S. E.; Tomlinson, T. 2017. Polycyclic aromatic hydrocarbons (PAH) and polychlorinated aromatic compounds in biochar, in B. Singh, M. Camps-Arbestain, J. Lehmann (Eds). *Biochar: a guide to analytical methods.* Melbourne: CSIRO Publishing, 625–654.

Cornelissen, G.; Rigterink, H.; ten Hulscher, D. E. M.; Vrind, B. A.; van Noort, P. C. M. 2001. A simple Tenax® extraction method to determine the availability of sediment-sorbed organic compounds, *Environmental Toxicology and Chemistry* 20(4): 706–711. https://doi.org/10.1002/etc.5620200403

Cui, L.; Li, L.; Zhang, A.; Pan, G.; Bao, D.; Chang, A. 2011. Biochar amendment greatly reduces rice Cd uptake in a contaminanted paddy soil: a two-year field experiment, *Bioresources* 6(3): 2605–2618.

Cui, L. Q.; Pan, G. X.; Li, L. Q.; Yan, J. L.; Zhang, A.; Bian, R. J.; Chang, A. 2012. The reduction of wheat Cd uptake in contaminated soil via biochar amendment: a two-year field experiment, *Bioresources* 7(4): 5666–5676. https://doi.org/10.15376/biores.7.4.5666-5676

Debela, F.; Thring, R. W.; Arocena, J. M. 2012. Immobilization of heavy metals by Co-pyrolysis of contaminated soil with woody biomass, *Water Air and Soil Pollution* 223(3): 1161–1170. https://doi.org/10.1007/s11270-011-0934-2

DeLuca, T. H.; MacKenzie, M. D.; Gundale, M. J.; Jones, D. L. 2015. Biochar effects on soil nutrient transformations, in J. Lehmann, S. Joseph (Eds). *Biochar for environmental management: science, technology and implementation.* London, UK: Routledge.

Denyes, M. J.; Rutter, A.; Zeeb, B. A. 2016. Bioavailability assessments following biochar and activated carbon amendment in DDT-contaminated soil, *Chemosphere* 144: 1428–1434. https://doi.org/10.1016/j.chemosphere.2015.10.029

Directive 2008/105/EC of the European parliament and the council of 16 December 2008 on the environmental quality standards in the field of water policy [online], [cited 20 September 2016]. Available from Internet: http://eur-lex.europa.eu/LexUriServ/LexUriServ.do?uri=OJ:L:2008:348:0084:0097:en:PDF

Domene, X.; Enders, A.; Hanley, K.; Lehmann, J. 2015a. Ecotoxicological characterization of biochars: role of feedstock and pyrolysis temperature, *Science of the Total Environment* 512–513: 552–561. https://doi.org/10.1016/j.scitotenv.2014.12.035

Domene, X.; Hanley, K.; Enders, A.; Lehmann, J. 2015b. Short-term mesofauna responses to soil additions of corn stover biochar and the role of microbial biomass, *Applied Soil Ecology* 89: 10–17. https://doi.org/10.1016/j.apsoil.2014.12.005

Domene, X.; Mattana, S.; Hanley, K.; Enders, A.; Lehmann, J. 2014. Medium-term effects of corn biochar addition on soil biota activities and functions in a temperate soil cropped to corn, *Soil Biology and Biochemistry* 72: 152–162. https://doi.org/10.1016/j.soilbio.2014.01.035

Dong, J.; Chi, Y.; Tang, Y.; Ni, M.; Nzihou, A.; Weiss-Hortala, E.; Huang, Q. 2015. Partitioning of heavy metals in municipal solid waste pyrolysis, gasification, and incineration, *Energy & Fuels* 29(11): 7516–7525. https://doi.org/10.1021/acs.energyfuels.5b01918

EBC. 2012. *European Biochar Certificate – Guidelines for a sustainable production of biochar* [online], [cited 20 September 2016]. Version 6.1 of June 19th, 2015. European Biochar Foundation (EBC), Arbaz, Switzerland. Available from Internet: http://www.european-biochar.org/biochar/media/doc/ebc-guidelines.pdf

Elad, Y.; Cytryn, E.; Meller Harel, Y.; Lew, B.; Graber, E. R. 2011. The biochar effect: plant resistance to biotic stresses, *Phytopathologia Mediterranea* 50(3): 335–349.

Enell, A.; Andersson-Skold, Y.; Vestin, J.; Wagelmans, M. 2016. Risk management and regeneration of brownfields using bioenergy crops, *Journal of Soils and Sediments* 16: 987–1000. https://doi.org/10.1007/s11368-015-1264-6

Fellet, G.; Marchiol, L.; Delle Vedove, G.; Peressotti, A. 2011. Application of biochar on mine tailings: effects and perspectives for land reclamation, *Chemosphere* 83(9): 1262–1267. https://doi.org/10.1016/j.chemosphere.2011.03.053

Freddo, A.; Cai, C.; Reid, B. J. 2012. Environmental contextualisation of potential toxic elements and polycyclic aromatic hydrocarbons in biochar, *Environmental Pollution* 171: 18–24. https://doi.org/10.1016/j.envpol.2012.07.009

Fristak, V.; Friesl-Hanl, W.; Pipiska, M.; Michalekova, B. R.; Soja, G. 2014. The response of artificial aging to sorption properties of biochar for potentially toxic heavy metals, *Nova Biotechnologica et Chimica* 13(2): 137–147. https://doi.org/10.1515/nbec-2015-0004

Gartler, J.; Robinson, B.; Burton, K.; Clucas, L. 2013. Carbonaceous soil amendments to biofortify crop plants with zinc, *Science of the Total Environment* 465: 308–313. https://doi.org/10.1016/j.scitotenv.2012.10.027

Genesio, L.; Vaccari, F. P.; Miglietta, F. 2016. Black carbon aerosol from biochar threats its negative emission potential, *Global Change Biology* 22(7): 2313–2314. https://doi.org/10.1111/gcb.13254

Glaser, B. 2007. Prehistorically modified soils of central Amazonia: a model for sustainable agriculture in the twenty-first century, *Philosophical Transactions of the Royal Society B-Biological Sciences* 362(1478): 187–196. https://doi.org/10.1098/rstb.2006.1978

Gomez-Eyles, J. L.; Beesley, L.; Moreno-Jimenez, E.; Ghosh, U.; Sizmur, T. 2013. The potential of biochar amendments to remediate contaminated soils, in N. Ladygina, F. Rineau (Eds). *Biochar and Soil Biota.* London, New York: CRC Press, 100–133. https://doi.org/10.1201/b14585-5

Gouliarmou, V.; Collins, C. D.; Christiansen, E.; Mayer, P. 2013. Sorptive physiologically based extraction of contaminated solid matrices: incorporating silicone rod as absorption sink for hydrophobic organic contaminants, *Environmental Science & Technology* 47(2): 941–948. https://doi.org/10.1021/es303165u

Gouliarmou, V.; Mayer, P. 2012. Sorptive bioaccessibility extraction (SBE) of soils: combining a mobilization medium with an absorption sink, *Environmental Science & Technology* 46(19): 10682–10689. https://doi.org/10.1021/es301515s

Graber, E. R.; Frenkel, O.; Jaiswal, A. K.; Elad, Y. 2014. How may biochar influence severity of diseases caused by soilborne pathogens?, *Carbon Management* 5(2): 169–183. https://doi.org/10.1080/17583004.2014.913360

Greger, M.; Malm, T.; Kautsky, L. 2007. Heavy metal transfer from composted macroalgae to crops, *European Journal of Agronomy* 26: 257–265. https://doi.org/10.1016/j.eja.2006.10.003

Hale, S. E.; Cornelissen, G.; Werner, D. 2015. Sorption and remedition of organic compounds in soils and sediments by (activated) biochar, in J. Lehmann, S. Joseph (Eds). *Biochar for environmental management: science, technology and implementation.* London, UK: Routledge.

Hale, S. E.; Hanley, K.; Lehmann, J.; Zimmerman, A. R.; Cornelissen, G. 2011. Effects of chemical, biological, and physical aging as well as soil addition on the sorption of pyrene to activated carbon and biochar, *Environmental Science & Technology* 45(24): 10445–10453.

Hale, S. E.; Jensen, J.; Jakob, L.; Oleszczuk, P.; Hartnik, T.; Henriksen, T.; Okkenhaug, G.; Martinsen, V.; Cornelissen, G. 2013. Short-term effect of the soil amendments activated carbon, biochar, and ferric oxyhydroxide on bacteria and invertebrates, *Environmental Science & Technology* 47(15): 8674–8683.

Hale, S. E.; Lehmann, J.; Rutherford, D.; Zimmerman, A. R.; Bachmann, R. T.; Shitumbanuma, V.; O'Toole, A.; Sundqvist, K. L.; Arp, H. P. H.; Cornelissen, G. 2012. Quantifying the total and bioavailable polycyclic aromatic hydrocarbons and dioxins in biochars, *Environmental Science & Technology* 46(5): 2830–2838.

Hamer, U.; Marschner, B.; Brodowski, S.; Amelung, W. 2004. Interactive priming of black carbon and glucose mineralisation, *Organic Geochemistry* 35(7): 823–830.

Hartley, W.; Dickinson, N. M.; Riby, P.; Lepp, N. W. 2009. Arsenic mobility in brownfield soils amended with green waste compost or biochar and planted with *Miscanthus*, *Environmental Pollution* 157(10): 2654–2662.

Hilber, I.; Bucheli, T. D.; Wyss, G.; Schulin, R. 2009. Assessing the phytoavailability of dieldrin residues in charcoal-amended soil using Tenax® extraction, *Journal of Agricultural and Food Chemistry* 57(10): 4293–4298.

Hilber, I.; Schmid, H. P.; Bucheli, T. D. 2017. Sampling, storage and preparation of biochar for laboratory analysis. *in* B. Singh, M. Camps-Arbestain, J. Lehmann (Eds). *Biochar: a guide to analytical methods.* CSIRO Publishing Melbourne.

Houben, D.; Evrard, L.; Sonnet, P. 2013a. Beneficial effects of biochar application to contaminated soils on the bioavailability of Cd, Pb and Zn and the biomass production of rapeseed (*Brassica napus L.*), *Biomass and Bioenergy* 57: 196–204.

Houben, D.; Evrard, L.; Sonnet, P. 2013b. Mobility, bioavailability and pH-dependent leaching of cadmium, zinc and lead in a contaminated soil amended with biochar, *Chemosphere* 92(11): 1450–1457.

IBI. 2015a. *Biochar Certification Program Manual. Requirements and Procedures for IBI Biochar Certification* [online], [cited 20 September 2016]. Available from Internet: http://www.biochar-international.org/sites/default/files/IBI_Biochar_Certification_Program_Manual_V2.1_Final.pdf

IBI. 2015b. *Standardized product definition and product testing guidelines for biochar that is used in soil (Version of Nov. 23, 2015)* [online], [cited 20 September 2016]. Available from Internet: http://www.biochar-international.org/sites/default/files/IBI_Biochar_Standards_V2.1_Final.pdf

Ippolito, J. A.; Spokas, K. A.; Novak, J. M.; Lentz, R. D.; Cantrell, K. B. 2015. Biochar elemental composition and factors influencing nutrient retention, in J. Lehmann, S. Joseph (Eds). *Biochar for environmental management: science, technology and implementation.* London, UK: Routledge.

ISO 10381-8. 2006. *Soil quality - Sampling - PArt 8: guidance on sampling of stockpiles.*

Jaafar, N. M.; Clode, P. L.; Abbott, L. K. 2014. Microscopy observations of habitable space in biochar for colonization by fungal hyphae from soil, *Journal of Integrative Agriculture* 13(3): 483–490. https://doi.org/10.1016/S2095-3119(13)60703-0

Jaafar, N. M.; Clode, P. L.; Abbott, L. K. 2015. Soil microbial responses to biochars varying in particle size, surface and pore properties, *Pedosphere* 25(5): 770–780. https://doi.org/10.1016/S1002-0160(15)30058-8

Jaffé, R.; Ding, Y.; Niggemann, J.; Vähätalo, A. V.; Stubbins, A.; Spencer, R. G. M.; Campbell, J.; Dittmar, T. 2013. Global charcoal mobilization from soils via dissolution and riverine transport to the oceans, *Science* 340(6130): 345–347. https://doi.org/10.1126/science.1231476

Janssen, R. P. T.; Posthuma, L.; Baerselman, R.; den Hollander, H. A. 1997. Equilibrium partitioning of heavy metals in Dutch field soils. II. Prediction of metal accumulation in earthworms, *Environmental Toxicology and Chemistry* 16: 2479–2488. https://doi.org/10.1002/etc.5620161207

Jeffery, S.; Abalos, D.; Spokas, K. A.; Verheijen, F. G. 2015. Biochar effects on crop yield., in J. Lehmann, S. Joseph (Eds). *Biochar for environmental management: science, technology and implementation.* London, UK: Routledge.

Jeffery, S.; Verheijen, F. G. A.; van der Velde, M.; Bastos, A. C. 2011. A quantitative review of the effects of biochar application to soils on crop productivity using meta-analysis, *Agriculture, Ecosystems & Environment* 144(1): 175–187. https://doi.org/10.1016/j.agee.2011.08.015

Jiang, J.; Xu, R. K.; Jiang, T.-Y.; Li, Z. 2012a. Immobilization of Cu(II), Pb(II) and Cd(II) by the addition of rice straw derived biochar to a simulated polluted Ultisol, *Journal of Hazardous Materials* 229–230: 145–150. https://doi.org/10.1016/j.jhazmat.2012.05.086

Jiang, J. Xu, R. K. 2013. Application of crop straw derived biochars to Cu(II) contaminated ultisol: evaluating role of alkali and organic functional groups in Cu(II) immobilization, *Bioresource Technology* 133: 537–545. https://doi.org/10.1016/j.biortech.2013.01.161

Jiang, T. Y.; Jiang, J.; Xu, R. K.; Li, Z. 2012b. Adsorption of Pb(II) on variable charge soils amended with rice-straw derived biochar, *Chemosphere* 89(3): 249–256. https://doi.org/10.1016/j.chemosphere.2012.04.028

Jonker, M. T. O. Koelmans, A. A. 2001. Polyoxymethylene solid phase extraction as a partitioning method for hydrophobic organic chemicals in sediment and soot, *Environmental Science & Technology* 35(18): 3742–3748. https://doi.org/10.1021/es0100470

Jonker, M. T. O. Koelmans, A. A. 2002. Sorption of polycyclic aromatic hydrocarbons and polychlorinated biphenyls to soot and soot-like materials in the aqueous environment mechanistic considerations, *Environmental Science & Technology* 36(17): 3725–3734. https://doi.org/10.1021/es020019x

Jung, C.; Park, J.; Lim, K. H.; Park, S.; Heo, J.; Her, N.; Oh, J.; Yun, S.; Yoon, Y. 2013. Adsorption of selected endocrine disrupting compounds and pharmaceuticals on activated biochars, *Journal of Hazardous Materials* 263: 702–710. https://doi.org/10.1016/j.jhazmat.2013.10.033

Karami, N.; Clemente, R.; Moreno-Jiménez, E.; Lepp, N. W.; Beesley, L. 2011. Efficiency of green waste compost and biochar soil amendments for reducing lead and copper mobility and uptake to ryegrass, *Journal of Hazardous Materials* 191(1–3): 41–48. https://doi.org/10.1016/j.jhazmat.2011.04.025

Karer, J.; Wawra, A.; Zehetner, F.; Dunst, G.; Wagner, M.; Pavel, P.-B.; Puschenreiter, M.; Friesl-Hanl, W.; Soja, G. 2015. Effects of biochars and compost mixtures and inorganic additives on immobilisation of heavy metals in contaminated soils, *Water, Air & Soil Pollution* 226(10): 1–12. https://doi.org/10.1007/s11270-015-2584-2

Kelly, C. N.; Peltz, C. D.; Stanton, M.; Rutherford, D. W.; Rostad, C. E. 2014. Biochar application to hardrock mine tailings: soil quality, microbial activity, and toxic element sorption, *Applied Geochemistry* 43: 35–48. https://doi.org/10.1016/j.apgeochem.2014.02.003

Khan, S.; Chao, C.; Waqas, M.; Arp, H. P. H.; Zhu, Y. G. 2013. Sewage Sludge Biochar Influence upon Rice (Oryza sativa L) Yield, Metal Bioaccumulation and Greenhouse Gas Emissions from Acidic Paddy Soil, *Environmental Science & Technology* 47(15): 8624–8632. https://doi.org/10.1021/es400554x

Kloss, S.; Zehetner, F.; Buecker, J.; Oburger, E.; Wenzel, W. W.; Enders, A.; Lehmann, J.; Soja, G. 2014a. Trace element biogeochemistry in the soil-water-plant system of a temperate agricultural soil amended with different biochars, *Environmental Science and Pollution Research* 22(6): 4513–4526. https://doi.org/10.1007/s11356-014-3685-y

Kloss, S.; Zehetner, F.; Oburger, E.; Buecker, J.; Kitzler, B.; Wenzel, W. W.; Wimmer, B.; Soja, G. 2014b. Trace element concentrations in leachates and mustard plant tissue (*Sinapis alba* L.) after biochar application to temperate soils, *Science of the Total Environment* 481: 498–508. https://doi.org/10.1016/j.scitotenv.2014.02.093

Kloss, S.; Zehetner, F.; Wimmer, B.; Buecker, J.; Rempt, F.; Soja, G. 2014c. Biochar application to temperate soils: effects on soil fertility and crop growth under greenhouse conditions, *Journal of Plant Nutrition and Soil Science* 177(1): 3–15. https://doi.org/10.1002/jpln.201200282

Kołtowski, M.; Hilber, I.; Bucheli, T. D.; Oleszczuk, P. 2016. Effect of activated carbon and biochars on the bioavailability of polycyclic aromatic hydrocarbons in different industrially contaminated soils, *Environmental Science and Pollution Research* 23(11): 11058–11068. https://doi.org/10.1007/s11356-016-6196-1

Kołtowski, M. Oleszczuk, P. 2015. Toxicity of biochars after polycyclic aromatic hydrocarbons removal by thermal treatment, *Ecological Engineering* 75(0): 79–85. https://doi.org/10.1016/j.ecoleng.2014.11.004

Kuppusamy, S.; Thavamani, P.; Megharaj, M.; Venkateswarlu, K.; Naidu, R. 2016. Agronomic and remedial benefits and risks of applying biochar to soil: current knowledge and future research directions, *Environment International* 87: 1–12. https://doi.org/10.1016/j.envint.2015.10.018

Kupryianchyk, D.; Hale, S. E.; Breedveld, G. D.; Cornelissen, G. 2016. Treatment of sites contaminated with perfluorinated compounds using biochar amendment, *Chemosphere* 142: 35–40. https://doi.org/10.1016/j.chemosphere.2015.04.085

Lang, S.-C.; Hursthouse, A.; Mayer, P.; Kötke, D.; Hand, I.; Schulz-Bull, D.; Witt, G. 2015. Equilibrium passive sampling as a tool to study polycyclic aromatic hydrocarbons in Baltic Sea sediment pore-water systems, *Marine Pollution Bulletin* 101(1): 296–303. https://doi.org/10.1016/j.marpolbul.2015.10.069

Lehmann, J.; Abiven, S.; Kleber, M.; Pan, G.; Singh, B. P.; Sohi, S. P.; Zimmerman, A. R.; Lehmann, J.; Joseph, S. 2015. Persistence of biochar in soil, in J. Lehmann, S. Joseph (Eds). *Biochar for environmental management: science, technology and implementation*. London, UK: Routledge.

Lehmann, J.; Rillig, M. C.; Thies, J.; Masiello, C. A.; Hockaday, W. C.; Crowley, D. 2011. Biochar effects on soil biota – a review, *Soil Biology and Biochemistry* 43(9): 1812–1836. https://doi.org/10.1016/j.soilbio.2011.04.022

Li, D.; Hockaday, W. C.; Masiello, C. A.; Alvarez, P. J. J. 2011. Earthworm avoidance of biochar can be mitigated by wetting, *Soil Biology and Biochemistry* 43(8): 1732–1737. https://doi.org/10.1016/j.soilbio.2011.04.019

Li, H.; Ye, X.; Geng, Z.; Zhou, H.; Guo, X.; Zhang, Y.; Zhao, H.; Wang, G. 2016. The influence of biochar type on long-term stabilization for Cd and Cu in contaminated paddy soils, *Journal of Hazardous Materials* 304: 40–48. https://doi.org/10.1016/j.jhazmat.2015.10.048

Li, R.; Zhao, W.; Li, Y.; Wang, W.; Zhu, X. 2015. Heavy metal removal and speciation transformation through the calcination treatment of phosphorus-enriched sewage sludge ash, *Journal of Hazardous Materials* 283: 423–431. https://doi.org/10.1016/j.jhazmat.2014.09.052

Lian, F.; Sun, B.; Song, Z.; Zhu, L.; Qi, X.; Xing, B. 2014. Physicochemical properties of herb-residue biochar and its sorption to ionizable antibiotic sulfamethoxazole, *Chemical Engineering Journal* 248: 128–134. https://doi.org/10.1016/j.cej.2014.03.021

Liang, Y.; Cao, X. D.; Zhao, L.; Arellano, E. 2014. Biochar- and phosphate-induced immobilization of heavy metals in contaminated soil and water: implication on simultaneous remediation of contaminated soil and groundwater, *Environmental Science and Pollution Research* 21(6): 4665–4674. https://doi.org/10.1007/s11356-013-2423-1

Liesch, A. M.; Weyers, S. L.; Gaskin, J. W.; Das, K. 2010. Impact of two different biochars on earthworm growth and survival, *Annals of Environmental Science* 4(1): 1–9.

Luan, J.; Li, R.; Zhang, Z.; Li, Y.; Zhao, Y. 2013. Influence of chlorine, sulfur and phosphorus on the volatilization behavior of heavy metals during sewage sludge thermal treatment, *Waste Management & Research* 31(10): 1012–1018. https://doi.org/10.1177/0734242X13493955

Lucchini, P.; Quilliam, R. S.; DeLuca, T. H.; Vamerali, T.; Jones, D. L. 2014a. Does biochar application alter heavy metal dynamics in agricultural soil?, *Agriculture, Ecosystems & Environment* 184: 149–157. https://doi.org/10.1016/j.agee.2013.11.018

Lucchini, P.; Quilliam, R. S.; Deluca, T. H.; Vamerali, T.; Jones, D. L. 2014b. Increased bioavailability of metals in two contrasting agricultural soils treated with waste wood-derived biochar and ash, *Environmental Science and Pollution Research* 21(5): 3230–3240. https://doi.org/10.1007/s11356-013-2272-y

Marks, E. A. N.; Mattana, S.; Alcañiz, J. M.; Domene, X. 2014. Biochars provoke diverse soil mesofauna reproductive responses in laboratory bioassays, *European Journal of Soil Biology* 60: 104–111. https://doi.org/10.1016/j.ejsobi.2013.12.002

Mayer, P.; Hilber, I.; Gouliarmou, V.; Hale, S. E.; Cornelissen, G.; Bucheli, T. D. 2016. How to determine the environmental exposure of PAHs originating from biochar, *Environmental Science & Technology* 50(4): 1941–1948. https://doi.org/10.1021/acs.est.5b05603

Mayer, P.; Olsen, J. L.; Gouliarmou, V.; Hasinger, M.; Kendler, R.; Loibner, A. P. 2011. A contaminant trap as a tool for isolating and measuring the desorption resistant fraction of soil pollutants, *Environmental Science & Technology* 45(7): 2932–2937. https://doi.org/10.1021/es1033124

McBeath, A. V.; Wurster, C. M.; Bird, M. I. 2015. Influence of feedstock properties and pyrolysis conditions on biochar carbon stability as determined by hydrogen pyrolysis, *Biomass and Bioenergy* 73: 155–173. https://doi.org/10.1016/j.biombioe.2014.12.022

Melo, L. C. A.; Coscione, A. R.; Abreu, C. A.; Puga, A. P.; Camargo, O. A. 2013. Influence of pyrolysis temperature on cadmium and zinc sorption capacity of sugar cane straw-derived biochar, *Bioresources* 8(4): 4992–5004. https://doi.org/10.15376/biores.8.4.4992-5004

Mendez, A.; Gomez, A.; Paz-Ferreiro, J.; Gasco, G. 2012. Effects of sewage sludge biochar on plant metal availability after application to a Mediterranean soil, *Chemosphere* 89(11): 1354–1359. https://doi.org/10.1016/j.chemosphere.2012.05.092

Moon, D. H.; Park, J. W.; Chang, Y. Y.; Ok, Y. S.; Lee, S. S.; Ahmad, M.; Koutsospyros, A.; Park, J. H.; Baek, K. 2013. Immobilization of lead in contaminated firing range soil using biochar, *Environmental Science and Pollution Research* 20(12): 8464–8471. https://doi.org/10.1007/s11356-013-1964-7

Moreno-Jiménez, E.; Esteban, E.; Carpena-Ruiz, R. O.; Lobo, M. C.; Peñalosa, J. M. 2012. Phytostabilisation with Mediterranean shrubs and liming improved soil quality in a pot experiment with a pyrite mine soil, *Journal of Hazardous Materials* 201: 52–59. https://doi.org/10.1016/j.jhazmat.2011.11.013

Morris, C. 1992. *Academic Press Dictionary of Science and Technology*. San Diego.

Mukherjee, A.; Zimmerman, A. R.; Hamdan, R.; Cooper, W. T. 2014. Physicochemical changes in pyrogenic organic matter (biochar) after 15 months of field aging, *Solid Earth* 5(2): 693–704. https://doi.org/10.5194/se-5-693-2014

Mukherjee, A.; Zimmerman, A. R.; Harris, W. 2011. Surface chemistry variations among a series of laboratory-produced biochars, *Geoderma* 163(3): 247–255. https://doi.org/10.1016/j.geoderma.2011.04.021

Mukome, F. N. D.; Kilcoyne, A. L. D.; Parikh, S. J. 2004. Alteration of biochar carbon chemistry during soil incubations: SR-FTIR and NEXAFS investigation, *Soil Science Society of America Journal* 78(5): 1632–1640. https://doi.org/10.2136/sssaj2014.05.0206

Namgay, T.; Singh, B.; Singh, B. P. 2010. Influence of biochar application to soil on the availability of As, Cd, Cu, Pb, and Zn to maize (*Zea mays L.*), *Australian Journal of Soil Research* 48(6–7): 638–647. https://doi.org/10.1071/SR10049

Oleszczuk, P.; Jośko, I.; Kuśmierz, M. 2013. Biochar properties regarding to contaminants content and ecotoxicological assessment, *Journal of Hazardous Materials* 260: 375–382. https://doi.org/10.1016/j.jhazmat.2013.05.044

Pan, J.-J.; Jiang, J.; Qian, W.; Xu, R. K. 2015. Arsenate adsorption from aqueous solution onto Fe (III)-modified crop straw biochars, *Environmental Engineering Science* 32(11): 922–929. https://doi.org/10.1089/ees.2014.0540

Park, J. H.; Choppala, G.; Lee, S. J.; Bolan, N.; Chung, J. W.; Edraki, M. 2013. Comparative sorption of Pb and Cd by biochars and its implication for metal immobilization in soils, *Water Air and Soil Pollution* 224(12): 1711–1711. https://doi.org/10.1007/s11270-013-1711-1

Park, J. H.; Choppala, G. K.; Bolan, N. S.; Chung, J. W.; Chuasavathi, T. 2011. Biochar reduces the bioavailability and phytotoxicity of heavy metals, *Plant and Soil* 348(1–2): 439–451. https://doi.org/10.1007/s11104-011-0948-y

Rees, F.; Simonnot, M. O.; Morel, J. L. 2014. Short-term effects of biochar on soil heavy metal mobility are controlled by intraparticle diffusion and soil pH increase, *European Journal of Soil Science* 65(1): 149–161. https://doi.org/10.1111/ejss.12107

Reichenberg, F.; Mayer, P. 2006. Two complementary sides of bioavailability: accessibility and chemical activity of organic contaminants in sediments and soils, *Environmental Toxicology and Chemistry* 25(5): 1239–1245. https://doi.org/10.1897/05-458R.1

Schimmelpfennig, S. Glaser, B. 2012. One step forward toward characterization: some important material properties to distinguish biochars, *Journal of Environmental Quality* 41(4): 1001–1013. https://doi.org/10.2134/jeq2011.0146

Schwarzenbach, R. P.; Gschwend, P. M.; Imboden, D. M. 2003. *Environmental organic chemistry*. 2nd ed. Wiley.

Semple, K. T.; Doick, K. J.; Jones, K. C.; Burauel, P.; Craven, A.; Harms, H. 2004. Peer deviewed: Defining bioavailability and bioaccessibility of contaminated soil and sediment is complicated, *Environmental Science & Technology* 38(12): 228A–231A. https://doi.org/10.1021/es040548w

Seth, R.; Mackay, D.; Muncke, J. 1999. Estimating the organic carbon partition coefficient and its variability for hydrophobic chemicals, *Environmental Science & Technology* 33(14): 2390–2394. https://doi.org/10.1021/es980893j

Shaheen, S. M. Rinklebe, J. 2015. Impact of emerging and low cost alternative amendments on the (im)mobilization and phytoavailability of Cd and Pb in a contaminated floodplain soil, *Ecological Engineering* 74: 319–326. https://doi.org/10.1016/j.ecoleng.2014.10.024

Shen, Z. T.; Som, A. M.; Wang, F.; Jin, F.; McMillan, O.; Al-Tabba, A. 2016. Long-term impact of biochar on the immobilisation of nickel (II) and zinc (II) and the revegetation of a contaminated site, *Science Of The Total Environment* 542: 771–776. https://doi.org/10.1016/j.scitotenv.2015.10.057

Silva, F. C.; Borrego, C.; Keizer, J. J.; Amorim, J. H.; Verheijen, F. G. A. 2015. Effects of moisture content on wind erosion thresholds of biochar, *Atmospheric Environment* 123 (Part A): 121–128.

Sizmur, T.; Wingate, J.; Hutchings, T.; Hodson, M. E. 2011. *Lumbricus terrestris L.* does not impact on the remediation efficiency of compost and biochar amendments, in *9th International Symposium on Earthworm EcologyXalapa*, 5–10 September 2010, Veracruz, Mexico, 54: S211–S216.

Smith, C. R.; Buzan, E. M.; Lee, J. W. 2013. Potential impact of biochar water-extractable substances on environmental sustainability, *ACS Sustainable Chemistry & Engineering* 1(1): 118–126. https://doi.org/10.1021/sc300063f

Sneath, H. E.; Hutchings, T. R.; de Leij, F. A. A. M. 2013. Assessment of biochar and iron filing amendments for the remediation of a metal, arsenic and phenanthrene co-contaminated spoil, *Environmental Pollution* 178: 361–366. https://doi.org/10.1016/j.envpol.2013.03.009

Sorrenti, G.; Masiello, C. A.; Dugan, B.; Toselli, M. 2016. Biochar physico-chemical properties as affected by environmental exposure, *Science of the Total Environment* 563–564: 237–246. https://doi.org/10.1016/j.scitotenv.2016.03.245

Spokas, K. A.; Novak, J. M.; Stewart, C. E.; Cantrell, K. B.; Uchimiya, M.; DuSaire, M. G.; Ro, K. S. 2011. Qualitative analysis of volatile organic compounds on biochar, *Chemosphere* 85(5): 869–882. https://doi.org/10.1016/j.chemosphere.2011.06.108

Tammeorg, P.; Parviainen, T.; Nuutinen, V.; Simojoki, A.; Vaara, E.; Helenius, J. 2014. Effects of biochar on earthworms in arable soil: avoidance test and field trial in boreal loamy sand, *Agriculture, Ecosystems & Environment* 191: 150–157. https://doi.org/10.1016/j.agee.2014.02.023

Tammeorg, P.; Bastos, A. C.; Jeffery, S.; Rees, F.; Kern, J.; Graber, E. R.; Ventura, M.; Kibblewhite, M.; Amaro, A.; Budai, A.; Cordovil, C. M. S.; Domene, X.; Gardi, C.; Gascó, G.; Horák, J.; Kammann, C.; Kondrlova, E.; Laird, D.; Loureiro, S.; Martins, M. A. S.; Panzacchi, P.; Prasad, M.; Prodana, M.; Puga, A. P.; Ruysschaert, G.; Sas-Paszt, L.; Silva, F. C.; Teixeira, W. G.; Tonon, G.; delle Vedove, G.; Zavalloni, C.; Glaser, B.; Verheijen, F. G. A. Biochars in soils: towards the required level of scientific understanding, *Journal of Environmental Engineering and Landscape management* (this issue).

Thies, J. E.; Rillig, M. C.; Graber, E. R. 2015. Biochar effects on the abundance, activity and diversity of the soil biota, in J. Lehmann, S. Joseph (Eds). *Biochar for environmental management: science, technology and implementation*. London, UK: Routledge.

Trakal, L.; Komarek, M.; Szakova, J.; Zemanova, V.; Tlustos, P. 2011. Biochar application to metal-contaminated soil: evaluating of Cd, Cu, Pb and Zn sorption behavior using single- and multi-element sorption experiment, *Plant Soil and Environment* 57(8): 372–380.

Uchimiya, M.; Bannon, D. I. 2013. Solubility of lead and copper in biochar-amended small arms range soils: influence of soil organic carbon and pH, *Journal of Agricultural and Food Chemistry* 61(32): 7679–7688. https://doi.org/10.1021/jf401481x

Uchimiya, M.; Bannon, D. I.; Wartelle, L. H. 2012a. Retention of heavy metals by carboxyl functional groups of biochars in small arms range soil, *Journal of Agricultural and Food Chemistry* 60(7): 1798–1809. https://doi.org/10.1021/jf2047898

Uchimiya, M.; Bannon, D. I.; Wartelle, L. H.; Lima, I. M.; Klasson, K. T. 2012b. Lead retention by broiler litter biochars in small arms range soil: impact of pyrolysis temperature, *Journal of Agricultural and Food Chemistry* 60(20): 5035–5044. https://doi.org/10.1021/jf300825n

Uchimiya, M.; Chang, S.; Klasson, K. T. 2011a. Screening biochars for heavy metal retention in soil: role of oxygen functional groups, *Journal of Hazardous Materials* 190(1–3): 432–441. https://doi.org/10.1016/j.jhazmat.2011.03.063

Uchimiya, M.; Klasson, K. T.; Wartelle, L. H.; Lima, I. M. 2011b. Influence of soil properties on heavy metal sequestration by biochar amendment: 1. copper sorption isotherms and the release of cations, *Chemosphere* 82(10): 1431–1437. https://doi.org/10.1016/j.chemosphere.2010.11.050

Uchimiya, M.; Klasson, K. T.; Wartelle, L. H.; Lima, I. M. 2011c. Influence of soil properties on heavy metal sequestration by biochar amendment: 2. copper desorption isotherms, *Chemosphere* 82(10): 1438–1447. https://doi.org/10.1016/j.chemosphere.2010.11.078

Uchimiya, M.; Lima, I. M.; Klasson, K. T.; Chang, S. C.; Wartelle, L. H.; Rodgers, J. E. 2010a. Immobilization of heavy metal ions (Cu(II), Cd(II), Ni(II), and Pb(II)) by broiler litter-derived biochars in water and soil, *Journal of Agricultural and Food Chemistry* 58(9): 5538–5544. https://doi.org/10.1021/jf9044217

Uchimiya, M.; Lima, I. M.; Klasson, K. T.; Wartelle, L. H. 2010b. Contaminant immobilization and nutrient release by biochar soil amendment: roles of natural organic matter, *Chemosphere* 80(8): 935–940. https://doi.org/10.1016/j.chemosphere.2010.05.020

Uchimiya, M.; Lima, I. M.; Thomas Klasson, K.; Chang, S.; Wartelle, L. H.; Rodgers, J. E. 2010c. Immobilization of heavy metal ions (CuII, CdII, NiII, and PbII) by broiler litter-derived biochars in water and soil, *Journal of Agricultural and Food Chemistry* 58(9): 5538–5544.

Uchimiya, M.; Wartelle, L. H.; Klasson, K. T.; Fortier, C. A.; Lima, I. M. 2011d. Influence of pyrolysis temperature on biochar property and function as a heavy metal sorbent in soil, *Journal of Agricultural and Food Chemistry* 59(6): 2501–2510. https://doi.org/10.1021/jf104206c

Van Zwieten, L.; Kammann, C.; Cayuela, M.; Singh, B.; Joseph, S.; Kimber, S.; Clough, T.; Spokas, K.; Lehmann, J. 2015. Biochar effects on nitrous oxide and methane emissions from soil, in J. Lehmann, S. Joseph (Eds). *Biochar for environmental management: science, technology and implementation*. London, UK: Routledge.

Van Zwieten, L.; Kimber, S.; Morris, S.; Chan, K. Y.; Downie, A.; Rust, J.; Joseph, S.; Cowie, A. 2010. Effects of biochar from slow pyrolysis of papermill waste on agronomic performance and soil fertility, *Plant and Soil* 327(1–2): 235–246. https://doi.org/10.1007/s11104-009-0050-x

Verheijen, F. G. A.; Mankasingh, U.; Penizek, V.; Panzacchi, P.; Glaser, B.; Jeffery, S.; Bastos, A. C.; Tammeorg, P.; Kern, J.; Zavalloni, C.; Zanchettin, G.; Sakrabani, R. Representativeness of European biochar research: part i – field experiments, *Journal of Environmental Engineering and Landscape Management* (this issue).

Verheijen, F.; Bastos, A. C.; Schmidt, H. P.; Brandão, M.; Jeffery, S. 2015a. Biochar sustainability and certification, in J. Lehmann, S. Joseph (Eds). *Biochar for Environmental Management: Science, Technology and Implementation*. London, UK: Routledge.

Verheijen, F. G. A.; Bastos, A. C.; Jeffery, S. 2015b. International policy: bury the idea that soils are a local issue, *Nature* 528(7583): 479–479. https://doi.org/10.1038/528479b

Verheijen, F. G. A.; Graber, E. R.; Ameloot, N.; Bastos, A. C.; Sohi, S.; Knicker, H. 2014. Biochars in soils: new insights and emerging research needs, *European Journal of Soil Science* 65(1): 22–27. https://doi.org/10.1111/ejss.12127

Verheijen, F. G. A.; Montanarella, L.; Bastos, A. C. 2012. Sustainability, certification, and regulation of biochar, *Pesquisa Agropecuária Brasileira* 47: 649–653. https://doi.org/10.1590/S0100-204X2012000500003

Wagner, A.; Kaupenjohann, M. 2014. Suitability of biochars (pyro- and hydrochars) for metal immobilization on former sewage-field soils, *European Journal of Soil Science* 65(1): 139–148. https://doi.org/10.1111/ejss.12090

Waqas, M.; Khan, S.; Qing, H.; Reid, B. J.; Chao, C. 2014. The effects of sewage sludge and sewage sludge biochar on PAHs and potentially toxic element bioaccumulation in *Cucumis sativa* L., *Chemosphere* 105: 53–61. https://doi.org/10.1016/j.chemosphere.2013.11.064

Werner, D.; Ghosh, U.; Luthy, R. G. 2006. Modeling polychlorinated biphenyl mass transfer after amendment of contaminated sediment with activated carbon, *Environmental Science & Technology* 40(13): 4211–4218. https://doi.org/10.1021/es052215k

Whitman, T.; Singh, B. P.; Zimmerman, A. R.; Lehmann, J.; Joseph, S. 2015. Priming effects in biochar-amended soils: implications of biochar-soil organic matter interactions for carbon storage, in J. Lehmann, S. Joseph (Eds). *Biochar for environmental management: science, technology and implementation*. London, UK: Routledge.

Xie, M.; Chen, W.; Xu, Z.; Zheng, S.; Zhu, D. 2014. Adsorption of sulfonamides to demineralized pine wood biochars prepared under different thermochemical conditions, *Environmental Pollution* 186: 187–194. https://doi.org/10.1016/j.envpol.2013.11.022

Xie, T.; Reddy, K. R.; Wang, C.; Yargicoglu, E.; Spokas, K. 2015. Characteristics and applications of biochar for environmental remediation: a review, *Critical Reviews in Environmental Science and Technology* 45(9): 939–969. https://doi.org/10.1080/10643389.2014.924180

Xu, R. K. Zhao, A. Z. 2013. Effect of biochars on adsorption of Cu(II), Pb(II) and Cd(II) by three variable charge soils from southern China, *Environmental Science and Pollution Research* 20(12): 8491–8501. https://doi.org/10.1007/s11356-013-1769-8

Yao, Y.; Gao, B.; Chen, H.; Jiang, L.; Inyang, M.; Zimmerman, A. R.; Cao, X.; Yang, L.; Xue, Y.; Li, H. 2012. Adsorption of sulfamethoxazole on biochar and its impact on reclaimed water irrigation, *Journal of Hazardous Materials* 209–210: 408–413. https://doi.org/10.1016/j.jhazmat.2012.01.046

Yargicoglu, E. N.; Sadasivam, B. Y.; Reddy, K. R.; Spokas, K. 2015. Physical and chemical characterization of waste wood derived biochars, *Waste Management* 36: 256–268. https://doi.org/10.1016/j.wasman.2014.10.029

Zhang, X.; Wang, H.; He, L.; Lu, K.; Sarmah, A.; Li, J.; Bolan, N. S.; Pei, J.; Huang, H. 2013. Using biochar for remediation of soils contaminated with heavy metals and organic pollutants, *Environmental Science and Pollution Research* 20(12): 8472–8483. https://doi.org/10.1007/s11356-013-1659-0

Zhang, Y.; Pignatello, J. J.; Tao, S.; Xing, B. 2015. Bioaccessibility of PAHs in fuel soot assessed by an *in vitro* digestive model with absorptive sink: effect of food ingestion, *Environmental Science & Technology* 49(24): 14641–14648. https://doi.org/10.1021/acs.est.5b04342

Zheng, H.; Wang, Z.; Zhao, J.; Herbert, S.; Xing, B. 2013a. Sorption of antibiotic sulfamethoxazole varies with biochars produced at different temperatures, *Environmental Pollution* 181: 60–67. https://doi.org/10.1016/j.envpol.2013.05.056

Zheng, R. L.; Cai, C.; Liang, J. H.; Huang, Q.; Chen, Z.; Huang, Y. Z.; Arp, H. P.; Sun, G. X. 2012. The effects of biochars from rice residue on the formation of iron plaque and the accumulation of Cd, Zn, Pb, As in rice (*Oryza sativa* L.) seedlings, *Chemosphere* 89(7): 856–862. https://doi.org/10.1016/j.chemosphere.2012.05.008

Zheng, R. L.; Chen, Z.; Cai, C.; Wang, X. H.; Huang, Y. Z.; Xiao, B.; Sun, G. X. 2013b. Effect of biochars from rice husk, bran, and straw on heavy metal uptake by pot-grown wheat seedling in a historically contaminated soil, *Bioresources* 8(4): 5965–5982. https://doi.org/10.15376/biores.8.4.5965-5982

Isabel HILBER studied environmental sciences at ETH Zurich from 1999–2005 and finished her PhD in 2008. The focus of her PhD was the immobilization of dieldrin, an organochlorine pesticide, by activated charcoal (AC) in horticultural soils. The binding of the pesticide to AC was observed via phytoavailability to cucumber fruits and via Tenax, a chemical proxy. After her PhD she worked in three international projects in collaboration with Pakistan, Poland and Cuba where the scientific focus was again on organic contaminants, biochar and/or AC, bioavailability, and soil. The administrative focus lay in the financial and time management and in the knowledge transfer of chemical and statistical analysis, paper writing and other important things a PhD student needs to know.

Ana Catarina BASTOS concluded her PhD at Cranfield University (CU, UK) in 2008, focused on soil biomonitoring and bioremediation. As a Post-Doc at CU, she combined her research activities with training in Environmental and Occupational Health and Management. In 2009, she initiated her research on biochar's impact on soil processes and functions, collaborating with the EC's Joint Research Centre on a scientific and technical report on that topic. She joined the Department of Biology and CESAM, University of Aveiro (UA, Portugal) in 2010. Since 2013, she is a Post-Doc fellow of the Portuguese Science Foundation (FCT) at the UA, where her research focuses on risk assessment of contaminants and emerging environmental management tools (incl. biochar) on biota-mediated soil functions, at laboratory and field scale.

Susana LOUREIRO, PhD in Biology, is a researcher and assistant professor at the University of Aveiro, Portugal. She is part of the applEE-Applied Ecology and Ecotoxicology research group and works for more than 15 years in soil and aquatic ecotoxicology, focusing on hazard and risk assessment of emergent chemicals and also agrochemicals. She has been actively participating in national (Portuguese) and international projects on chemical mixtures assessment and nanoecotoxicology.

Gerhard SOJA received his Master's and PhD degrees from the University of Natural Resources and Life Sciences Vienna, Austria, in 1983 and 1988, respectively. Trained as a crop physiologist, he habilitated in the subject "Applied Plant Physiology" in 2004. He is employed at AIT Austrian Institute of Technology GmbH since 1984 and has contributed to or coordinated national or international research projects in environmental sciences, crop sciences, hydrology and soil science. G. Soja is an IPMA-certified project manager (level C). From 2003–2008 he was head of the business unit "Environmental Research" and currently is a senior scientist in "Environmental Resources and Technologies", leading a research group on biochar for environmental and agricultural applications.

Aleksandra MARSZ graduated from Warsaw University of Life Sciences as Environmental Protection Engineer in 2011. She then obtained her Master's Degree in Environmental Sciences from the University of Natural Resources and Life Sciences in Vienna and from the Swedish University of Agricultural Sciences in Uppsala as a part of the European Double Degree Master Program in Environmental Sciences EnvEuro – Soil, Water, Biodiversity in 2014. In 2014–2015 she has been working at AIT in the Department of Environmental Resources and Technologies. She had conducted a major meta-analysis research about the influence of biochar on pollutants (heavy-metals and organic pollutants) in soil.

Gerard CORNELISSEN is technical expert at the Norwegian Geotechnical Institute (NGI; 80%) and professor (20%) at the Norwegian University of Life Sciences (NMBU). GC's current research is geared towards two aspects of black carbon: its strong binding of organic compounds and application for in-situ remediation, and its potential to improve soil quality of weathered tropical soil through biochar amendment, leading projects on these themes in Zambia, Indonesia, Nepal and Norway.

Thomas D. BUCHELI studied Environmental Sciences at the ETH Zürich (1987–1992). He conducted his PhD at Eawag Dübendorf and received his degree in Natural Sciences from the ETH in 1997 (awarded with the ETH's Otto Jaag-water protection prize). From 1997 until 2001, he was a Postdoc at ETH Zürich, Department of Chemistry, and at the Institute of Applied Environmental Research (ITM), Stockholm University. He has been Head of the Organic Trace Analysis Group at Agroscope ISS since 2001, and leads the Research Group Environmental Analytics at the same institute since 2014. He investigates the occurrence, fate and behaviour of organic chemicals in the agro-environment. His current research focus is on natural toxins, PAHs, carbonaceous nanoparticles, and related sorption and bioavailability studies. Since 2012, he is a (guest) lecturer at ETH.

10

QUATERNARY INTERGLACIAL SEDIMENTS AS POSSIBLE NATURAL SOURCES OF ARSENIC AND MOLYBDENUM ANOMALIES IN STREAM SEDIMENTS IN LITHUANIA

Rimantė ZINKUTĖ, Valentinas BALTRŪNAS, Ričardas TARAŠKEVIČIUS, Bronislavas KARMAZA, Rimutė STAKĖNIENĖ, Vaida ŠEIRIENĖ, Dalia KISIELIENĖ

Nature Research Centre Institute of Geology and Geography, T. Ševčenkos g. 13, 03223 Vilnius, Lithuania

Abstract. Geochemical investigations were conducted on Quaternary interglacial sediments in order to reveal if they can be a natural source of 10 potentially harmful chemical elements: As, Ba, Cr, Cu, Mn, Mo, Ni, Pb, V and Zn. Determination of the total contents of elements in 680 samples was performed by EDXRF analysis. The content of Ni, Cu, Pb, Ba in all analysed samples was lower than maximum permissible concentrations (MPC) for soil, the content of Zn, V, Cr, Mn exceeded MPC values only in several samples (≤4%). The highest percentage of anomalous samples where MPC was exceeded was characteristic of Mo (21%) and As (12.6%). Therefore, interglacial sediments, especially enriched in organic matter and other sorbents, comprise one of the possible natural sources of Mo and As. It is probable that they contribute to As and Mo anomalies in stream sediments in Lithuania.

Keywords: natural sources, interglacial sediments, arsenic, molybdenum, total organic carbon, total inorganic carbon, iron.

Introduction

Elevated contents of harmful chemical elements can originate not only from anthropogenic activity, but also from natural sources (Nagajyoti *et al.* 2010). Volcanoes (Ernst 2012; Ayris, Delmelle 2012; Calvo *et al.* 2013) and forest fires (Pereira, Ubeda 2010) are the main natural sources of air pollution. Leaching from deposits is the main natural source of groundwater and surface water pollution (Falk *et al.* 2006; Baba, Tayfur 2011). Ore deposits are the most obvious natural sources of harmful chemical elements (Von der Heyden, New 2004).

In Lithuania insufficient attention has been paid to various possible natural sources of anomalies of harmful chemical elements, because the main part of geochemical investigations has been concentrated in urban territories or near, where anthropogenic sources are abundant (Vasarevičius *et al.* 2010; Taraškevičius, Zinkutė 2011; Kumpienė *et al.* 2011; Pundytė *et al.* 2011; Stakėnienė *et al.* 2011; Zinkutė *et al.* 2011; Galkus *et al.* 2012). Research was done also on military grounds (Vasarevičius, Greičiūtė 2004; Idzelis *et al.* 2006), landfills (Jaskelevičius, Lynikienė 2009) or near highways (Butkus, Šalčiūnienė 2011). The focus on urban areas is understandable, because of their high heavy metal fluxes (Milukaitė *et al.* 2008). Elevated contents of harmful chemical elements in topsoil of the country also often have been explained by anthropogenic reasons (e.g. the influence of distant pollution sources or diffuse pollution by agriculture). The influence of forest fires in Lithuania, which can be both natural and anthropogenic, has also been analysed and slight increases of Cd, Cr, Cu, Mn, Zn and Fe in the mineral topsoil of Scots pine forests has been found (Marozas *et al.* 2013). Growth of Cu, Pb, Zn in stream water has been correlated with land fires (Ignatavičius *et al.* 2006).

Natural geochemical anomalies have been analysed only in surface sediments of the country (i.e. topsoil and stream sediments) (Kadūnas *et al.* 1999). From natural reasons, the unequal amount of soil clayey component has been often mentioned in Lithuania (Kadūnas *et al.* 1999). It depends on different origin or age of soil parent materials. Soil of most regions of the country is less weathered and contains higher contents of most chemical elements than soil of Ašmena-Medininkai Heights (Kadūnas *et al.* 1999). The latter is ~10000 years older and was formed

Corresponding author: Rimantė Zinkutė
E-mail: zinkute@geo.lt

on glaciogenic deposits of the Medininkai glaciation of Middle Pleistocene. Similar explanations have been given for elevated contents of harmful chemical elements in stream sediments.

Stream sediments are not only influenced by surface run-off, but also by groundwater discharge (Lachmar *et al.* 2006). So part of the content of chemical elements can be leached from underlying layers. Parent material or rock outcrops comprise the most important natural source of heavy metals; besides, the contents of the elements depend on both rock type and environmental conditions activating their weathering (Nagajyoti *et al.* 2010). So search for rock outcrops with elevated concentrations of harmful chemical elements is very important, especially in Lithuania where there is lack of geochemical analysis of upper Quaternary sediments. Part of these sediments may be potential sources of anomalies in stream sediments.

Pre-Quaternary sedimentary rocks in Lithuania are covered by a thick Quaternary cover: average thickness is 130 m (Kadūnas *et al.* 1999) with increased thickness in palaeo-incisions of the sub-Quaternary surface (Baltrūnas, Gaigalas 2004). Subdivision of the Quaternary is based on climate, i.e. glacials and interglacials. Lithuanian Quaternary sediments were formed during seven sedimentary cycles (Fig. 1): Daumantai pre-glacial, Kalviai glacial and Vindžiūnai interglacial, Dzūkija glacial and Turgeliai interglacial, Dainava glacial and Butėnai interglacial, Žemaitija glacial and Snaigupėlė interglacial, Medininkai glacial and Merkinė interglacial, Nemunas glacial (with Grūda and Baltija stages) and the Holocene (Kondratienė 1996; Baltrūnas *et al.* 2013a).

The greatest number of the Pleistocene global coolings and warmings is recorded in Poland, Belarus and Ukraine (Lindner *et al.* 2004). Thus, correlation of Pleistocene deposits in Poland and in neighbouring countries is very important (Ber 2006; Lindner *et al.* 2006). Some interglacial sediments are composite with interglacial-rank warmings separated by non-glacial coolings (Lindner *et al.* 2013). Investigations of Quaternary interglacial sediments started in Lithuania, and the geochemical database was compiled in 2012 and anomalous contents of some harmful chemical elements were revealed (Baltrūnas *et al.* 2013b).

Quaternary interglacial sediments were chosen as the study object in the search for possible natural sources of harmful chemical elements, due to several reasons: 1) they have undoubtfully natural origin; 2) they were usually formed in palaeolakes where high accumulations of most chemical elements (including potentially harmful) with fine clay particles, often with increased content of organic matter were possible; 3) the outcrops of these sediments are usually on the banks of streams, so during flood events harmful chemical elements can be washed out and migrate downstream causing natural anomalies in stream sediments. Of course, the real possibility of migration of harmful chemical elements from interglacial Quaternary sediments to contemporary stream sediments is not analysed here.

The aim of this research is to demonstrate that there are many more samples with anomalous contents of As and Mo compared with the other harmful chemical elements Ba, Cr, Cu, Mn, Ni, Pb, V and Zn in Quaternary interglacial sediments and to give possible explanation of As and Mo enrichment in some intervals. The findings of this investigation might be useful for researchers of stream water and sediments, because As and Mo anomalies might be related to one of the factors affecting water quality in Lithuanian rivers (Gudas, Povilaitis 2013).

Division	Subdivision	Step formation
Pleistocene	Upper	Nemunas (Weichselian), glacial Merkinė (Eemian), interglacial
	Middle	Medininkai (Warthe), glacial Snaigupėlė (Karlich), interglacial Žemaitija (Drenthe), glacial Butėnai (Holsteinian), interglacial Dainava (Elsterian 2), glacial Turgeliai (Voigstedt), interglacial Dzūkija (Elsterian 1), glacial Vindžiūnai (Bavelian), interglacial Kalviai (Nidanian), glacial
	Lower	Daumantai complex (preglacial)

Fig. 1. Quaternary stratigraphic scheme of Lithuania

1. Methods

Samples of interglacial sediments were collected at five outcrops and four boreholes with the sampling interval of 5–10 or 15–20 cm. Vertical sections of Daumantai (dm) preglacial, Vindžiūnai (vn), Turgeliai (tr) and Butėnai (bt) interglacials are near Šventoji and Šlavė stream (Anykščiai District), of Snaigupėlė interglacial (sn) on the bank of the Snaigupėlė stream near Druskininkai, and of Merkinė interglacial (mr) on the bank of the River Nemunas (Varėna District) (Fig. 2).

Diatom analysis and plant macrofossil survey was performed in all sections. Respective methods are described in Baltrūnas *et al.* (2013c). Geochemical analysis was performed in the Nature Research Centre Institute of Geology and Geography. Energy-dispersive x-ray fluorescence equipment SPECTRO XEPOS with TurboQuant for pressed pellets calibration method was used for the determination of total contents of chemical elements in 680 samples. Sample preparation included air-drying, crushing and homogenization in agate mortar with agate pestle, then milling of sub-sample by MM 400 mixer mill with zirconium oxide grinding jars and balls and mixing with binder (4 g of material and 0.9 g of Licowax, dilution factor (DF) was 0.82).

Pressed pellets of 32 mm diameter were prepared by PP15 press. Quality control is assured by the participation of the laboratory in the "International Soil-analytical Exchange" (ISE) program of Wageningen Evaluating Programs for Analytical Laboratories (WEPAL) (Taraškevičius *et al.* 2013). ISE samples can be used for re-calibration. The contents of 10 harmful chemical elements As, Ba, Cr, Cu, Mn, Mo, Ni, Pb, V, Zn were compared with their maximum permitted concentrations (MPC) in soil (Taraškevičius *et al.* 2013).

Fig. 2. Location of study objects of interglacial sediments

Notes: Interglacials: dm – Daumantai, vn – Vindžiūnai, tr – Turgeliai, bt – Butėnai, sn – Snaigupėlė, mr – Merkinė. Sampled depth intervals (in parentheses) of vertical sections: Daumantai (2.2–19.2 m), Vindžiūnai-136 (146–160 m), Šlavė (2.4–4.8 m), Kudrė-915 (75.8–91.8 m), Jononys-938 (20.2–39.2 m), Butėnai (0.5–1.85 m), Snaigupėlė-705 (16.0–40.2 m), Snaigupėlė (4.65–6.20 m), Netiesos (4.05–17.3 m).

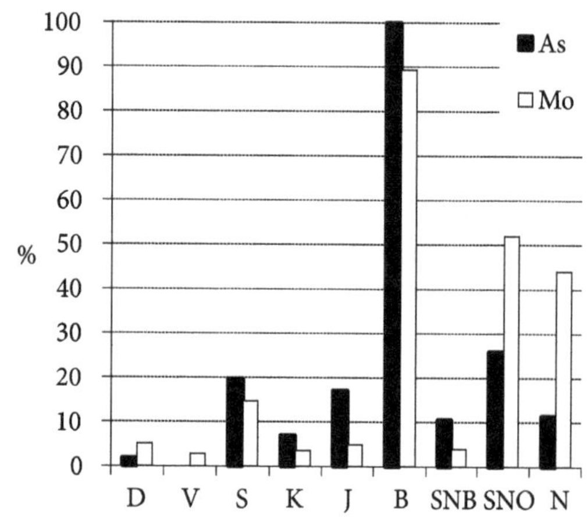

Fig. 3. Percentage of samples where MPC of As and Mo is exceeded in different sections. Outcrops: D - Daumantai, S - Šlavė, B - Butėnai, SNO - Snaigupėlė, N - Netiesos. Boreholes: V - Vindžiūnai-136, K - Kudrė-915, J - Jononys-938, SNB - Snaigupėlė-705

Total carbon (TC) was determined in Šlavė, Vindžiūnai and Kudrė sections by elemental analyzer liquiTOC (Elementar analysensysteme GmbH, Hanau-Germany). The procedure involves heating samples at 950 °C and measuring the combustion products by infrared energy detector with a precision of ±1%. The total organic carbon (TOC) was measured in the same way in carbonate-free subsamples, which were obtained by removing the carbonates with HCl and washing with deionized water. The difference between the two carbon measurements (TC–TOC) gives the total inorganic carbon (TIC) content. In part of the Netiesos section, the percentages of organic matter (OM) and total carbonates (TCB), calculated as $CaCO_3$, were determined by loss on ignition (LOI) at 550 °C and 950 °C, respectively (Santisteban *et al.* 2004). The amount of inorganic non-carbonate minerals (NCM) was calculated as 100%–OM%–TCB%. The main components of sediments (OM, TCB and NCM) were used for characterization of lithological units in the Netiesos section. The percentages of OM were recalculated to TOC, and TCB to TIC.

2. Results

The content of Ni, Cu, Pb, Ba in all samples was lower than respective MPC (75 mg/kg for Ni, 100 mg/kg for Cu and Pb, 600 mg/kg for Ba), indicating that interglacial sediments cannot be natural sources of these harmful chemical elements (Table 1). The contents of Zn, V, Cr, Mn exceeded MPC values (300, 150, 100, 1500 mg/kg) only in several samples and only in some sections. Such anomalous values comprise ≤4% from the total number of samples. The highest number of samples where element content exceeds MPC is characteristic of Mo and As. For Mo this comprises 21% and for As 12.6% from the total number of samples. MPC of Mo (5 mg/kg) is exceeded in all sections and As (10 mg/kg) in almost all. Butėnai, Snaigupėlė and Netiesos outcrop sections are most abundant in As and Mo anomalies (Fig. 3).

Butėnai section (0.5–1.85 m) has the highest percentage of anomalous samples: MPC of As is exceeded in all samples, of Mo in 89.3% of samples and of Mn in 42.9% of samples. There is also one sample where MPC of Zn is exceeded. The highest values of As, Mn, Zn are also in this section which consists of organogenic sediments. Snaigupėlė section (4.65–6.20 m) with organogenic sediments is also characterized by high percentages of Mo and As

Table 1. Maximum content of harmful chemical elements (mg/kg) in sections of interglacial sediments compared with soil maximum permitted concentrations (mg/kg)

Element	MPC	D(156)	S(41)	V(70)	K(57)	J(41)	B(28)	SNB(76)	SNO(27)	N(184)
		dm	vn	vn	tr	bt	bt	sn	sn	mr
As(86)	10	43.6(3)	59.0(8)	9.4	14.3(4)	61.9(7)	**89.4(28)**	23.4(8)	88.8(7)	44.1(21)
Mo(143)	5	6.7(8)	14.0(6)	6.3(2)	6.7(2)	7.7(2)	21.6(25)	6.9(3)	**27.5(14)**	19.6(81)
Mn(26)	1500	137	249	1714(1)	1686(1)	2594(1)	**8847(12)**	2412(8)	580	1589(3)
Cr(5)	100	103(1)	152(2)	78.0	74.5	69.3	71.2	84.2	72.4	**392(2)**
V(3)	150	101	**209(3)**	100	85.4	80.9	60.5	115	88.6	92.9
Zn(1)	300	56.4	119	54.0	70.4	198	**303(1)**	189	174	211
Ba	600	530	**580**	507	418	560	370	497	466	504
Cu	100	19.5	**52.6**	22.3	38.7	21.7	48.7	26.1	22.6	24.2
Ni	75	46.6	**67.0**	26.9	28.1	28.0	31.4	42.7	38.9	26.4
Pb	100	17.0	**20.0**	19.0	15.8	19.6	18.0	19.6	16.3	16.2

Notes: Interglacials: dm – Daumantai, vn – Vindžiūnai, tr – Turgeliai, bt – Butėnai, sn – Snaigupėlė, mr – Merkinė. Outcrops: D – Daumantai, S – Šlavė, B – Butėnai, SNO – Snaigupėlė, N – Netiesos. Boreholes: V – Vindžiūnai-136, K – Kudrė-915, J – Jononys-938, SNB – Snaigupėlė-705. Total number of samples analysed in each section is given in parentheses after abbreviation. The greatest maximum value in all 9 sections is in bold. Maximum content of element in the section is shaded if it exceeds MPC and the number of samples of the section where MPC is exceeded is given in parentheses. Total number of samples where MPC is exceeded is in parentheses after element. The table is based on TurboQuant for pressed pellets calibration method results.

anomalies: 51.9 and 25.9%, respectively. Netiesos section (4.05–17.3 m) is on the third place with 44.0% of Mo and 11.4% of As anomalies (Fig. 3). Unlike Butėnai and Snaigupėlė, the lithology of the section is not uniform: the lowermost layer of silt is overlain by gyttja, the latter is covered by peat and the upper part consists of sand. Median content of As in gyttja is 9.7 mg/kg, of Mo is 9.5 mg/kg, and in the silt layer the respective median values are 8.7 and 7.4 mg/kg. The re-calibrated (using ISE samples) median values of As and Mo in gyttja, silt and peat from Netiesos section exceed not only the respective median values in stream sediments of Lithuania, Šventoji and Middle Nemunas basin, but also the respective medians in different types of Lithuanian soil (Table 2). They are also higher than median values in European stream sediments and subsoil. In gyttja and silt, median values of As and Mo are especially high: they exceed respective median values in European topsoil and MPC of soil. Median content of Mo in gyttja, silt and peat is higher even than Mo maximum in Lithuanian stream sediments and Mo maximum in Lithuanian mineral soil. Median content of As in gyttja also exceeds As maximum in Lithuanian mineral soil and is close to As maximum in Lithuanian stream sediments.

In section of Snaigupėlė outcrop the median recalibrated values of As and Mo also exceed median values in Lithuanian stream sediments. They are especially high in the Butėnai section: re-calibrated As median is 59.7 mg/kg, Mo is 7.93 mg/kg.

All these facts confirm that interglacial Quaternary sediments comprise one of the possible natural sources of As and Mo anomalies. Netiesos section, which has been analysed in more detail by Baltrūnas *et al.* (2013c), can help to find out the peculiarities of layers enriched in As and Mo. The highest percentage of As anomalies is in gyttja layer (41.7% of its samples), it is followed by silt (22.2%); the same lithological units are even more abundant in Mo anomalies: 100 and 97.2%, respectively (Fig. 4).

Table 2. Arsenic and molybdenum contents (mg/kg) in interglacial sediments compared to soil and sediments

Contents (mg/kg)	As	Mo
Median in gyttja of Netiesos section	11.1	9.23
Median in silt of Netiesos section	10.2	7.24
Median in peat of Netiesos section	6.59	5.06
MPC in soil (1)*	10	5
Median in Lithuanian stream sediments (2)*	3.0	0.8
Median in Middle Nemunas basin sediments (2)*	3.3	0.9
Median in Šventoji basin sediments (2)*	2.9	0.7
Maximum in Lithuanian stream sediments (2)*	11.7	4.95
Median in sandy soil of Lithuania (2)*	1.8	0.62
Median in sandy loamy soil of Lithuania (2)*	2.9	0.67
Median in loamy soil of Lithuania (2)*	3.7	0.70
Median in organic soil of Lithuania (2)*	1.7	0.68
Median in all soil types of Lithuania (2)*	2.4	0.67
Maximum in mineral soil of Lithuania (2)*	7.9	2.55
Median in European topsoil (3)*	7.03	0.62
Median in European subsoil (3)*	6.02	0.52
Median in European stream sediments (3)*	6.00	0.63

*Sources: 1: Taraškevičius *et al.* (2013); 2: Kadūnas *et al.* (1999); 3: Salminen *et al.* (2005).

Plant macrofossil and pollen analysis shows that they were formed during the Merkinė (Eemian) interglacial climatic optimum when palaeoflora became most thermophilic. The intervals of increase of As and Mo correspond to two subzones of the thermal optimum. Comparison of median percentage of main components in four lithological units of Netiesos section reveals the differences among them (Fig. 5).

Sand unit predominantly consists of non-carbonate minerals (median NCM is 91.6%), median value of carbonates is much lower (TCB = 8.0%) and organic part (OM) makes up only ~0.4%. In peat, the percentage of NCM is still the highest, but decreases to 50.0%. Meanwhile OM increases to 36.7%, the part of carbonates is the lowest (TCB = 2.6%). Both gyttja and silt have the highest part of carbonates (40.3% and 51.7%, respectively), but also high percentage of organics, especially gyttja (OM = 29.2%). Meanwhile NCM percentage is much lower than in sand and peat. Elevated contents of As (Fig. 6b) and Mo (Fig. 6c) in the Netiesos section are obviously related to gyttja and silt, partly to peat where distribution pattern of main components shows higher percentage of organic matter and carbonates (Fig. 6a). The anomalies of both chemical elements are obviously not related to non-carbonate minerals: significant ($p < 0.0001$, number of observations $n = 70$) negative Pearson correlation coefficients of As ($r = -0.57$) and Mo ($r = -0.78$) with NCM confirm this statement. Since gyttja and silt are abundant in As and Mo anomalies and at the same time rich in carbonates and organics, the next task is to find out which of two components has more influence on the formation of the above-mentioned anomalies.

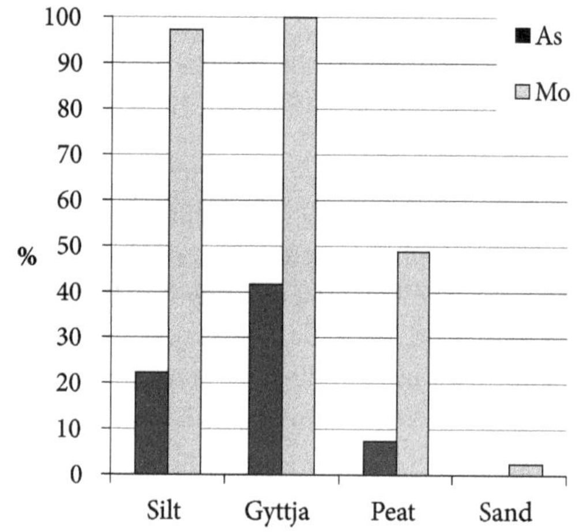

Fig. 4. Percentage of samples where MPC of As and Mo is exceeded in different layers of Netiesos section

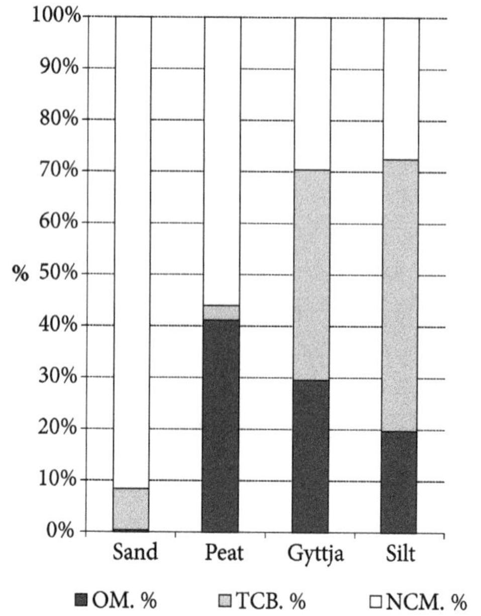

Fig. 5. Median percentage of main components in Netiesos. NCM: non-carbonate minerals, TCB: total carbonates, OM: organic matter

In the Netiesos section, As and Mo have significant positive correlation both with TOC and with TIC. However, correlation of As is higher and more significant with TOC than with TIC, and correlation of Mo is higher with TIC than with TOC (Table 3).

Correlation of As and Mo with TOC and TIC has also been analysed in Šlavė outcrop and Kudrė-915 borehole sections. The percentage of As and Mo anomalies in these sections is lower than in Butėnai, Snaigupėlė and Netiesos sections (Fig. 3). Arsenic in both these sections has the strongest correlation coefficient with TOC compared to other chemical elements (including Mo); meanwhile its correlation with TIC is lower, as well as the significance level (Table 3). Molybdenum in Šlavė section is significantly positively correlated both with TOC (higher p) and with TIC (lower p) (Table 3). As in the Netiesos section, the Mo correlation coefficient with TOC is lower than correlation of As with TOC. Meanwhile correlation of Mo with TIC is stronger than correlation of As with TIC. Although in Kudrė-915 section correlation of Mo with TOC and TIC is insignificant, the coefficient of Mo with TIC is positive, meanwhile with TOC it is negative. The above-mentioned correlations indicate that As anomalies are more related to OM and partly to carbonates, meanwhile Mo anomalies, on the contrary, are more related to carbonates than to OM.

Two dendrograms (Fig. 7a and b) were obtained for two different subsets of interglacial sediments: one of them included 86 samples with anomalous (>10 mg/kg) As contents, another one 143 samples with anomalous (>5 mg/kg) Mo contents. In both dendrograms As and Mo are in the same branch with eight other chemical elements: S, Br, Zn, Mn, Fe, P, Ca and Sr. These elements are proxies of organic matter (S, Br, Zn), iron and manganese hydroxides (Fe, Mn), carbonates (Ca, Sr) and phosphates (P). Arsenic and molybdenum can be related to these components, most of which are good sorbents.

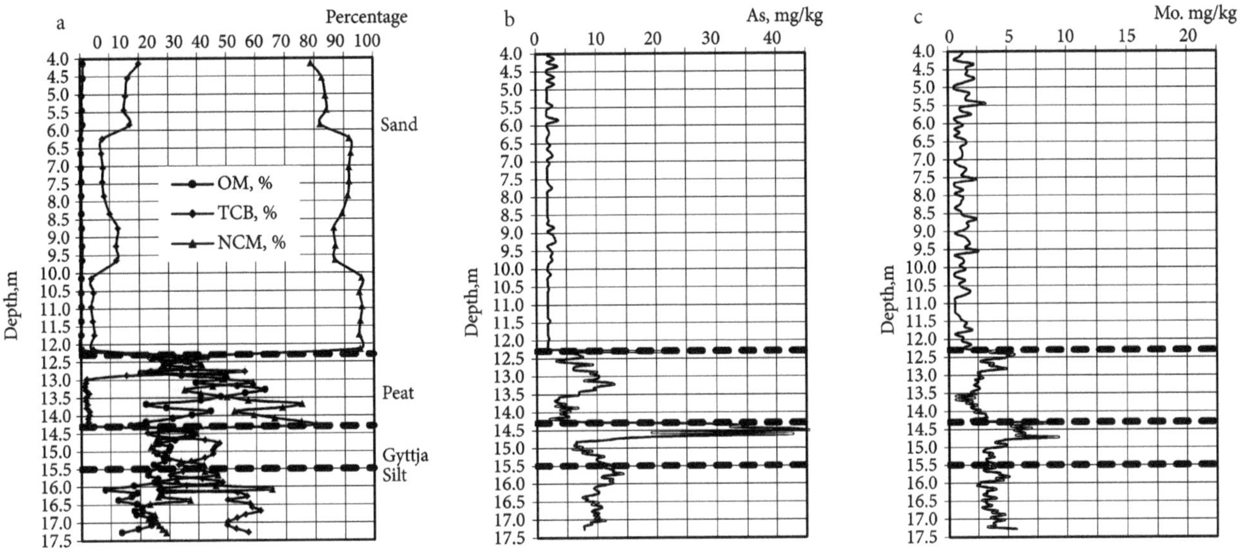

Fig. 6. Relationship of As and Mo to main components of palaeolake Netiesos sediments. NCM: non-carbonate minerals, TCB: total carbonates, OM: organic matter.

Table 3. Correlation of chemical elements with total organic carbon and total inorganic carbon in three sections

Netiesos (n = 70)				Šlavė (n = 41)				Kudrė-915 (n = 57)			
El	C(TOC)	El	C(TIC)	El	C(TOC)	El	C(TIC)	El	C(TOC)	El	C(TIC)
S	0.834***	Ca	0.986***	As	0.974***	Fe	0.472*	As	0.851***	Ca	0.547***
Cu	0.808***	Sr	0.875***	Br	0.963***	Mo	0.388x	S	0.839***	Br	0.510***
Zn	0.748***	Mn	0.772***	S	0.917***	V	0.373x	Br	0.821***	S	0.483**
Ni	0.709***	Br	0.713***	Cu	0.905***	Ca	0.372x	Zn	0.573***	Sr	0.339*
V	0.600***	Mo	0.641***	Mo	0.856***	S	0.321x	Fe	0.467**	Fe	0.336x
Br	0.579***	Fe	0.608***	Fe	0.812***	As	0.318x	P	0.437**	Mn	0.334x
As	0.509***	S	0.367*	Ni	0.779***	Cr	0.288	Mn	0.331x	As	0.323x
Fe	0.506***	As	0.340*	V	0.771***	Pb	0.282	Ni	0.253	Zn	0.280x
Mo	0.484***	Mg	−0.053	Cr	0.767***	Ni	0.249	Cr	0.222	Ni	0.222
Ga	0.144	Ba	−0.149	Ca	0.612***	Br	0.235	V	0.084	Cr	0.221
Ca	0.118	Cu	−0.185	Y	0.525**	Cu	0.229	Ti	0.076	P	0.185
Cr	0.088	V	−0.200	Pb	0.427*	Ga	0.226	Ca	0.032	Mo	0.135
Y	−0.057	Ni	−0.240x	Ga	0.163	Zn	0.160	Nb	0.030	V	0.120
Mn	−0.063	Zn	−0.373*	Ba	0.063	Mg	0.106	Cu	−0.040	Cu	0.097
Al	−0.065	P	−0.445**	Sr	0.006	Al	0.053	Mo	−0.065	Nb	0.085
Nb	−0.086	Cr	−0.517***	Zn	−0.028	Rb	0.027	Pb	−0.082	Ti	0.078
Sr	−0.121	Nb	−0.577***	Mn	−0.134	Y	0.023	Hf	−0.086	Pb	0.013
Pb	−0.151	Zr	−0.610***	Na	−0.150	Ba	−0.072	Y	−0.095	Ga	0.008
Ti	−0.165	Ga	−0.663***	P	−0.178	P	−0.130	Ga	−0.098	Rb	−0.032
Rb	−0.199	Ti	−0.675***	Rb	−0.286	Mn	−0.139	Rb	−0.174	Y	−0.102
P	−0.366*	Hf	−0.686***	Hf	−0.301	Sr	−0.147	Ba	−0.239	K	−0.168
Hf	−0.391**	Pb	−0.697***	Al	−0.325x	K	−0.202	Al	−0.255	Hf	−0.170
K	−0.449***	Na	−0.703***	K	−0.379x	Nb	−0.203	Sr	−0.320x	Al	−0.181
Mg	−0.457***	Si	−0.711***	Zr	−0.403*	Na	−0.282	Zr	−0.344*	Mg	−0.214
Zr	−0.565***	Al	−0.780***	Mg	−0.560**	Ti	−0.325x	K	−0.399*	Ba	−0.247
Na	−0.567***	Y	−0.782***	Nb	−0.611***	Hf	−0.343x	Mg	−0.642***	Zr	−0.479**
Ba	−0.573***	K	−0.795***	Ti	−0.782***	Zr	−0.366x	Na	−0.662***	Na	−0.528***
Si	−0.697***	Rb	−0.805***	Si	−0.903***	Si	−0.412*	Si	−0.735***	Si	−0.645***

Notes: El – chemical element, n – number of observations. Sorted Pearson correlation coefficients of chemical element contents: C(TOC) – with TOC, C(TIC) – with TIC. Significance level p of correlation coefficients: x $p < 0.05$, * $p < 0.01$, ** $p < 0.001$, *** $p < 0.0001$.

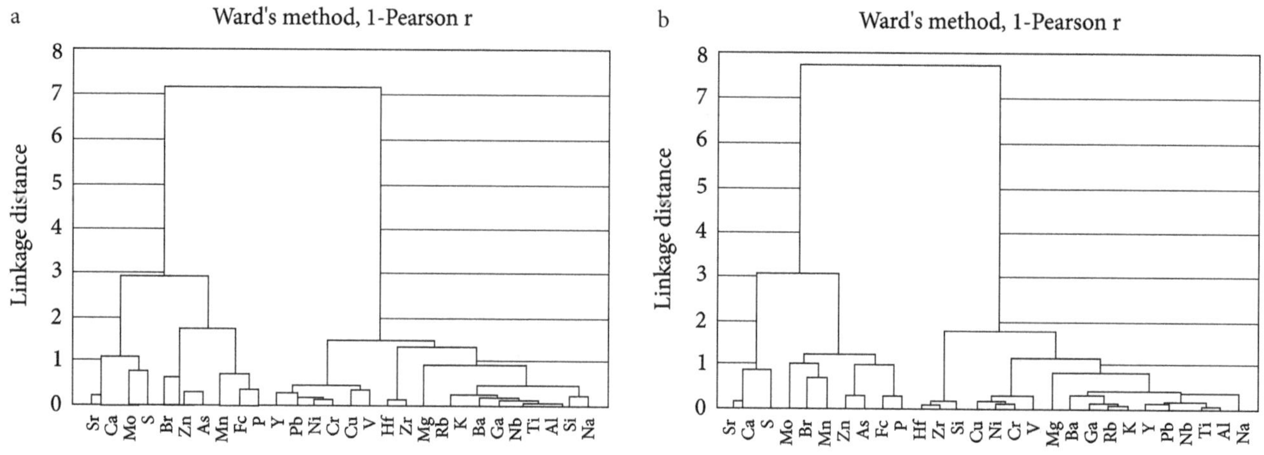

Fig. 7. Dendrograms of anomalous As (a) and anomalous Mo (b) data sets (Fc means Fe)

Such ideally coinciding groups of chemical elements indicate that the sorbents of As and Mo can be the same and that the difference between main sorbents of As and main sorbents of Mo is only relative. Anomalous As ($n = 86$) has significant positive Pearson correlation coefficients with Zn ($r = 0.69$, $p < 0.0001$), Br ($r = 0.46$, $p < 0.0001$), Mo ($r = 0.40$, $p < 0.001$), Fe ($r = 0.25$, $p < 0.05$), meanwhile anomalous Mo ($n = 143$) with As ($r = 0.42$, $p < 0.0001$), Mn ($r = 0.23$, $p < 0.01$), S ($r = 0.19$, $p < 0.05$), Fe ($r = 0.16$, $p < 0.05$). Significant correlation between As and Mo in both datasets can be explained by significant correlation of both elements with Fe. In the data set of 170 anomalous samples, where at least one of two elements (As or Mo) exceeds MPC, there are even 59 samples (34.7%) where both elements have anomalous contents. This fact once again confirms that the reasons for high content of As and Mo are very similar. The above-mentioned eight chemical elements are related to organic, carbonate or other authigenic components of lacustrine sediments. So biogenic and authigenic sedimentation processes were the main reason for As and Mo anomalies.

3. Discussion

The fact that part of Quaternary interglacial sediments is enriched in As and Mo adds information to the findings of researchers who emphasise the role of natural sources of contamination and at the same time stimulates further research. Contamination by As from natural sources has been investigated much more compared with Mo.

The geochemical atlas of Europe does not show large-scale anthropogenic contamination by As, but "there are significant areas with naturally occurring high As levels, many of which are above currently used national acceptable guideline values" (Salminen *et al.* 2005). According to Wang and Mulligan (2006a), the main natural As sources in Canada are weathering and erosion of arsenic-containing rocks and soil. Not only volcanic rocks, their weathering products and ash, hydrothermal ore deposits and geothermal waters can be sources of As, but also oceanic sedimentary rocks (e.g. phosphorites) (Baturin 2008), or fossil fuels (e.g. coals) (Yudovich, Ketris 2005). Black shales deposited in stagnant aquatic environments with high organic productivity and oxygen deficiency at the bottom contain especially high As contents (Paikaray 2012).

Natural sources of Mo are known to form (e.g. in pelagic zones of oceanic sediments which contain ferromanganese nodules) (Dubinin *et al.* 2008). Black shales can be enriched in Mo reaching 70 mg/kg (Salminen *et al.* 2005). Coal often has rather high Mo content (Sun *et al.* 2007).

The causes of As enrichment in sedimentary rocks can be different, because despite solubility of its minerals and compounds, As migration is greatly limited due to its intensive sorption (Reimann *et al.* 2003). Adsorption and co-precipitation are the main processes controlling As mobility; this element can be sorbed onto different mineral surfaces: clay, oxyhydroxides of Fe, Mn, Al and calcite (Romero *et al.* 2004), and it can be associated with organic matter (Fu *et al.* 2013). Such variety of components to which As can be related explains the significant positive correlation of As with both OM and carbonates in interglacial sediments. Besides, our investigation has shown that accumulation of OM and carbonates in interglacial sediments often occurs simultaneously. Higher correlation of As with TOC than with TIC in Netiesos, Šlavė and Kudrė-915 sections corresponds to the statement of Reimann *et al.* (2003) that organic matter may have a marked influence on As concentration and to the findings of Meharg *et al.* (2006) and Zahid *et al.* (2009). Besides, high correlation of As with organic matter in these sections corresponds to previous results from Daumantai section (Baltrūnas *et al.* 2013a) where correlation of As with TOC ($r = 0.21$) was significant ($n = 156$, $p < 0.01$), meanwhile with TIC ($r = 0.14$) it was insignificant. Daumantai section consists of sand but some silt interlayers are enriched in OM, S, Fe, TIC and Ca.

Adsorption and co-precipitation processes also control Mo mobility. Molybdenum is well adsorbed by Fe, Al and Mn oxyhydroxides (Reimann *et al.* 2003). Since Mo is a biologically essential element and serves various biochemical functions (Brucker *et al.* 2011), its significant correlation with organic matter in interglacial sediments is understandable. Correlation of authigenic Mo with TOC has been observed in different objects, e.g. Tanganyika lake sediments (Brucker *et al.* 2011), marine oil shale from northern Tibet (Fu *et al.* 2013). Higher correlation of Mo with TIC than with TOC in interglacial sediments can be explained by the fact that molybdate oxyanions are readily co-precipitated not only by OM, but also by $CaCO_3$ and several cations: Pb^{2+}, Cu^{+}, Zn^{2+}, $Mn^{2+}$and Ca^{2+} (Salminen *et al.* 2005).

Not only interglacial sediments of Lithuania are enriched in As and Mo. Both As and Mo are also among the most enriched trace elements in marine oil shale (Fu *et al.* 2013). The relationship of As and Mo and their correlation with Fe in interglacial sediments enables us to presume that their enrichment can be related either to sulphides or to Fe oxyhydroxides, because the experience from Latin America has shown the importance of these minerals for As distribution (Bundschuh *et al.* 2012). According to Paikaray (2012) who studied black shales, iron sulphidization occurs in relatively deeper depositional environments. These conditions favour a stagnant euxinic layer for OM deposition, besides, lack of clastic dilution enhances OM deposition. Microbial sulphate reduction to hydrogen sulphide and subsequent precipitation as metal sulphides takes place there (Paikaray 2012). Fu *et al.* (2013) explained As and Mo anomalies in marine oil shale by sulphide minerals and organic matter. Arsenic in coal is also either in sulphides or in organic compounds, both being authigenic (Yudovich, Ketris 2005). However, since correlation of As with S in anomalous interglacial sediments is not significant, the second presumption about relationship of As and Mo with Fe oxyhydroxides seems to us more reasonable. The relationship of As with Fe oxides and hydroxides has been proven by SEM image analysis of some stream sediments from Greece (Alexakis 2011). Increased contents of Fe, Mn, Mo and P were observed in $FeOOH_x$ layers; the explanation was that Fe and Mn oxides strongly absorb oxyanions such as phosphate, arsenate and molybdate (Zahid *et al.* 2009).

The possibility of As and Mo release from interglacial sediments is still not estimated. About 18% of the total content of As and Mo may not be extracted from soil even by *aqua regia*, because their median degree of extractability in *aqua regia* from some European soil samples analysed by ISE participants is only 82% (Taraškevičius *et al.* 2013). In agricultural soils of 10 European countries the respective index for As is even lower: from 28% in Russia to 77% in Lithuania and Poland (Reimann *et al.* 2003).

On the other hand, obvious release of As from Holocene aquifers to groundwater where As concentration reached toxic levels has been observed in 61 of 64 districts of Bangladesh (Zahid *et al.* 2009). Groundwater in Bangladesh when used for irrigation of rice paddies also causes problems, though the exposure of population to As in rice is of less concern compared to As in untreated groundwater (Van Geen *et al.* 2006).

Release of As to ground- and surface water greatly depends on pH, redox potential, contents of OM, ions present and ionic strength (Bundschuh *et al.* 2012). Natural organic matter not only reduces, but often enhances, As mobility (Wang, Mulligan 2006b). According to Bundschuh *et al.* (2012), remobilisation of As from metal oxides and oxyhydroxides can be either by their dissolution under very acid conditions or by desorption of As from them at high pH (>8) and oxidising conditions. The mechanisms of As release from black shales have been investigated by Paikaray (2012); the conclusion was that in suitable geochemical conditions these deposits can be important source of As in the environment.

Conclusions

1. Interglacial sediments, especially enriched in organic matter, carbonates and other sorbents, comprise one of the possible natural sources of As and Mo anomalies in stream sediments of Lithuania.
2. Biogenic and authigenic sedimentation processes often take place simultaneously and were the main reason for As and Mo anomalies.
3. Anomalous contents of As and Mo correlate with the content of Fe, indicating iron minerals as probable sources and sinks. Anomalies of As are related to organic matter and partly to carbonates. Mo is more related to carbonates than to organic matter.

Acknowledgements

The authors are grateful to the Research Council of Lithuania for financial support (project registration No LEK-12007).

References

Alexakis, D. 2011. Diagnosis of stream sediment quality and assessment of toxic element contamination sources in East Attica, Greece, *Environmental Earth Sciences* 63(6): 1369–1383. http://dx.doi.org/10.1007/s12665-010-0807-9

Ayris, P. M.; Delmelle, P. 2012. The immediate environmental effects of tephra emission, *Bulletin of Volcanology* 74(9): 1905–1936. http://dx.doi.org/10.1007/s00445-012-0654-5

Baba, A.; Tayfur, G. 2011. Groundwater contamination and its effect on health in Turkey, *Environmental Monitoring and Assessment* 183(1–4): 77–94. http://dx.doi.org/10.1007/s10661-011-1907-z

Baltrūnas, V.; Gaigalas, A. 2004. Entropy of Pleistocene till composition as an indicator of sedimentation conditions in southern Lithuania [online], [cited 17 April 2014], *Geological Quarterly* 48(2): 115–122. Available from Internet: https://gq.pgi.gov.pl/article/view/7337

Baltrūnas, V.; Zinkutė, R.; Šeirienė, V.; Katinas, V.; Karmaza, B.; Kisielienė, D.; Taraškevičius, R.; Lagunavičienė, L. 2013a. Sedimentary environment changes during the Early-Middle Pleistocene transition as recorded by the Daumantai sections in Lithuania, *Geological Quarterly* 57(1): 45–59. http://dx.doi.org/10.7306/gq.1076

Baltrūnas, V.; Zinkutė, R.; Taraškevičius, R.; Karmaza, B.; Šeirienė, V.; Kisielienė, D. 2013b. Interglacial Quaternary sediments as possible natural source of anomalies of harmful chemical elements in Lithuania, *E3S Web of Conferences* 1: 16010. http://dx.doi.org/10.1051/e3sconf/20130116010

Baltrūnas, V.; Šeirienė, V.; Molodkov, A.; Zinkutė, R.; Katinas, V.; Karmaza, B.; Kisielienė, D.; Petrošius, R.; Taraškevičius, R.; Piličiauskas, G.; Schmolcke, U.; Heinrich, D. 2013c. Depositional environment and climate changes during the late Pleistocene as recorded by the Netiesos section in southern Lithuania, *Quaternary International* 292: 136–149. http://dx.doi.org/10.1016/j.quaint.2012.11.038

Baturin, G. N. 2008. Arsenic, antimony, and bismuth in oceanic phosphorites, *Doklady Earth Sciences* 419(2): 266–269. http://dx.doi.org/10.1134/S1028334X08020177

Ber, A. 2006. Pleistocene interglacials and glaciations of north-eastern Poland compared to neighbouring areas, *Quaternary International* 149(1): 12–23. http://dx.doi.org/10.1016/j.quaint.2005.11.014

Brucker, R. P.; McManus, J.; Severmann, S; Owens, J.; Lyons, T. W. 2011. Trace element enrichments in Lake Tanganyika sediments: controls on trace metal burial in lacustrine systems, *Geochimica et Cosmochimica Acta* 75(2): 483–499. http://dx.doi.org/10.1016/j.gca.2010.09.041

Bundschuh, J.; Litter, M. I.; Parvez, F.; Román-Ross, G.; Nicolli, H. B.; Jean, J.-S.; Liu, C.-W.; López, D.; Armienta, M. A.; Guilherme, L. R. G.; Gomez Cuevas, A.; Cornejo, L.; Cumbal, L.; Toujaguez, R. 2012. One century of arsenic exposure in Latin America: a review of history and occurrence from 14 countries, *Science of The Total Environment* 429: 2–35. http://dx.doi.org/10.1016/j.scitotenv.2011.06.024

Butkus, D.; Šalciūnienė, K. 2011. Investigation of heavy metal and radionuclide distribution in silt of Lake Didžiulis, *Journal of Environmental Engineering and Landscape Management* 19(3): 215–224. http://dx.doi.org/10.3846/16486897.2011.602554

Calvo, A. I.; Alves, C.; Castro, A.; Pont, V.; Vicente, A. M.; Fraile, R. 2013. Research on aerosol sources and chemical composition: past, current and emerging issues, *Atmospheric Research* 120–121: 1–28. http://dx.doi.org/10.1016/j.atmosres.2012.09.021

Dubinin, A. V.; Sval'nov, V. N.; Uspenskaya, T. Yu. 2008. Geochemistry of the authigenic ferromanganese ore formation in sediments of the Northeast Pacific Basin, *Lithology and Mineral Resources* 43(2): 99–110. http://dx.doi.org/10.1134/S0024490208020016

Ernst, W. G. 2012. Overview of naturally occurring Earth materials and human health concerns, *Journal of Asian Earth Sciences* 59: 108–126. http://dx.doi.org/10.1016/j.jseaes.2012.05.030

Falk, H.; Lavergren, U.; Bergbäck, B. 2006. Metal mobility in alum shale from Öland, Sweden, *Journal of Geochemical Exploration* 90(3): 157–165. http://dx.doi.org/10.1144/1467-7873/08-188

Fu, X.; Wang, J.; Zeng, Y.; Tan, F.; Feng, X. 2013. Trace elements and their behaviour during the combustion of marine oil shale from Changliang Mountain, northern Tibet, China, *Environmental Earth Sciences* 70: 1125–1134. http://dx.doi.org/10.1007/s12665-012-2199-5

Galkus, A.; Stakėnienė, R.; Jokšas, K.; Lagunavičienė, L. 2012. Assessing bottom sediment contamination with heavy metals in the Klaipėda port semi-closed bays, *Journal of Environmental Engineering and Landscape Management* 20(4): 307–315. http://dx.doi.org/10.3846/16486897.2012.660883

Gudas, M.; Povilaitis, A. 2013. Factors affecting seasonal and spatial patterns of water quality in Lithuanian rivers, *Journal of Environmental Engineering and Landscape Management* 21(1): 26–35. http://dx.doi.org/10.3846/16486897.2012.696058

Idzelis, R. L.; Greičiūtė, K.; Paliulis, D. 2006. Investigation and evaluation of surface water pollution with heavy metals and oil products in Kairiai military ground territory, *Journal of Environmental Engineering and Landscape Management* 14(4): 183–190. http://dx.doi.org/10.1080/16486897.2006.9636896

Ignatavičius, G.; Sakalauskienė, G; Oškinis, V. 2006. Influence of land fires on increase of heavy metal concentrations in river waters of Lithuania, *Journal of Environmental Engineering and Landscape Management* 14(1): 46–51. http://dx.doi.org/10.1080/16486897.2006.9636878

Jaskelevičius, B.; Lynikienė, V. 2009. Investigation of influence of Lapes landfill leachate on ground and surface water pollution with heavy metals, *Journal of Environmental Engineering and Landscape Management* 17(3): 131–139. http://dx.doi.org/10.3846/1648-6897.2009.17.131-139

Kadūnas, V.; Budavičius, R.; Gregorauskienė, V.; Katinas, V.; Kliaugienė, E.; Radzevičius, A.; Taraškevičius, R. 1999. *Lietuvos geocheminis atlasas* [Geochemical Atlas of Lithuania]. Vilnius (in Lithuanian and English).

Kondratienė, O. 1996. *Stratigrafija i paleogeografija kvartera Litvy po paleobotanicheskim dannym* [Stratigraphy and paleogeography of Quaternary in Lithuania according to palaeobotanic data]. Academia, Vilnius (in Russian, with English summary).

Kumpienė, J.; Brännvall, E.; Taraškevičius, R.; Aksamitauskas, Č.; Zinkutė, R. 2011. Spatial variability of topsoil contamination with trace elements in preschools in Vilnius, Lithuania, *Journal of Geochemical Exploration* 108(1): 15–20. http://dx.doi.org/10.1016/j.gexplo.2010.08.003

Lachmar, T. E.; Burk, N. I.; Kolesar, P. T. 2006. Groundwater contribution of metals from an abandoned mine to the north fork of the American Fork River, Utah, *Water, Air, and Soil Pollution* 173: 103–120. http://dx.doi.org/10.1007/s11270-005-9031-8

Lindner, L.; Gozhik, P.; Marciniak, B.; Marks, L.; Yelovicheva, Y. 2004. Main climatic changes in the Quaternary of Poland, Belarus and Ukraine [online], [cited 17 April 2014], *Geological Quarterly* 48(2): 97–114. Available from Internet: https://gq.pgi.gov.pl/article/view/7336

Lindner, L.; Bogutsky, A.; Gozhik, P.; Marks, L.; Łanczont, M.; Wojtanowicz, J. 2006. Correlation of Pleistocene deposits in the area between the Baltic and Black Sea, Central Europe [online], [cited April 17, 2014], *Geological Quarterly* 50(1): 195–210. Available from Internet: https://gq.pgi.gov.pl/article/view/7405

Lindner, L.; Marks, L.; Nita, M. 2013. Climatostratigraphy of interglacials in Poland: Middle and Upper Pleistocene lower boundaries from Polish perspective, *Quaternary International* 292: 113–123. http://dx.doi.org/10.1016/j.quaint.2012.11.018

Marozas, V.; Armolaitis, K.; Aleinikovienė, J. 2013. Changes of ground vegetation, soil chemical properties and microbiota following the surface fires in Scots pine forests, *Journal of Environmental Engineering and Landscape Management* 21(1): 67–75. http://dx.doi.org/10.3846/16486897.2012.663087

Meharg, A. A.; Scrimgeour, C.; Hossain, S. A.; Fuller, K.; Cruickshank, K.; Williams, P. N., *et al.* 2006. Codeposition of organic carbon and arsenic in Bengal Delta aquifers, *Environmental Science and Technology* 40: 4928–4935. http://dx.doi.org/10.1021/es060722b

Milukaitė, A.; Valiulis, D.; Šakalys, J. 2008. Fluxes and loading of heavy metals, benzo[*a*]pyrene and oil products in Vilnius City, *Water, Air and Soil Pollution: Focus* 8: 485–494. http://dx.doi.org/10.1007/s11267-008-9181-y

Nagajyoti, P. C.; Lee, K. D.; Sreekanth, T. V. M. 2010. Heavy metals, occurrence and toxicity for plants: a review, *Environmental Chemistry Letters* 8: 199–216. http://dx.doi.org/10.1007/s10311-010-0297-8

Paikaray, S. 2012. Environmental hazards of arsenic associated with black shales: a review on geochemistry, enrichment and leaching mechanism, *Reviews in Environmental Science and Biotechnology* 11: 289–303. http://dx.doi.org/10.1007/s11157-012-9281-z

Pereira, P.; Ubeda, X. 2010. Spatial distribution of heavy metals released from ashes after a wildfire, *Journal of Environmental Engineering and Landscape Management* 18(1): 13–22. http://dx.doi.org/10.3846/jeelm.2010.02

Pundytė, N.; Baltrėnaitė, E.; Pereira, P.; Paliulis, D. 2011. Anthropogenic effects on heavy metals and macronutrients accumulation in soil and wood of *Pinus sylvestris* L., *Journal of Environmental Engineering and Landscape Management* 11(1): 34–43. http://dx.doi.org/10.3846/16486897.2011.557473

Reimann, C.; Siewers, U.; Tarvainen, T.; Bityukova, L.; Eriksson, J.; Gilucis, A.; Gregorauskiene, V.; Lukashev, V. K.; Matinian, N. N.; Pasieczna, A. 2003. *Agricultural soils in Northern Europe: A Geochemical Atlas.* E. Schweizerbart'sche Verlagsbuchhandlung, Stuttgart.

Romero, F. M.; Armienta, M. A.; Carrillo-Chavez, A. 2004. Arsenic sorption by carbonate-rich aquifer material, a control on arsenic mobility at Zimapán, México, *Archives of Environmental Contamination and Toxicology* 47(1): 1–13. http://dx.doi.org/10.1007/s00244-004-3009-1

Salminen, R.; Plant, J.; Reeder, S. (Eds.) 2005. Geochemical Atlas of Europe. *Part 1, Background information, methodology and maps.* Espoo, Finland, Geological Survey of Finland.

Santisteban, J. I.; Mediavilla, R.; López-Pamo, E.; Dabrio, C. J.; Ruiz Zapata, M. B.; Gil Garcia, M. J.; Castano, S.; Martinez-Alfaro, P. E. 2004. Loss on ignition: a qualitative or quantitative method for organic matter and carbonate mineral content in sediments?, *Journal of Paleolimnology* 32(3): 287–299. http://dx.doi.org/10.1023/B:JOPL.0000042999.30131.5b

Sun, Y.; Lin, M.; Qin, P.; Zhao, C.; Jin, K. 2007. Geochemistry of the barkinite liptobiolith (Late Permian) from the Jinshan mine, Anhui Province, China, *Environmental Geochemistry and Health* 29: 33–44. http://dx.doi.org/10.1007/s10653-006-9059-8

Stakėnienė, R.; Galkus, A.; Jokšas, K. 2011. Pollution of Klaipeda Port waters [online], [cited 17 April 2014], *Polish Journal of Environmental Studies* 20(2): 445-459. Available from Internet: http://www.pjoes.com/pdf/20.2/Pol.J.Environ.Stud.Vol.20.No.2.445-459.pdf

Taraškevičius, R.; Zinkutė, R. 2011. Lietuvos miestų geocheminės anomalijos ir jų sklaida [Urban geochemical anomalies of Lithuania and their spread] [online], [cited 17 April 2014], *Baltica 24, Special Issue "Geosciences in Lithuania: challenges and perspectives"*: 163–168 (in Lithuanian). Available from Internet: http://www.geo.lt/geo/index.php?id=1375

Taraškevičius, R.; Zinkutė, R.; Stakėnienė, R.; Radavičius, M. 2013. Case study of relationship between *aqua regia* and real total contents of harmful trace elements in some European soils, *Journal of Chemistry* 2013(2013): 15. http://dx.doi.org/10.1155/2013/678140

Van Geen, A.; Zheng, Y.; Cheng, Z.; He, Y.; Dhar, R. K.; Garnier, J. M.; Rose, J.; Seddique, A.; Hoque, M. A.; Ahmed, K. M. 2006. Impact of irrigating rice paddies with groundwater containing arsenic in Bangladesh, *Science of The Total Environment* 367(2–3): 769–777. http://dx.doi.org/10.1016/j.scitotenv.2006.01.030

Vasarevičius, S.; Greičiūtė, K. 2004. Investigation of soil pollution with heavy metals in Lithuanian military grounds, *Journal of Environmental Engineering and Landscape Management* 12(4): 132–137. http://dx.doi.org/10.1080/16486897.2004.9636834

Vasarevičius, S.; Mineikaitė, A.; Vaitiekūnas, P. 2010. Investigation into heavy metals in storm wastewater from Vilnius Žirmūnai District and pollutants spread model in Neris River, *Journal of Environmental Engineering and Landscape Management* 18(3): 242–249. http://dx.doi.org/10.3846/jeelm.2010.28

Von der Heyden, C. J.; New, M. G. 2004. Groundwater pollution on the Zambian Copperbelt: deciphering the source and the risk, *Science of The Total Environment* 327(1–3): 17–30. http://dx.doi.org/10.1016/j.scitotenv.2003.08.028

Wang, S.; Mulligan, C. N. 2006a. Occurrence of arsenic contamination in Canada: Sources, behaviour and distribution, *Science of The Total Environment* 366: 701–721. http://dx.doi.org/10.1016/j.scitotenv.2005.09.005

Wang, S.; Mulligan, C. N. 2006b. Effect of natural organic matter on arsenic release from soils and sediments into groundwater, *Environmental Geochemistry and Health* 28: 197–214. http://dx.doi.org/10.1007/s10653-005-9032-y

Yudovich, Ya. E.; Ketris, M. P. 2005. Arsenic in coal: a review, *International Journal of Coal Geology* 61(3–4): 141–196. http://dx.doi.org/10.1016/j.coal.2004.09.003

Zahid, A.; Hassan, M. Q.; Breit, G. N.; Balke, K.-D.; Flegr, M. 2009. Accumulation of iron and arsenic in the Chandina alluvium of the lower delta plain, Southeastern Bangladesh, *Environmental Geochemistry and Health* 31: 69–84. http://dx.doi.org/10.1007/s10653-008-9226-1

Zinkutė, R.; Taraškevičius, R.; Želvys, T. 2011. Major elements as possible factors of trace element urban pedochemical anomalies, *Central European Journal of Chemistry* 9(2): 337–347. http://dx.doi.org/10.2478/s11532-011-0012-z

Rimantė ZINKUTĖ. Senior Researcher at the Nature Research Centre Institute of Geology and Geography. Doctor of Physical (Geology) Sciences (Vilnius University and Institute of Geology, 1998). In 1998–2013 published 52 scientific articles and one monograph. Co-author of two atlases and two monographs. Scientific interests: major and trace elements in contemporary and Quaternary sediments, geochemical associations, assessment of contamination, application of mathematical-statistical methods for treatment of geochemical data.

Valentinas BALTRŪNAS. Head of the project at the Nature Research Centre Institute of Geology and Geography. Dr Habil. of Physical (Geology) Sciences (Vilnius University and Institute of Geology, 1998), Professor of Lithuanian University of Educational Sciences. In 1998–2013 published 74 scientific articles and monographs. Scientific interests: sedimentology (glacial and interglacial formations of the Pleistocene), environmental geology, methodology of environmental research.

Ričardas TARAŠKEVIČIUS. Head of Geochemistry Division and Senior Researcher at Nature Research Centre Institute of Geology and Geography. Doctor of Physical (Geochemistry) Sciences (Institute of Geochemistry and Geophysics, Minsk, 1992). Co-author of two atlases, three monographs and over 70 scientific articles. Scientific interests: analytical methods, quality control, geochemical anomalies, their determinant factors, reflection and influence on biota, health effects, ecogeochemical evaluation, visualization and mathematical processing of geochemical data.

Bronislavas KARMAZA. Senior Researcher at the Nature Research Centre Institute of Geology and Geography. Doctor of Physical (Geology) Sciences (Vilnius University and Institute of Geology, 1995). Scientific interests: Lithuanian Pleistocene glaciolacustrine, glaciofluvial sediments, their structure and composition, geological heritage, application of distant methods for the research of Quaternary sediments, archaeological camps and settlements. Author/co-author of over 30 scientific articles. Research interests: geochemistry, land degradation, soil pollution.

Rimutė STAKĖNIENĖ. Researcher at the Nature Research Centre Institute of Geology and Geography. Doctor of Physical (Geography) Sciences (Vilnius University and Institute of Geography, 1999). Author and co-author of over 40 scientific papers and one monograph. Scientific interests: organic geochemistry of sediments and water column; environmental pollution and its reduction.

Vaida ŠEIRIENĖ. Head of Quaternary Research Laboratory and Senior Researcher at the Nature Research Centre Institute of Geology and Geography. Doctor of Physical (Geology) Sciences (Vilnius University and Institute of Geology, 1996). Author and co-author of over 50 scientific papers and five monographs. Scientific interests: Quaternary stratigraphy and geology, diatom flora, interglacial palaeoecology and palaeogeography.

Dalia KISIELIENĖ. Senior researcher at Nature Research Centre Institute of Geology and Geography. Doctor of Physical (Geology) Sciences (Vilnius University and Institute of Geology and Geography, 2002). Author and co-author of more than 30 scientific papers and five monographs. Scientific interests: Quaternary stratigraphy and geology, palaeobotany, palaeoecology and palaeogeography, human impact on palaeoecosystems.

11

DEPOSITION OF PETROLEUM HYDROCARBONS WITH SEDIMENT TRAPPED IN SNOW IN ROADSIDE AREAS

Larysa Mykhailova[a, b], Thomas Fischer[b], Valentina Iurchenko[a]

[a]*Kharkiv National Automobile and Highway University, 25 Petrovskoho str., Kharkiv, Ukraine*
[b]*Brandenburg University of Technology Cottbus-Senftenberg, Central Analytical Laboratory, Konrad-Wachsmann-Allee 6, 03046 Cottbus, Germany*

Abstract. Total petroleum hydrocarbon (TPH) deposition rates were determined along various roads using the natural snow cover as deposition trap. Daily deposition rates decreased with distance from the roads, which coincided with long-term TPH accumulation in roadside soils. Scanning electron microscopy (SEM-EDX) of the snow meltwater sediment revealed occurrence of carbon-rich plaques, which were identified as hydrocarbons using FTIR-microscopy. GC-MS revealed that the compounds extracted from the sediment consisted of an unresolved complex hydrocarbon mixture (UCM). Individual n-alkanes could not be resolved in the sediment extract, whereas TPHs extracted from soils contained a series of *n*-alkanes peaking at C25-C27. The proportion of UCM compounds from TPHs decreased with distance from road. We conclude that high-boiling hydrocarbons bind to coarse mineral dust and/or to splash water and vehicle spray, which preferentially deposit within a 10 m roadside strip.

Keywords: vehicle spray, unresolved complex hydrocarbon mixture, *n*-alkanes, dust deposition, soil contamination.

Introduction

The main sources of traffic pollution in roadside soils are leakage and combustion of oil and fuel, wear from tires and from road furniture, de-icing agents as well as accidental spillages (DWW 2003). The widespread use of hydrocarbons in fuels causes their predominance among organic atmospheric pollutants, and petroleum products are the major source of anthropogenic hydrocarbon pollutants found in the atmosphere (Mykhailova *et al.* 2013; Adeniyi, Owoade 2010; Okonkwo *et al.* 2006; Manahan 1993). Total petroleum hydrocarbons (TPH) from fossil sources are not readily biodegradable as compared to biomass or soil organic matter, which have been demonstrated to be consumed during hydrocarbon degradation (Mykhailova *et al.* 2013; Zyakun *et al.* 2011; Dilly *et al.* 2011). However, n-alkanes have been reported to degrade more rapidly in soil and water than components of the unresolved complex mixture (UCM) contained in mineral oils (Reddy *et al.* 2002). Once deposited on the surface, hydrocarbons may persist and bioaccumulate in environmental media (Alloway 1992), infiltrate into groundwater aquifers via leaching or into surface aquifers by runoff with severe effects on plants (Adam, Duncan 2002; Masakorala *et al.* 2013), humans and animals (Amadi *et al.* 1996; Albers 1995). Organic contaminants in roadside soils have been receiving considerable attention as a result of traffic intensity (Mykhailova *et al.* 2013; Adeniyi, Owoade 2010; Biasioli, Ajmone-Marsan 2007; Okonkwo *et al.* 2006; Unger, Prinz 1992).

Car emissions to the air, mainly incompletely burnt gases and dust particles, contribute to the background concentration in the atmosphere and are deposited over large distances from the road (Schipper *et al.* 2007). However, elevated amounts of petroleum hydrocarbons were reported from within a narrow strip ranging from 10–25 m (Unger, Prinz 1992; Iurchenko *et al.* 2013; Mykhailova *et al.* 2013) up to 50–100 m, which were attributed to the influence of vehicle spray (Schipper *et al.* 2007).

It was the aim of this study to determine deposition rates of total petroleum hydrocarbons (TPH) on roadside soils in Kharkov, Ukraine, and to elucidate the role of near-road sediment deposition under no-runoff conditions in a situation of a considerable age of the vehicle fleet (UNECE Transport Division Database, 31.12.2008),

Corresponding author: Thomas Fischer
E-mail: thomas.fischer@tu-cottbus.de

which typically is characterized by elevated oil leakage and tailpipe emissions.

The natural snow cover was used as deposition trap, which offers two main advantages: (1) Using meteorological snowfall data, the time period of deposition can be determined exactly; (2) Snow with its maximal temperature of 0 °C minimizes evaporative hydrocarbon losses from the surface. Long-term accumulation of TPH was further investigated in soils, and the sediment trapped in snow was investigated using scanning electron microscopy, equipped with X-ray detection (SEM-EDX) and Fourier transformed infrared (FTIR) microscopy.

1. Material and methods

1.1. Sampling sites

Soil samples were collected according to ISO 10381-2 (2003) at distances of 1, 6–8, 15, 40, 60 and 100 m in October 2010 from various asphalt roads in Kharkov, Ukraine (mean annual temperature: 7.5 °C, mean annual precipitation: 520 mm). According to AG Boden (2005), the soils were characterized as chernozems developed on loess, the soil texture was sandy to clayey silt. The background organic C concentration of the soils ranged from 22 to 27 g kg^{-1}, the content of carbonates was negligible (Mykhailova *et al.* 2013). Five replicates along a 100 m section at the windward (Akademika Pavlova and Traktorostroiteley av. only) and the leeward sides of the roads, as referred to the long-term wind direction, were sampled. The ground vegetation cover consisted of ruderal grass species. No road runoff could be observed, because all roads investigated were confined by curbs (Fig. 1). Instead, road traffic on ponding water caused appearance of splash water and vehicle spray during and after rain events, or after application of de-icing agents.

a)
b)

Fig. 1. Roadside of Akademika Pavlova street (a) and Traktorostroiteley avenue (b). Surface runoff can be excluded the given study, because all roads were confined by curbs

Snow samples, residing in the roadside area for a time period of 19 days, were collected from the windward and leeward sides at 5 points from areas of 20×20 cm^2 at distances of 2 and 8 m from the roads to the total depth of the snow cover following the general rules of volume-based sampling. To avoid hydrocarbon adsorption to the walls of the sampling containers as well as gaseous hydrocarbon losses, samples were sealed in glass containers immediately after collection and transported to the lab. Snow sediment was extracted from fresh meltwater after melting at 4 °C using 0.2 μm membrane filters, and was air-dried at ambient room temperature.

1.2. Analytical procedures

TPH concentrations in soil and snow were determined gravimetrically according to DIN 38409-56 (2009) after triple extraction of 20 g soil or in the entire amount of fresh meltwater from the snow samples, respectively, with 50 mL hexane p.a., clean up over a short aluminium oxide column, and solvent evaporation to constant weight at 40 °C.

TPH and fractional composition of petroleum hydrocarbons of selected soil and snow meltwater samples were further determined using GC-MS (Carlo Erba, Fisons) after triple extraction of 2 g soil with 5 mL hexane for residue analysis, clean up over a short fluorosil column, and solvent evaporation under hydrocarbon-free nitrogen at 20 °C to a sample volume of 10–100 μL, depending on the results of the gravimetric hydrocarbon determination. The analytical procedure corresponded to EN ISO 9377-2 (2001). A SGE HT8 column at 2 mL min^{-1} linear Helium flow with 1 μL split injection at a split ratio of 20:1 was used in GC-MS. Injection temperature was 300 °C. The column temperature was 70 °C during injection, was increased after 1 min with a heating rate of 30 K min^{-1} and held constant for 20 min after reaching 320 °C. For detection, positive electron impact ionization (EI+) at 70 eV and MS full scan mode with m/z ranging from 50 to 250 was used. Commercial diesel fuel and Hewlett Packard alkane standard Part. No. 18710-60170 were used for quantification and identification of total hydrocarbons and n-alkanes, respectively (Mykhailova *et al.* 2013).

Volatile organic compounds (VOC, boiling point <250 °C, alkane chainlength <C15), semi-volatile organic compounds (SVOC, boiling point 250–380 °C, alkane chainlength C15–C23) and organic compounds associated with particulate matter (PM) or particulate organic matter

(POM, boiling point >380 °C, alkane chainlength >C23) were identified in the chromatograms using the total ionic current (TIC) after elimination of siloxane signals resulting from column bleeding. The proportion of UCM components in soil extracts was determined by subtraction of the peak area of resolved peaks from the total chromatogram peak area, related to the total chromatogram area.

Scanning electron microscopy (Zeiss DSM 962), equipped with energy dispersive X-ray fluorescence detection (EDXRF, Oxford Instruments), as well as FTIR microscopy (Thermo Centaurµs) were used for microstructural analysis of the meltwater sediment. FTIR measurements were performed in diffuse reflectance mode with Kubelka-Munk correction and with a spectral resolution of 4 cm^{-1} against a gold mirror as background. In FTIR spectroscopy, net peak heights were determined to account for the intense OH vibration peaking at 3400 cm^{-1} and for baseline drift in the C = O region. Soil carbon was determined by dry combustion using a CNS Analyser (Elementar, Germany).

1.3. Statistics

Statistical procedures included analysis of variance (ANOVA) and the post-hoc Tukey-test after Bartletts test for homogeneity of variance. The Wilcoxon signed rank test was used if homogeneity of variance was not met. The R software suite was used for all statistical calculations.

Mattsons Win1st FTIR software was used for processing the FTIR spectra. Box-whisker-plots were used to depict median values (bold lines) and variation (upper and lower quartiles in boxes, data ranges as whiskers, outliers as circles) of CH/C = O signal ratios and the proportion of the chromatogram area of UCM compounds on the total chromatogram area.

2. Results

2.1. Soil

Soil TPH concentrations are given Table 1. Maximal values were observed in 1 m distance from the roads and amounted up to >3000 mg kg^{-1} at a sampling depth of 0–0.5 cm. TPH concentrations decreased more or less exponentially with distance from road and with sampling depth (data not shown), where background values amounted to approximately 100 mg kg^{-1}, which were reached in 100 m distance on the windward sides of the roads.

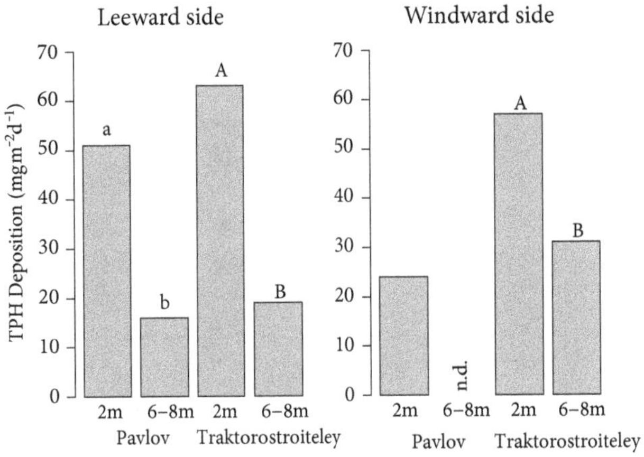

Fig. 2. Depositional flow of total petroleum hydrocarbons on snow. Same letters indicate non-significant ranges (Tukey-test, $p = 0.05$), n.d. – not determined.

Clearly elevated above background values were observed on the leeward sides, reaching amounts of 390–540 mg kg^{-1} in 100 m distance from Pushkinskaya and Bluechera streets, respectively. Analysis of variance, where logarithmic distance and depth data were processed, revealed a significant influence of traffic intensity (F-Test, $p = 0.013$), distance from road (F-test, $p < 0.001$), sampling depth (F-test, $p < 0.001$), exposition to wind direction (F-test, $p < 0.001$) and period of road exploitation (F-test, $p < 0.001$) on soil TPH concentrations.

2.1. Meltwater sediment

Depositional flow of total petroleum hydrocarbons to snow amounted up to 63 mg m^{-2} d^{-1} in 2 m distance, which was significantly (Tukey-test, $p = 0.05$) higher than values obtained in 8 m distance regardless of the road investigated, and which amounted up to 31 mg m^{-2} d^{-1} (Fig. 2). Values tended to be higher on the leeward side,

Table 1. Soil total petroleum hydrocarbon (TPH) concentrations (mg kg^{-1}), sampling depth 0–0.5 cm, n.d. – not determined

Road	Period of exploitation (yrs.)	Traffic Intensity (veh. h^{-1})	Wind exposition	Distance from road (m)					
				1 m	6–8 m	15 m	40 m	60 m	100 m
Akademika Pavlova	40	2050	Leeward	3149	n.d.	347	233	225	125
			Windward	785	n.d.	361	151	146	113
Pushkinskaya	20	960	Leeward	1900	n.d.	n.d.	n.d.	n.d.	390
Bluechera	40	1430	Leeward	2800	n.d.	n.d.	n.d.	n.d.	540
Traktorostroiteley	40	1120	Leeward	2766	2712	302	121	116	98
			Windward	661	462	248	229	194	176

a)

b)

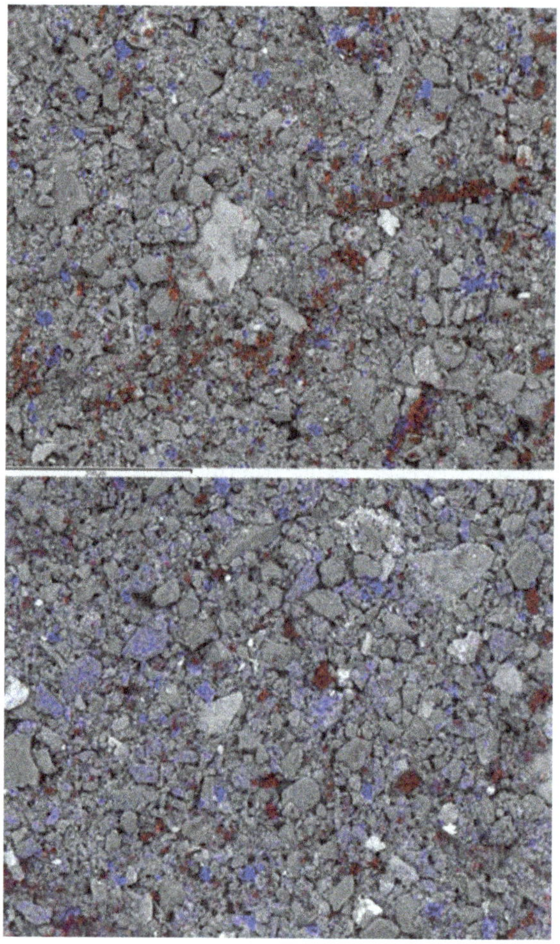

Fig. 3. SEM image of snow meltwater filter sediments: (a) sampling distance 1 m; (b) sampling distance 20 m. Carbon is mapped in red, Ca is mapped in blue colour. Ca may be geogenic (from loess) or anthropogenic (from de-icing agent)

but did not differ significantly from the samples taken from the windward side.

The meltwater sediment mainly consisted of mineral fine sand and silt (mean particle diameter < 200 µm, AG Boden (2005)). SEM-EDX images of snow meltwater filter sediments, including Ca and C element maps, are depicted in Figure 3. Carbon-rich plaques on the surface of the sediment (mapped in red) were more pronounced in 1 m distance from the road compared to 8 m.

A microscopic image of snow meltwater filter sediment sampled at 1 m distance from road is shown Figure 4, which exemplifies a dark-coloured measuring field

Fig. 4. Microscopic image of snow meltwater filter sediment sampled at 1 m distance from road: (1) dark-coloured measuring field; and (2) background measuring field for FTIR microscopy

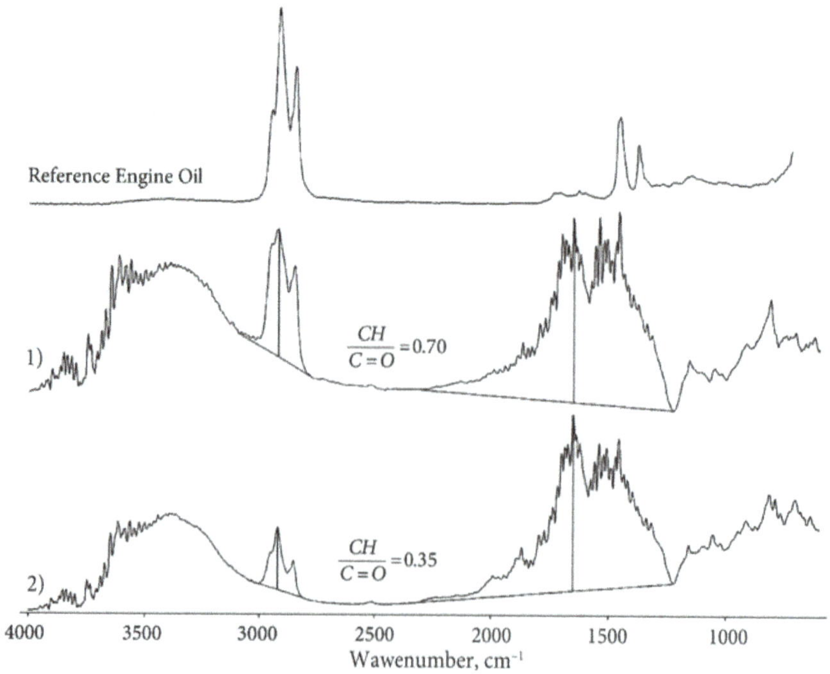

Fig. 5. FTIR spectra of: (1) the dark-coloured; and (2) the background measuring fields depicted in Figure 4. Engine oil is provided as analytical reference

and its respective background for FTIR microscopy. The respective FTIR spectra are shown Figure 5.

FTIR spectra were generally characterized by a broad band peaking at 3400 cm^{-1}, which can be attributed to the OH vibration originating from soil organic matter (alcoholic, phenolic OH), from residual moisture (H_2O), but also from OH bound to phyllosilicates. Individual peaks on the left shoulder of the broad OH-band (3500–4000 cm^{-1}) most likely derive from phyllosilicatic inner surface Al-OH vibrations (mica, kaolinite, smectites). A sharp and intense band between 2800 and 3000 cm^{-1} characterizes the CH stretching vibration of saturated C-structures, which is very typical for organic matter. Due to missing sharps peak between 3000 and 3100 cm^{-1} the presence of unsaturated aliphatic structures could not be confirmed. Peaks in the 1700–1750 cm^{-1} range typically characterize organic C = O double bonds, like carboxyl-, carbonyl- and ester-C, which are typical for biogenic structures like biomass or soil organic matter, but absent in petroleum products (cf. reference Fig. 5).

The median values of the CH/C = O signal intensity ratios amounted to 0.57 and 0.36 for the dark coloured hot spots on snow meltwater filter sediment in 1 and 20 m distance from the roads, respectively, and significantly (paired Wilcoxon signed rank test, $p < 0.01$) exceeded the respective neighbouring background values, which amounted to 0.34 and 0.15 for the reference measuring fields in 1 and 20 m distance from the roads, respectively (Fig. 6).

GC-MS revealed that hydrocarbons belonged mainly to the VOC and SVOC fraction in commercial diesel fuel, peaking at C18 (Fig. 7a), to the SVOC and PM/POM fractions in soil extracts (Fig. 7b–d), but mainly to the fraction bound to particulate matter or POM in the meltwater filter extract. Typical series of n-alkane peaks could not be observed in the chromatograms of the meltwater filter extracts (Fig. 7e). Soil extracts had an n-alkane pattern resembling that of diesel fuel and consisting of a series of even- and odd-numbered chainlengths, but at a higher boiling range, peaking at C25–C27. On the other hand, filter extracts mainly contained an unresolved complex mixture (UCM) forming broad humps. The proportion of the chromatogram area of UCM components on the total chromatogram area for samples taken in 0–15 m distance from the roads significantly exceeded that of samples taken in 40–100 m distance (Wilcoxon rank sum test with continuity correction, $p < 0.001$) and amounted to median values of 88% and 83%, respectively (Fig. 7f).

3. Discussion

The TPH concentration in soils depends on traffic intensity (Adeniyi, Owoade 2010; Biasioli, Ajmone-Marsan 2007; Okonkwo *et al.* 2006; Unger, Prinz 1992), as well as on meteorological and on further peripheral conditions,

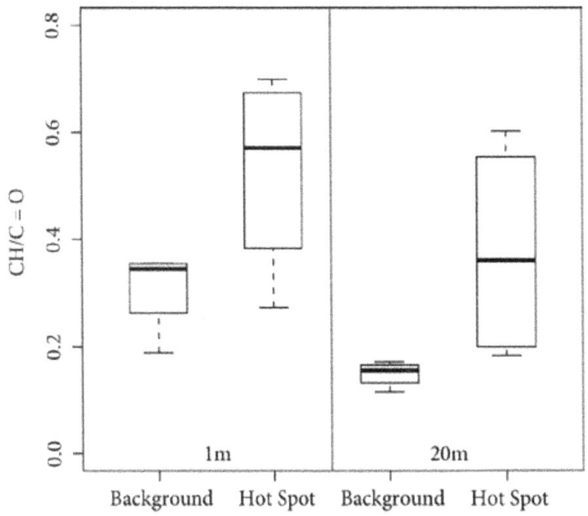

Fig. 6. Boxplot of the CH/C = O ratios for dark coloured hot spots and neighbouring background of snow meltwater filter sediment sampled at 1 m (left) and 20 m distance (right) from road. $n = 5$

like wind, geomorphology, road construction, buildings, or on vegetation. Hence, TPH amounts along roadside soils reported in the literature vary widely, reaching maximum levels of approximately 1100 mg kg^{-1} soil in 1 m distance from a German freeway (50000 vehicles per day, Unger, Prinz 1992), and up to approximately 19500 mg kg^{-1} soil in Nigeria (Mykhailova *et al.* 2013; Adeniyi, Owoade 2010). The maximum levels found in this study amounted to >3000 mg kg^{-1} soil at a traffic intensity of up to 25000 vehicles per day. In particular, for perpendicular approaching wind, markedly higher atmospheric pollutant concentrations at the leeward side and slightly lower concentrations at the windward side were observed by Gromke and Ruck (2009), which may explore the higher soil TPH concentrations at the leeward side here. Unger, Prinz (1992) reported that background levels were reached at a distance of 10–25 m from the road, which could be confirmed by the findings of this study with elevated concentrations at 15 m distance from road (Table 1, Mykhailova *et al.* 2013).

SEM-EDX revealed carbon-rich plaques on the surface of the sediment, which were more pronounced in 1 m distance from the road compared to 8 m. The XRF C signal (red) did not coincide with the Ca signal (blue), indicating the organic nature of the plaques instead of carbonates (Fig. 3). This observation confirms the findings of snow meltwater extraction that TPH deposition decreases with distance (Fig. 2). Stewart *et al.* (2013) suggested that available heavy hydrocarbon contaminants tended to sorb onto sediment as snow melts rather than partitioning into the aqueous phase. Consequently, hydrocarbons do not need to be adsorbed primarily to sediment, because dissolved or dispersed in splash water or vehicle spray hydrocarbons accumulate in the solid phase while water infiltrates into the soil (this process was simulated in our experiment by membrane filtration).

FTIR microscopy (Fig. 4) revealed a wider CH/C = O ratio for the dark-spotted areas, indicating accumulation of compounds poor in carboxyl, carbonyl or ester bonds. Absence of narrow bands in the 3000–3100 cm^{-1} range further indicates the aliphatic nature of the respective compounds (Figs 5 and 6).

Diesel fuel typically contains large amounts of n-alkanes, which form a characteristic pattern in chromatograms (Fig. 7a) and which significantly exceed the signal intensity of an unresolved complex mixture (UCM). Contrarily, components of high-boiling oils or tar cannot be resolved sufficiently in one-dimensional GC and form a typical UCM hump (Wang *et al.* 2012). Thus, the appearance of diesel-like n-alkane patterns in roadside soil extracts points to incompletely combusted high-boiling components of fuel as the origin of soil hydrocarbons, which clearly can be distinguished in all soils regardless of the distance from the roads (Fig. 7b–e), and, hence, translocate over larger distances. It is unlikely that the series of n-alkanes found in soil extracts originates from biomass, because fatty acid biosynthesis commonly generates even-numbered alkyl chains by subsequently adding C2-units deriving from acetyl CoA. In contrast, mineral oil contains series of both even- and odd-numbered n-alkanes (Fig. 7a–d).

UCM predominantly found in near-road snow sediment also dominated in extracts of soils sampled within a 10 m range, whereas background soils contained smaller proportions of UCM compounds (Fig. 7f). It should be considered here that laboratory-produced meltwater samples typically suffer from heating during the melting

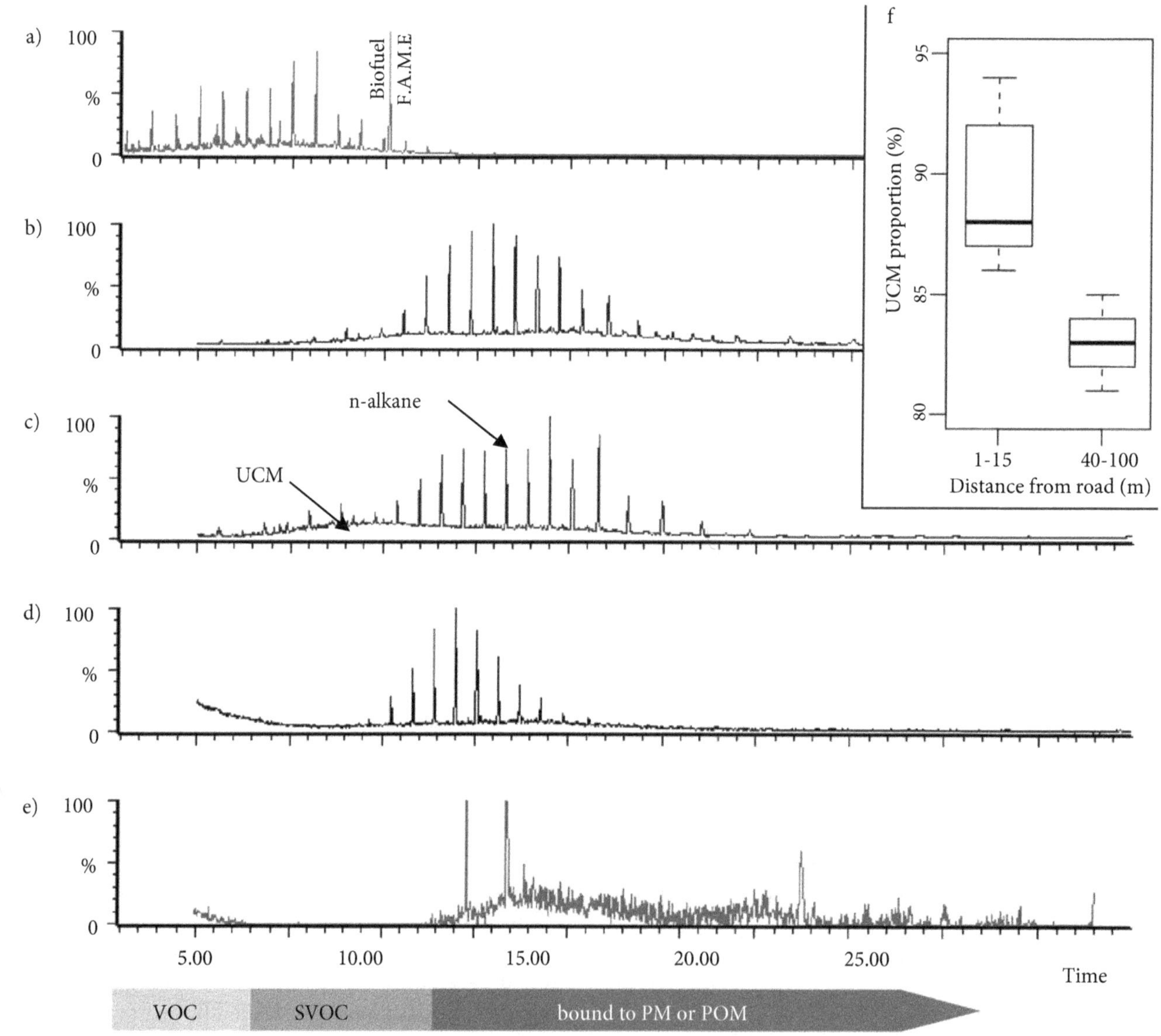

Fig. 7. GC-MS chromatograms of: (a) commercial diesel fuel, containing 5% fatty acid methyl esters (F.A.M.E.), purified hexane extract of roadside soils, distance; (b) 1 m ; (c) 40 m; and (d) 100 m from Pushkinskaya street; (e) purified hexane filter extract; (f) boxplot of the percentage of unresolved complex mixture (UCM) compounds from total, distance classes 1–15 and 40–100 m, sampling depth 0–0.5 cm

process; thus most volatile compounds likely volatilised prior or during this experiment (Stewart *et al.* 2013). However, our hypothesis of predominance of UCM meltwater sediment compared to soil is supported by the fact that roadside soils are subjected to higher temperatures: Soils would contain more light hydrocarbons if they would not heat up during summer, which is not possible for snow sediment prepared at a maximal temperature of 4 °C.

High-boiling organic tailpipe emissions condense in the air forming secondary organic aerosol (SOA, "micro tarballs"). Gentner *et al.* (2012) reported that chainlengths in SOA formed from diesel car emissions range from C6 to >C25 and peak at around C20, whereas our results indicate a shifted range towards longer chains for soil extracts, peaking at C25–C27 (Fig. 7b–e). Car emissions to the air, mainly incompletely burnt gases and dust particles, are deposited over larger distances from the road (Schipper *et al.* 2007). Due to elevated TPH concentrations within a 10 m strip (Table 1), due to elevated hydrocarbon deposition rates close to the roads (Fig. 2), and due to the depositional patterns of vehicle spray reported by Schipper *et al.* (2007), we hypothesize that dry deposition of hydrocarbons bound to mineral soil particles, or wet deposition in conjunction with splashwater or vehicle spray would be the preferable pathways of hydrocarbon deposition on roadside soil. In contrast, direct adsorption of VOCs from the gas phase onto the soil surface and dry deposition of hydrocarbon carrying (fine) aerosol, including secondary organic aerosol (SOA) formed from tailpipe emissions, are considered to be less significant for elevated roadside TPH concentrations, because these fractions are known to be subjected to long-distance transport. While preferred sedimentation of these fractions within a couple of meters from their emission source cannot be expected, they well may contribute to elevated large-scale background levels. Further, typical VOCs (like low boiling alkanes, cycloalkanes, benzene, toluene, xylenes, napthalenes, etc. Gentner *et al.* 2012; Sheng *et al.* 2006) could not be extracted from any of the soils of this study. Due to their high vapour pressure typical VOCs likely prevail in the gas phase and adsorb little to the soil surface.

The decomposition of n-alkanes in soil is preferred compared to the decomposition of UCM hump forming iso-alkanes (Reddy *et al.* 2002). Hence, the presence of n-alkane patterns in soil extracts after decades of road exploitation indicates the establishment of a dynamic equilibrium state, pointing to ongoing deposition of SVOC hydrocarbons into, as well as their ongoing decay in soils.

Conclusions

Summarizing the FTIR findings that snow sediment carries aliphatic substances poor in C = O, the GC-MS findings that predominantly UCM compounds could be extracted from this sediment, the spatial TPH depositional patterns on snow and soil, as well as the SEM finding that high-boiling petroleum hydrocarbons like engine oils or tar are associated with mineral particles belonging to the fine sand and silt fraction, it can be concluded that high boiling UCM hydrocarbons preferably deposit within a 10 m strip along roads. Under wet road conditions particles carrying hydrocarbons are likely bound to or washed out by splash water or vehicle spray, resulting in accumulation of UCM compounds near the road, whereas n-alkanes prevail at larger distances. A dynamic equilibrium between deposition, degradation and leaching of hydrocarbon fractions can be proposed. Further investigation of this equilibrium promises to utilize natural hydrocarbon degradation for the reclamation of contaminated sites. Elevated depositional hydrocarbon flow corresponded to long-term TPH accumulation in roadside soils.

Acknowledgments

We thank Wolfgang Wiehe (Brandenburg University of Technology Cottbus-Senftenberg, Central Analytical Laboratory) for technical assistance with SEM-EDX and Wolfgang Wieprecht (Brandenburg University of Technology Cottbus-Senftenberg, Chair of Atmospheric Chemistry and Air Pollution Prevention) for fruitful discussions. The authors are grateful to the Ministry of Education and Science of the Ukraine and to the German Academic Exchange Service (DAAD) for financial support.

References

Adam, G.; Duncan, H. 2002. Influence of diesel fuel on seed germination, *Environmental Pollution* 120(2): 363–370. http://dx.doi.org/10.1016/S0269-7491(02)00119-7

Adeniyi, A. A; Owoade, O. J. 2010. Total petroleum hydrocarbons and trace heavy metals roadside soils along the Lagos–Badagry expressway, Nigeria, *Environmental Monitoring and Assessment* 167(1–4): 625–630. http://dx.doi.org/10.1007/s10661-009-1078-3

Albers, P. H. 1995. Petroleum and individual polycyclic aromatic hydrocarbons, in Haffman, D. T.; Rattner, B. A.; Burton, G. A.; Cairns, J. (Eds.). *Handbook of ecotoxicology*. London: Lewis, 330–335.

AG Boden. 2005. *Bodenkundliche Kartieranleitung, 5. Verbesserte und erweiterte Auflage*. Schweizerbarthsche Verlagsbuchhandlung, Stuttgart, 142–147.

Alloway, B. J. 1992. Land contamination and reclamation, in Harrison, R. M. (Ed.). *Understanding our environment. An introduction to environmental chemistry and pollution*. Cambridge: Royal Society of Chemistry, 144–163.

Amadi, A.; Abbey, S. D.; Nma, A. 1996. Chronic effects of oil spill on soil properties and microflora of a rainforest ecosystem Nigeria, *Water, Air, and Soil Pollution* 86(1–4): 1–11. http://dx.doi.org/10.1007/BF00279142

Biasioli, M.; Ajmone-Marsan, F. 2007. Organic and inorganic diffuse contamination urban soils: the case of Torino (Italy), *Journal of Environmental Monitoring* 9(8): 862–868. http://dx.doi.org/10.1039/B705285E

Dilly, O.; Nii-Annang, S.; Franke, G.; Fischer, T.; Buegger, F.; Zyakun, A. 2011. Resilience of microbial respiration, respiratory quotient and stable isotope characteristics to soil hydrocarbon addition, *Soil Biology & Biochemistry* 43(9): 1809–1812. http://dx.doi.org/10.1016/j.soilbio.2010.09.026

DIN 38409-56. 2009. *German standard methods for the examination of water, waste water and sludge – Parameters characterizing effects and substances (group H) – Part 56: Gravimetric determination of low volatile lipophilic substances after solvent extraction (H 56).* 11 p.

DWW. 2003. *The POLMIT Handbook, A practical guideline on causes and remedies of the pollution of groundwater and soil by road and traffic sources* [online], [cited 05 August 2013]. Available from Internet: http://de.scribd.com/doc/78999080/The-POLMIT-handbook-a-practical-guideline-on-causes-and-remedies-of-the-pollution-of-groundwater-and-soil-by-road-and-traffic-sources

EN ISO 9377-2: 2001. *Water quality – Determination of hydrocarbon oil index – Part 2: Method using solvent extraction and gas chromatography.* 22 p.

Gentner, D. R.; Isaacman, G.; Worton, D. R.; Chan, A. W. H.; Dallmann, T. R.; Davis, L.; Liud, S.; Dayd, D. A.; Russel, L. M.; Wilson, K. R.; Weber, R.; Guha, A.; Harley, R. A.; Goldstein, A. H. 2012. Elucidating secondary organic aerosol from diesel and gasoline vehicles through detailed characterization of organic carbon emissions, *PNAS* 109(45): 18318–18323. http://dx.doi.org/10.1073/iti4512109

Gromke, C.; Ruck, B. 2009. On the impact of trees on dispersion processes of traffic emissions street canyons, *Boundary-Layer Meteorology* 131(1): 19–34.
http://dx.doi.org/10.1007/s10546-008-9301-2

Iurchenko, V.; Mykhailova, L.; Fischer, T. 2013. Kinetic characteristics of petrochemicals transposition and accumulation soils of roadside area, in *Transbaltica 2013 Conference Proceedings,* 09–10 May, 2013, Vilnius, Lithuania.
http://dx.doi.org/10.3846/transbaltica2013.016

ISO 10381-2: 2003. *Soil quality - Sampling - Part 2: Guidance on sampling techniques.* 23 p.

Manahan, E. S. 1993. *Fundamentals of environmental chemistry.* 4th ed. Chelsea: Lewis, 183–199.

Masakorala, K.; Yao, J.; Guo, H.; Chandankere, R.; Wang, J.; Cai, M.; Liu, H.; Choi, M. M. F. 2013. Phytotoxicity of long-term total petroleum hydrocarbon-contaminated soil – a comparative and combined approach, *Water, Air & Soil Pollution* 224: 1553. http://dx.doi.org/10.1007/s11270-013-1553-x

Mykhailova, L.; Fischer, T.; Iurchenko, V. 2013. Distribution and fractional composition of petroleum hydrocarbons roadside soils, *Applied and Environmental Soil Science* 2013: 6 p. http://dx.doi.org/10.1155/2013/938703

Okonkwo, J. O.; Awofolu, O. R.; Moja, S. J.; Forbes, P. C. B.; Senwo, Z. N. 2006. Total petroleum hydrocarbons and trace metals street dusts from Tshwane metropolitan area, South Africa, *Journal of Environmental Science and Health, Part A* 41(12): 2789–2798. http://dx.doi.org/10.1080/10934520600966920

Reddy, C. M.; Eglinton, T. I.; Hounshell, A.; White, H. K.; Xu, L.; Gaines, R. B.; Frysinger, G. S. 2002. The West Falmouth oil spill after thirty years: the persistence of petroleum hydrocarbons marsh sediments, *Environmental Science & Technology* 36(22): 4754–4760. http://dx.doi.org/10.1021/es020656n

Sheng, G. J.; Goldberg, J. M.; Harigaya, R.; Ty, J. C. 2006. GS-MS determination of volatile organic compounds gasoline and diesel emissions, *Dartmouth Undergraduate Journal of Science* Spring issue: 47–53.

Schipper, P. N. M.; Comans, R. N. J.; Dijkstra, J. J.; Vergouwen, L. 2007. Runoff and windblown vehicle spray from road surfaces, risks and measures for soil and water, *Water Science and Technology* 55(3): 87–96.
http://dx.doi.org/10.2166/wst.2007.076

Stewart, M.; Holden, A.; Haveroen, M.; Ulrich, A. 2013. Quantification of hydrocarbon contaminants meltwater and sediment a city snow pile, *Journal of Environmental Engineering* 139(2): 295–301.
http://dx.doi.org/10.1061/(ASCE)EE.1943-7870.0000606

UNECE. 2008. *United Nations Economic Commission for Europe Transport Division Database* [online], [cited 08 August 2013]. Available from Internet: http://w3.unece.org/pxweb/database/STAT/40-TRTRANS/02-TRRoadFleet/?lang=1

Unger, H.-J.; Prinz, D. 1992. *Verkehrsbedingte Immissionen Baden-Württemberg – Schwermetalle und organische Fremdstoffe straßennahen Böden und Aufwuchs. Ministerium für Umwelt Baden-Württenberg,* Luft, Boden, Abfall, Heft 19.

Wang, Z.; Yang, C.; Yang, Z.; Hollebone, B.; Brown, C. E.; Landriault, M.; Sun, S.; Mudge, S. M.; Kelly-Hoopere, F.; Dixon, D. G. 2012. Fingerprinting of petroleum hydrocarbons (PHC) and other biogenic organic compounds (BOC) oil-contaminated and background soil samples, *Journal of Environmental Monitoring* 14(9): 2367–2381.
http://dx.doi.org/10.1039/C2EM30339F

Zyakun, A.; Nii-Annang, S.; Franke, G.; Fischer, T.; Buegger, F.; Dilly, O. 2011. Microbial activity and 13C/12C ratio as evidence of N-Hexadecane and N-Hexadecanoic acid biodegradation agricultural and forest soils, *Geomicrobiology Journal* 28(7): 632–647.
http://dx.doi.org/10.1080/01490451.2010.489922

Larysa MYKHAILOVA is a PhD student of the Chair of Ecology of the Kharkiv National Automobile and Highway University.

Thomas FISCHER is a Doctor of Natural Sciences, Soil Scientist focussing on ecohydrology, microbial ecology, initial ecosystem development, as well as analytical and forensic chemistry. He is the Director of the Central Analytical Laboratory of the Brandenburg University of Technology Cottbus-Senftenberg.

Valentina IURCHENKO is a Doctor of Technical Sciences, Microbiologist, Hydro-Biologist and Ecologist, Professor of the Chair of Ecology of the Kharkiv National Automobile and Highway University, Head of the Chair of Life Safety & Environmental Engineering of the Kharkiv National University of Construction and Architecture.

12

THE EFFECT OF BIOCHAR ON PLANT DISEASES: WHAT SHOULD WE LEARN WHILE DESIGNING BIOCHAR SUBSTRATES?

Omer FRENKEL[a], Amit K. JAISWAL[a, b], Yigal ELAD[a], Beni LEW[d], Claudia KAMMANN[c], Ellen R. GRABER[d]

[a]*Department of Plant Pathology, The Volcani Center - Agricultural Research Organization, Rishon Lezion, Israel*

[b]*Department of Plant Pathology and Microbiology, Faculty of Agriculture, Food and Environment, The Hebrew University of Jerusalem, Rehovot, Israel*

[c]*Hochschule Geisenheim University, Geisenheim, Germany*

[d]*Soil Chemistry, Plant Nutrition and Microbiology, The Volcani Center - Agricultural Research Organization, Rishon Lezion, Israel*

Abstract. The increasing demand for soilless substrates and rising environmental concerns about the use of non-renewable resources such as peat has led to the search for alternative constituents of growing mixtures for containerized plants. In this report we reviewed the works concerning biochar as constituent of growing media, targeting its influence on plant growth and plant disease. Biochar mostly has positive or neutral influences on plant growth compared with peat media when present in concentrations higher than 25% (v:v). However, studies of biochar influence on plant disease reveals that while lower concentrations (≤1%) of biochar often suppressed several diseases, higher concentrations (≥3%) were mostly ineffective or induced plant disease. For use as horticultural peat replacement, it is recommended that biochar feedstocks and concentrations be standardized and the potential effect of biochar on plant disease be considered, so that growers can rely on consistent and reproducible biochars for desired effects.

Keywords: *Clavibacter michiganensis*, damping-off, disease control, foliar pathogen, hormesis effect, organic amendments, plant nurseries, peat replacement, root rot, soilborne pathogens.

Introduction

Research and development into biochar, the application of charcoal to soil, has been increasing substantially over the last decade. The major driver of interest is biochar's longevity in the soil, with many biochar types having estimated half-lives of 100s to 1000s of years or more. This means that by adding biochar to soil, carbon that originated in the atmosphere as CO_2 can be sequestered, leading to a desirable carbon removal from the atmosphere. In addition to this benefit, an increasing number of studies point to positive influences of biochar addition to soil, such as increased cation exchange capacity in organic matter-poor soils (Silber *et al.* 2010; Cornelissen *et al.* 2013; Obia *et al.* 2015), increased nutrient and water retention in light soils (Glaser *et al.* 2002; Haider *et al.* 2016), improvement in pH of acidic soils (Yamato *et al.* 2006; Chan *et al.* 2007), enrichment in beneficial soil microorganisms (Kolb *et al.* 2009; Kolton *et al.* 2011), and reductions in greenhouse gas emissions (Kammann et al., this issue). Several pot and field trials have shown that adding different biochars to soils can enhance productivity and performance of crops such as wheat, maize, cucumber, bean, tomato, strawberry, and sweet pepper (Graber *et al.* 2010; Meller Harel *et al.* 2012; Cornelissen *et al.* 2013; Jaiswal *et al.* 2014; Jaiswal *et al.* 2015; De Tender *et al.* 2016), including when the biochar is used as a fertilizer carrier (Joseph *et al.* 2013) or has been previously loaded with nutrients e.g. by co-composting (Kammann *et al.* 2015). In contrast, a number

Corresponding author: Omer Frenkel
E-mail: omerf@volcani.agri.gov.il

of studies have shown no effect of biochar on crop yields (Jeffery *et al.* 2011; Ruysschaert *et al.* 2016), and some have also shown that biochar additions to soil can have unwanted effects on crop productivity, often due to tying up of nitrogen (Kammann *et al.* 2015; Haider *et al.* 2016), particularly in clay-rich fertile soils. Other works have pointed to biochar inactivation of soil-applied pesticides (Graber, Kookana 2015) and to soil contamination arising from biochar addition. Seeing that biochar has such a long half-life and that cannot be removed once it has been applied, it seems prudent to take small measured steps in its deployment, before widespread use.

A convenient "pre-soil" stage and research playground may be the addition of biochar to detached soilless growth media for partial peat replacement or peat-free growth media improvement. This idea has a number of advantages over the approach of widespread application to field soils, particularly in the initial stages:

1. Horticultural media can be used more easily as a model system for studying different aspects of biochar types and doses, including very general phenomena but also for fine tuning;
2. Horticultural crops are generally high return crops, which can justify investment in biochar more readily than extensive, low return field crops;
3. It is low risk and although not always strictly mimicking field studies, it still may assist us to detect some general phenomena and to outline the biochar positive borders. In addition, if there are any unexpected negative impacts of the added biochar on plant performance and health, no permanent damage to non-renewable soil resources has been done;
4. Usually, the culture turnover time is faster than the field and can comprise several cultures per year, speeding up the gain of knowledge and experience;
5. Economically successful implementation of biochar in horticultural detached media can have a positive impact on both industries;
6. The pyrolysis/biochar platform may be a welcome green solution for local nursery wastes;
7. Successful implementation of biochar in nursery media will encourage more extensive use of biochar in the future, and its benefits will be more easily quantified after gaining valuable experience in the nursery setting;
8. Peat deposits, which are non-renewable resources, are over-exploited. The growth media industry and various R&D initiatives have been trying to replace peat for a couple of decades, with limited success. Biochar as a partial peat replacement, or tool for improving the performance of peat-free alternative growth media, may help achieve this goal.

This last point, biochar as a potential peat replacement, has been examined in detail in a companion paper in this volume (Kern et al., this issue) and will not be examined here. The emphasis of the current communication is the effect of biochar additions to detached soilless media and the unrecognized influence it may have on plant susceptibility to diseases caused by plant pathogens, as plant protection is an economically very important aspect of modern horticulture.

1. Current state of the art

Biochar as a component of soilless substrates has been tested in several experimental systems; data are summarized in Table 1. Studies were conducted with various types of biochars, and several works involved combinations of biochar with other supplements such as mycorrhiza and fertilizers (Conversa *et al.* 2015), and humic acid products

Table 1. List of studies that involve biochar as a peat replacement in soilless media. Data include the range of concentrations in which biochar was found to be beneficial or non-harmful to plant growth are presented in the 2nd column. Concentrations at which negative effects on plant growth were detected are presented in the 3rd column

Crop	Beneficial impact (% biochar)	Negative impact (% biochar)	Reference
Blanket flower (*Gaillardia* spp.)	25	50	Dumroese *et al.* (2011)
Calathea (*Calathea insignis*)	20–35		Zhang *et al.* (2014)
Calathea (*Calathea rotundifola*)	50	>50	Tian *et al.* (2012)
Crown of Thorns (*Euphorbia x Lomi*)	15–60		Fascella (2015)
Horse-shoe pelargonium (*Pelargonium zonale*)	30	70	Conversa *et al.* (2015)
Kale (*Brassica oleracea L.* var. *acephala*)	1–5		Kim *et al.* (2016)
Lettuce (*Lactuca sativa*)	38–70		Mendez *et al.* (2015)
Lettuce (*L. sativa*)	50–75		Nieto *et al.* (2016)
Sunflower (*Helianthus annuus*)	25–75	100	Steiner and Harttung (2014)
Tomato (*Solanum lycopersicum*) & Marigold (*Tagetes erecta*)	5		Vaughn *et al.* (2013)

(Zhang *et al.* 2014). The studies tested elevated ratios of biochar: growth media (mostly peat) occasionally reaching very high biochar percentages (>60%) (Steiner, Harttung 2014; Mendez *et al.* 2015; Nieto *et al.* 2016). Analyses included chemical properties and various parameters of plant growth and other measurements such as chlorophyll (Fascella 2015). In most cases, biochar had a neutral or positive influence on plant growth compared with peat media when present in concentrations lower than 30% (v:v) (Dumroese *et al.* 2011; Conversa *et al.* 2015), and in some works even a much higher concentration was found to be not harmful (Fascella 2015; Mendez *et al.* 2015; Nieto *et al.* 2016). However, all of these studies focused on plant survival and several growth parameters, and did not consider the influence on plant diseases.

A positive influence of biochar on reducing plant diseases such as rust in wheat and mildew in other crops was first reported some 170 years ago (Allen 1847) and drew attention in the last decade where several pathosystems were studied by different groups worldwide (Elad *et al.* 2010; Elmer, Pignatello 2011; Jaiswal *et al.* 2014; Copley *et al.* 2015; Jaiswal *et al.* 2015). Pathosystems included both foliar pathogens and soil borne pathogens (Elad *et al.* 2011; Graber *et al.* 2014a). Lately, Bonanomi *et al.* (2015) reviewed and summarized the data from 13 pathosystems that tested the effect of biochar on plant disease. In their analysis, they reported that 85% of the studies showed a positive influence of biochar in reducing plant disease severity, 12% had no effect, and only 3% showed that biochar additions were conducive to plant disease. However, their analysis did not consider the fact that many of these studies revealed that plant susceptibility/resistance to disease was dependent on the crucial factor of the biochar dose.

In Table 2, we inspected the systems reviewed in Bonanomi *et al.* (2015) and additional newer studies. We summarized the data from 15 pathogens (fungi, oomycetes and nematodes) in 30 different pathosystems (i.e. plant/pathogen system). We then compared the effect of the highest tested biochar concentration on disease severity to the effect of the unamended control and also to the biochar concentration which had the most positive effect on disease suppression. The data show that for 60% of the pathogens and in 70% of the pathosystems, the highest tested biochar concentrations (ranging in most cases between 0.5 to 5%) had neutral to negative effects on plant disease, compared with the control or with the maximally effective biochar concentration, thus demonstrating the start of an inverted U-shaped biochar dose/response curve (that may also be described as negative quadratic relationship curve) in the majority of systems (Jaiswal *et al.* 2014, 2015). Moreover, several studies on biochar in detached growing media reported that while relatively low concentrations (≤1%) of biochar suppressed the diseases (Jaiswal *et al.* 2014, 2015; Huang *et al.* 2015), higher concentrations (3%) were mostly ineffective or even accelerated plant disease (Jaiswal *et al.* 2014, 2015; Copley *et al.* 2015; Huang *et al.* 2015). These studies currently encompass 12 different hosts and 5 different biochars including biochars with very different characters in terms of mineral content, alkalinity, and carbon content, suggesting that most, if not all, plant disease responses would eventually show such inverted U-shaped dose/response relationships if biochar dose would be further increased.

Interestingly, this effect was less obvious in foliar pathogens where only 2 out of 6 pathosystems (33%) demonstrated potential U-shaped relationships (Elad *et al.* 2010, 2011; Meller Harel *et al.* 2012) but was very common in the soilborne pathogen pathosystems (20 out of 24 pathosystems, 83%). As the experiments with the foliar pathogens showed milder susceptibility response (and for only one pathogen (B. cinerea)) and were not tested at concentrations relevant to partial peat replacements (>10%), a nursery tray experiment was conducted with the foliar bacterial pathogen *Clavibacter michiganesis* subsp *michiganesis*, the causal agent of bacterial canker of tomato plants. The following experiment was conducted twice: tomato seeds (Cv. *Fantasia*) were grown in a commercial nursery growing mixture with added biochar concentrations (0, 3, 6 and 18% w/w). Biochar was produced from greenhouse wastes of pepper plants (chemical and physical property of the biochar are described by Graber *et al.* (2014b)). Seedlings were grown under commercial condition with foliar fertigation. Twenty-one days after sowing, seedlings at the second leaf stage were transferred from the nursery to the laboratory and sprayed with 10^8/ml of the bacterial pathogen *C. michiganensis* and placed in 25 °C growth chambers. After 8 days, disease symptoms appeared as white blisters on the leaves and disease severity was evaluated as the number of blisters on each plant which has been shown to be a reliable disease severity assessment method (Frenkel *et al.* 2016). While no differences were detected in plant biomass, the disease symptoms in the unamended control was 8.6 lesions/plants, and were not significantly lower at 3% and 6% biochar (6.4 and 9.4 lesions/plant). However, the disease symptoms were significantly more severe at a higher biochar concentration of 18% when compared with the unamended control (15.1 *versus* 8.6 lesions/plants, respectively; Fig. 1). In this experiment, not only did the biochar have no positive effect at any tested concentration, but it had a negative effect at the highest one, demonstrating additional evidence for a U shape dose response for foliar pathogens and strengthening the point that high concentrations of biochar in growth media might be problematic also in the presence of a foliar pathogen.

Table 2. List of the studies that tested biochar in growth media on soilborne and foliar plant diseases. The last column gives the biochar concentration which exhibits negative influence on plant health compared with the control (no biochar) or compared with the biochar concentration having the most positive effect on disease suppression treatment, therefore representing the phenomena of the U-shape dose response

Pathogen	Host plant	Feedstock of biochar	Reference	Biochar % needs attention
Botrytis cinerea (F)[1]	*Lycopersicum esculatum*	Citrus wood (3, 5%) Eucalyptus wood (0.5, 1, 3%) Olive pomice (0.5, 1, 3%) Greenhouse waste (0.5, 1, 3%)	Elad *et al.* (2010) Elad *et al.* (2011) Mehari *et al.* (2015)	N[2] 3% N
	Capsicum annum	Greenhouse waste (0.5, 1, 3%)	Elad *et al.* (2010)	N
	Fragia x ananassa	Citrus wood, Greenhouse waste (3, 5%) Holm oak (1, 3%)	Meller Harel *et al.* (2012) De Tender *et al.* (2016)	N N
Colletotrichum acutatum (F)	*Fragia x ananassa*	Greenhouse (1, 3%)	Meller Harel *et al.* (2012)	N
Leveilulla taurica (F)	*C. annum*	Citrus wood (1, 3%)	Elad *et al.* (2010)	3%
Podosphora aphanis (F)	*Fragia x ananassa*	Citrus wood, Greenhouse waste (1, 3%)	Meller Harel *et al.* (2012)	N
Fusarium oxysporum f.sp asparagi (S)	*Asparagus sp.*	Commercial Quest biochar (0.5, 1.5, 3%)	Elmer and Pignatello (2011)	N
		Coconut charcoal (10, 30%)	Matsubara *et al.* (2002)	N
Fusarium oxysporum f.sp lycopersici (S)	*L. esculatum*	Wood & Green waste biochar (3% v/v)	Akhter *et al.* (2015) Akhter *et al.* (2016)	3% (v/v)[3]
Meloidogyne graminicola (Root knot nematodes; (S)	*Oryza sativa*	Oak wood (0.6, 1.2, 2.5, 5%)	Huang *et al.* (2015)	5%
Phytophthora cactorum (S)	*Acer rubrum*	Wood (0.5, 10, 20%)	Zwart and Kim (2012)	>5%
Phytophthora cinnanomi (S)	*Quercus rubra*	Wood (0.5, 10, 20%)	Zwart and Kim (2012)	>5%
Pythium ultimum (S)	*C. annum*, *Ocimum basilicum*	Spruce bark (50% v/v)	Gravel *et al.* (2013)	50% (v/v)
Plasmodiophora brassica (S)	*Brassica rapa*	Miscanthus (0.4, 0.8%)	Knox *et al.* (2015)	0.5%
Pratylenchus penetrans (root lesion nematodes; (S))	*Daucus carota*	Pine wood (0.8%) Pine bark (0.92) Wood (1.24%) Spelt husks (0.64%)	George *et al.* (2016)	0.8%
Ralstonia solanacearum (S)	*L. esculatum*	Municipal waste (0, 20%) Peanuts shells (2%)	Nerome *et al.* (2005) Lu *et al.* (2016)	N N
Replant disease (S)	*Prunus persica*	Wood 10, 20% (v/v)	Atucha and Litus (2015)	N
Rhizoctonia solani (S)	*Cucumis sativus* *Phaseolus vulgaris*	Greenhouse waste and Eucalyptus wood (0.5, 1, 3%)	Jaiswal *et al.* (2014) Jaiswal *et al.* (2015)	1% 3%
	Glycine max, *Pisum sativum*, *Beta vulgaris*, *Medicago sativa*, *C. annum*, *L. esculatum*, *C. sativus*, *Raphanus sativus*, *D. carota*, *Allium ampeloprasum*	Maple bark (1, 3, 5%)	Copley *et al.* (2015)	3%

[1] The letter F relates to foliar pathogens and the letter S to soil borne pathogens.
[2] The letter N represents studies where no significant negative effect was detected in any concentration compare with the control or with the biochar concentration with the most positive effect on disease control treatment.
[3] Concentration of biochar was determined as wt:wt unless otherwise stated as v:v.

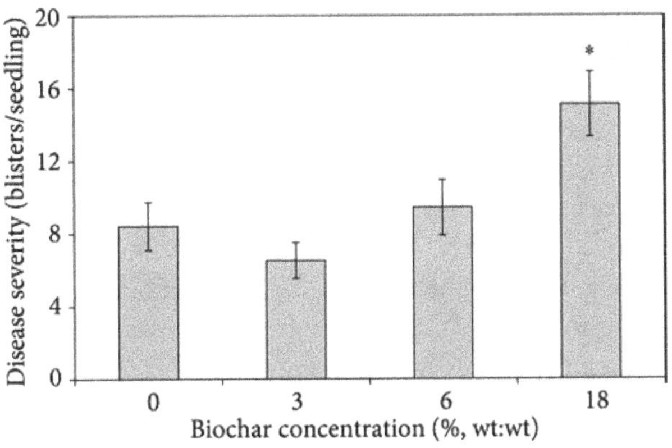

Fig. 1. The influence of increasing greenhouse waste biochar concentrations (w/w) on the disease severity caused by *Clavibacter michiganensis* subsp. *michiganensis* on tomato seedlings. Tomato cv. *Fantasia* was grown from seeds in commercial nursery. 17 days after sawing, seedlings were inoculated until runoff with 10^8 CFU/ml. Disease severities as white blisters on leaves were counted on each plant 8 days after inoculation. Vertical lines = Standard error. Asterisk represents treatment which is significantly different from the rest of the treatments by using Tukey HSD test ($\alpha = 0.05$)

2. The shifted dose response between plant growth and disease response

Jaiswal *et al.* (2015) noted that not only is plant resistance to disease frequently biochar-concentration-dependent, but also that in many instances, an unwanted effect on plant growth and biomass occurred at relatively higher biochar doses (Rondon *et al.* 2007; Rajkovich *et al.* 2012; Spokas *et al.* 2012). Moreover, even when high biochar content has positive impacts on plant growth, there will be potentially irreversible damage if plant disease emerges and spreads. It has been noted that biochar dose-response curves for plant growth and plant disease are frequently shifted relative to each other along the biochar dose axis, with diseased plants being much more sensitive to higher biochar doses than healthy plants. Jaiswal *et al.* (2015) described this phenomenon while testing the impacts of biochars produced from greenhouse wastes and from eucalyptus wood chips on beans and cucumbers infected with the soilborne pathogen *R. solani.* In that study, biochar concentrations up to 3% by weight resulted in higher plant biomass in healthy plants, but in diseased plants, the biochar was beneficial only up to 1 wt %. Jaiswal *et al.* (2015) termed this phenomenon the "Shifted R_{max}-Effect", where R_{max} refers to the biochar dose at which there is a maximum growth response (G–R_{max}) and maximum disease reduction (D–R_{max}). This finding was later corroborated in additional reports. Akhter *et al.* (2015) described a similar phenomenon using 3% green waste biochar, wherein tomato root and shoot length were both increased in disease-free growing medium while the same biochar concentration was found to be conducive to diseases caused by the pathogen *F. oxysporum lycopersici.*

We postulate that the "Shifted R_{max}-Effect" underlies the reports (Viger *et al.* 2014) of positive effects of high biochar doses (4.2 and 8.4%) on Arabidopsis growth and concomitant down-regulation of defense-related genes. Viger *et al.* (2014) found that when *Arabidopsis thaliana* plants were subjected to increasing amounts of biochar mixed with soil, the leaf area, rosette diameter and root length were stimulated between 50 to 150 percent. By checking the response of more than 10,000 genes of *A. thaliana*, they suggested that up-regulation of genes involved in the brassinosteroids and auxin pathways, both known as growth-promoting hormones and signaling molecules, was a key to the growth stimulation observed in biochar treatments. However, the positive impacts of biochar were coupled with negative findings for a suite of genes that are known to determine the ability of a plant to withstand attack from pests and pathogens. These defense-related genes were consistently down-regulated following biochar application to the soil, including genes that regulate jasmonic acid, salicylic acid and ethylene pathways. The shifted R_{max} paradigm also could resolve apparent contradictions between Viger *et al.* (2014) and an earlier report by Meller Harel *et al.* (2012), which showed improved strawberry plant growth, decreased disease susceptibility, and up-regulation of defense-related genes in systems at relatively lower biochar doses (1 and 3 wt %; (Meller Harel *et al.* 2012).

These examples may all be related to wide range of materials that stimulate positive effects at low concentrations but cause toxicity and inhibition at higher doses (hormesis effect) (Jaiswal *et al.* 2014; Kammann, Graber 2015). As an example, low doses of an herbicide such as glyphosate might be used to beneficially modulate plant growth, development, or composition. At higher concentrations, glyphosate is a highly effective herbicide (Kortekamp 2011). Biochars can contain a plethora of small and large organic molecules that may individually or in combination have hormone-like or phytotoxic activities that are dose dependent. One example is the emission of ethylene from some biochars (Spokas *et al.* 2010). The plant hormone ethylene plays a significant role at low doses both as a plant growth promotor and in promoting plant defenses against various stressors. At higher concentrations, its influence on those aspects is negative (Kammann, Graber 2015). Different dose/response curves to ethylene concentrations for plant growth and plant defense could potentially be one of the mechanisms that explain the shifted R_{max} effect. The hormesis effect might also be related to macro and micro nutrients that are well known to influence disease severity along with plant growth, but are not necessarily correlated to the same positive ranges. The hormesis effect on plant disease is still widely unknown and needs to be further investigated. Hormesis might be caused by organic compounds that are abundant in some biochars, as a function of the biochar feedstock and preparation temperatures (Spokas *et al.* 2012). Some compounds may cause minor stresses at low concentrations and hence prime systemic resistance defenses, but at higher concentrations may be toxic or directly weaken the root and collar region of the plant. Such stresses and minor wounding may be the weak point that provides soil borne pathogens with favorable conditions to successfully attack and penetrate the mechanical defense layer of the root surface (Jaiswal *et al.* 2014, 2015). This possibility should be tested as one hypothesis for why foliar pathogens seem to be less sensitive to higher biochar concentrations than soilborne pathogens.

3. The challenge of the U shape effect: how to use biochar in soilless media

Considering the frequent U-shaped biochar dose/response curves, we suggest that caution should be taken before using biochar as an extensive peat replacement at high concentrations. We suggest that at the current stage, biochar be considered a supplemental additive at lower doses, where in many cases it improves plant performance against pathogens. This approach is at odds with the idea that biochar can make up a substantial proportion of the soilless media and serve to replace peat on a large scale.

While it is true that plant nurseries take strict sanitation and control measures and therefore disease risks may be considered low, there are several bacterial and fungal diseases that are highly problematic during the nursery stage, including *Xanthomonas campestris* (Krauthausen *et al.* 2011); *C. michiganesis* (Frenkel *et al.* 2016) and oomycetes pathogens such as *Pythium aphanidermatum* (Hendrix, Campbell 1973). Moreover, biochar positive or negative effects on plant responses may be long-lived, as suggested by Elmer and Pignatello (2011) and Elmer (2016), who reported a decrease in asparagus plant size in the second year of their experiment.

Moreover, as previously found (Jaiswal *et al.* 2014, 2015), different biochars have different optimum application rates for disease suppression. There is not yet any "one concentration fits all" paradigm that can be adopted for biochar application in soilless culture. This requires more studies of individual crop-pathogen-biochar systems. More importantly, it is essential to understand the mechanisms that are involved in disease suppression and promotion under biochar addition, in order to develop methods that will encompass a range of optimum application protocols. Some or all of the following mechanisms may be operating: i) induced resistance; ii) alteration of beneficial microbial communities; iii) nutrient content and supply; and iv) hormesis effects of biochar-derived phytohormones. In the meanwhile, care should be taken to standardize biochar feedstocks, production conditions, and concentrations, so that growers can rely on consistent and reproducible biochars for the desired effects.

Last, we believe that an additional approach should involve testing a range of treatments that might stretch the positive boundaries of biochar application in substrates. These means may include pre-treatment of biochar substrates with physical and biological enhancers and combining biochar with other growing media alternatives such as compost (Akhter *et al.* 2015; Kammann *et al.* 2015). The benefits of such combinations have yet to be shown in rigorous scientific studies and indeed provide a big challenge to the desired standardization process. This approach may also include addition of beneficial microorganisms: biological control agents and plant growth promoting bacteria. There are already several works which investigated the role of arbuscular mycorrhizal fungi combinations with biochar plant pathogens, and usually reported that biochar increased the density of arbusular mycorrhiza (Elmer, Pignatello 2011; Lecroy *et al.* 2013; Akhter *et al.* 2015) and plant growth promoting bacteria (Saxena *et al.* 2013). However, very few works have included biological control agents (Postma *et al.* 2013), such that additional known agents like *Trichoderma harzianum* and commercial bacillus strains can be tried. Another aspect worthy of research involves whether, in the presence of biochar, doses of fungicides and other pesticides in plant nurseries can be reduced. It may also be that damage caused by soil fauna such as nematodes may be reduced when biochar is added to the potting medium (George *et al.* 2016); this should be further explored.

Conclusions

In summary, biochar as a peat replacement in high concentrations may pose a hidden risk by weakening the plants' defenses or predisposing the roots to pathogen attack if applied in concentrations that are too high. Determining the safe limits of biochar concentrations in pathosystems that are common in nurseries and soilless culture systems may help stretch limits of beneficial (i.e., economically meaningful) biochar use in horticulture. Such benefits are needed if biochar is to have a role as partial peat replacement or as improver of peat-less alternative growth media.

Acknowledgements

We thank Mrs. Dalia Rav David, Mrs. Indira Paudel, Mr. Menahem Borenshtein and Mrs. Ludmilla Tsechansky, for their assistance during experiments. The authors are grateful for the EU COST Action TD1107 (Biochar as option for sustainable resource management). Contribution from the Agricultural Research Organization, The Volcani Center, Bet Dagan, Israel, No: 571/17.

Funding

The research was funded by The Chief Scientist of the Ministry of Agriculture and Rural Development, Israel [Grant number 1321826].

Disclosure statement

The authors declare no competing financial, professional, or personal interests from other parties.

References

Akhter, A.; Hage-Ahmed, K.; Soja, G.; Steinkellner, S. 2015. Compost and biochar alter mycorrhization, tomato root exudation, and development of *Fusarium oxysporum* f. sp *lycopersici*, *Frontiers in Plant Science* 6: 529. https://doi.org/10.3389/fpls.2015.00529

Akhter, A.; Hage-Ahmed, K.; Soja, G.; Steinkellner, S. 2016. Potential of Fusarium wilt-inducing chlamydospores, in vitro behaviour in root exudates and physiology of tomato in biochar and compost amended soil, *Plant and Soil* 406: 425–440. https://doi.org/10.1007/s11104-016-2948-4

Allen, R. L. 1847. *A Brief Compend of American Agriculture*. 2nd ed. New York: C. M. Saxton. https://doi.org/10.5962/bhl.title.20151

Atucha, A.; Litus, G. 2015. Effect of Biochar Amendments Peach Replant Disease, *Hortscience* 50: 863–868.

Bonanomi, G.; Ippolito, F.; Scala, F. 2015. A "black" future for plant pathology? Biochar as a new soil amendment for controlling plant diseases, *Journal of Plant Pathology* 97: 223–234.

Chan, K. Y.; Van Zwieten, L.; Meszaros, I.; Downie, A.; Joseph, S. 2007. Agronomic values of greenwaste biochar as a soil amendment, *Australian Journal of Soil Research* 45: 629–634. https://doi.org/10.1071/SR07109

Conversa, G.; Bonasia, A.; Lazzizera, C.; Elia, A. 2015. Influence of biochar, mycorrhizal inoculation, and fertilizer rate on growth and flowering of Pelargonium (*Pelargonium zonale* L.) plants, *Frontiers in Plant Science* 6: 429. https://doi.org/10.3389/fpls.2015.00429

Copley, T. R.; Aliferis, K. A.; Jabaji, S. 2015. Maple bark biochar affects *Rhizoctonia solani* metabolism and increases damping-off severity, *Phytopathology* 105: 1334–1346. https://doi.org/10.1094/PHYTO-08-14-0231-R

Cornelissen, G.; Martinsen, V.; Shitumbanuma, V.; Alling, V.; Breedveld, G. D.; Rutherford, D. W.; Sparrevik, M.; Hale, S. E.; Obia, A.; Mulder, J. 2013. Biochar effect on maize yield and soil characteristics in five conservation farming sites in Zambia, *Agronomy* 3: 256–274. https://doi.org/10.3390/agronomy3020256

De Tender, C. A.; Debode, J.; Vandecasteele, B.; D'Hose, T.; Cremelie, P.; Haegeman, A.; Ruttink, T.; Dawyndt, P.; Maes, M. 2016. Biological, physicochemical and plant health responses in lettuce and strawberry in soil or peat amended with biochar, *Applied Soil Ecology* 107: 1–12. https://doi.org/10.1016/j.apsoil.2016.05.001

Dumroese, R. K.; Heiskanen, J.; Englund, K.; Tervahauta, A. 2011. Pelleted biochar: chemical and physical properties show potential use as a substrate in container nurseries, *Biomass & Bioenergy* 35: 2018–2027. https://doi.org/10.1016/j.biombioe.2011.01.053

Elad, Y.; Cytryn, E.; Meller Harel, Y.; Lew, B.; Graber, E. R. 2011. The biochar effect: plant resistance to biotic stresses, *Phytopathologia Mediterranea* 50: 335–349.

Elad, Y.; David, D. R.; Meller Harel, Y.; Borenshtein, M.; Ben Kalifa, H.; Silber, A.; Graber, E. R. 2010. Induction of systemic resistance in plants by biochar, a soil-applied carbon sequestering agent, *Phytopathology* 100: 913–921. https://doi.org/10.1094/PHYTO-100-9-0913

Elmer, W. H.; Pignatello, J. J. 2011. Effect of biochar amendments on mycorrhizal associations and Fusarium crown and root rot of asparagus in replant soils, *Plant Disease* 95: 960–966. https://doi.org/10.1094/PDIS-10-10-0741

Elmer, W. H. 2016. Effect of leaf mold mulch, biochar, and earthworms on mycorrhizal colonization and yield of asparagus affected by Fusarium crown and root rot, *Plant Disease* 100: 2507–2512. https://doi.org/10.1094/PDIS-10-15-1196-RE

Fascella, G. 2015. Growing substrates alternative to peat for ornamental plants, in Md. Asaduzzaman (Eds.). *Soilless Culture-Use of Substrates for the Production of Quality Horticultural Crops*. InTech, 47–67.

Frenkel, O.; Bornestein, M.; Shulhani, R.; Sharabani, G.; Sofer, M.; Abo-Moch, F.; Lofthouse, M.; Manulis-Sasson, S.; Shtienberg, D. 2016. Secondary spread of *Clavibacter michiganensis* subsp. *michiganensis* in nurseries and the conditions leading to infection of tomato seedlings, *European Journal of Plant Pathology* 144: 569–579. https://doi.org/10.1007/s10658-015-0795-4

George, C.; Kohler, J.; Rillig, M. C. 2016. Biochars reduce infection rates of the root-lesion nematode Pratylenchus penetrans and associated biomass loss in carrot, *Soil Biology & Biochemistry* 95 11–18. https://doi.org/10.1016/j.soilbio.2015.12.003

Glaser, B.; Lehmann, J.; Zech, W. 2002. Ameliorating physical and chemical properties of highly weathered soils in the tropics with charcoal – a review, *Biology and Fertility of Soils* 35: 219–230. https://doi.org/10.1007/s00374-002-0466-4

Graber, E. R.; Frenkel, O.; Jaiswal, A. K.; Elad, Y. 2014a. How may biochar influence severity of diseases caused by soilborne pathogens?, *Carbon Management* 5: 169–183. https://doi.org/10.1080/17583004.2014.913360

Graber, E. R.; Kookana, R. S. 2015. Biochar and retention/efficacy of pesticides, in J. Lehmann (Eds.). *Biochar for environmental management: science and technology*. London: Earthscan Books Ltd, 655–678.

Graber, E. R.; Meller Harel, Y.; Kolton, M.; Cytryn, E.; Silber, A.; David, D. R.; Tsechansky, L.; Borenshtein, M.; Elad, Y. 2010. Biochar impact on development and productivity of pepper and tomato grown in fertigated soilless media, *Plant and Soil* 337: 481–496. https://doi.org/10.1007/s11104-010-0544-6

Graber, E. R.; Tsechansky, L.; Lew, B.; Cohen, E. 2014b. Reducing capacity of water extracts of biochars and their solubilization of soil Mn and Fe, *European Journal of Soil Science* 65: 162–172. https://doi.org/10.1111/ejss.12071

Gravel, V.; Dorais, M.; Menard, C. 2013. Organic potted plants amended with biochar: its effect on growth and Pythium colonization, *Canadian Journal of Plant Science* 93: 1217–1227. https://doi.org/10.4141/cjps2013-315

Haider, G.; Steffens, D.; Müller, C.; Kammann, C. I. 2016. Standard extraction methods may underestimate nitrate stocks captured by field-aged biochar, *Journal of Environmental Quality* 45(4): 1196–1204. https://doi.org/10.2134/jeq2015.10.0529

Hendrix, F.; Campbell, W. 1973. Pythiums as plant pathogens, *Annual Review of Phytopathology* 11: 77–98. https://doi.org/10.1146/annurev.py.11.090173.000453

Huang, W. K.; Ji, H. L.; Gheysen, G.; Debode, J.; Kyndt, T. 2015. Biochar-amended potting medium reduces the susceptibility of rice to root-knot nematode infections, *BMC Plant Biology* 15: 267. https://doi.org/10.1186/s12870-015-0654-7

Jaiswal, A. K.; Elad, Y.; Graber, E. R.; Frenkel, O. 2014. Rhizoctonia solani suppression and plant growth promotion in cucumber as affected by biochar pyrolysis temperature, feedstock and concentration, *Soil Biology & Biochemistry* 69: 110–118. https://doi.org/10.1016/j.soilbio.2013.10.051

Jaiswal, A. K.; Frenkel, O.; Elad, Y.; Lew, B.; Graber, E. R. 2015. Non-monotonic influence of biochar dose on bean seedling growth and susceptibility to *Rhizoctonia solani*: the "Shifted Rmax-Effect", *Plant and Soil* 395: 1–16. https://doi.org/10.1007/s11104-014-2331-2

Jeffery, S.; Verheijen, F. G.; Van Der Velde, M.; Bastos, A. C. 2011. A quantitative review of the effects of biochar application to soils on crop productivity using meta-analysis, *Agriculture, Ecosystems & Environment* 144: 175–187. https://doi.org/10.1016/j.agee.2011.08.015

Joseph, S.; Graber, E. R.; Chia, C.; Munroe, P.; Donne, S.; Thomas, T.; Nielsen, S.; Marjo, C.; Rutlidge, H.; Pan, G. X.; Li, L.; Taylor, P.; Rawal, A.; Hook, J. 2013. Shifting paradigms: development of high-efficiency biochar fertilizers based on nano-structures and soluble components, *Carbon Management* 4: 323–343. https://doi.org/10.4155/cmt.13.23

Kammann, C.; Graber, E. R. 2015. Biochar effects on plant ecophysiology, in J. Lehmann, S. Joseph. *Biochar for environmental management: science and technology*. London: Earthscan Books Ltd, 391–419.

Kammann, C. I.; Schmidt, H. P.; Messerschmidt, N.; Linsel, S.; Steffens, D.; Muller, C.; Koyro, H. W.; Conte, P.; Stephen, J. 2015. Plant growth improvement mediated by nitrate capture in co-composted biochar, *Scientific Reports* 5: 12. https://doi.org/10.1038/srep11080

Kammann, C.; Ippolito, J.; Hagemann, N.; Borchard, N.; Cayuela, M. L.; Estavillo, J. M.; Fuertes-Mendizabal, T.; Jeffery, S.; Kern, J.; Novak, J.; Rasse, D.; Saarnio, S.; Schmidt, H. P.; Spokas, K.; Wrage-Mönnig, N. Biochar as a tool to reduce the agricultural greenhouse-gas burden – knowns, unknowns and future research needs, *Journal of Environmental Engineering and Landscape Management* (this issue).

Kern, J.; Tammeorg, P.; Shanskiy, M.; Sakrabani, R.; Knicker, H.; Kammann, C.; Tuhkanen, E.-M.; Smidt, G.; Prasad, M.; Tiilikkala, K.; Sohi, S.; Gascó, G.; Steiner, C.; Glaser, B. Synergistic use of peat and charred material in growing media – an option to reduce the pressure on peatlands?, *Journal of Environmental Engineering and Landscape Management* (this issue).

Kim, H. S.; Kim, K. R.; Yang, J. E.; Ok, Y. S.; Kim, W. I.; Kunhikrishnan, A.; Kim, K. H. 2016. Amelioration of horticultural growing media properties through rice hull biochar incorporation, *Waste and Biomass Valorization* 7(3): 1–10.

Knox, O. G. G.; Oghoro, C. O.; Burnett, F. J.; Fountaine, J. M. 2015. Biochar increases soil pH, but is as ineffective as liming at controlling clubroot, *Journal of Plant Pathology* 97: 149–152.

Kolb, S. E.; Fermanich, K. J.; Dornbush, M. E. 2009. Effect of charcoal quantity on microbial biomass and activity in temperate soils, *Soil Science Society of America Journal* 73: 1173–1181. https://doi.org/10.2136/sssaj2008.0232

Kolton, M.; Harel, Y. M.; Pasternak, Z.; Graber, E. R.; Elad, Y.; Cytryn, E. 2011. Impact of biochar application to soil on the root-associated bacterial community structure of fully developed greenhouse pepper plants, *Applied and Environmental Microbiology* 77: 4924–4930. https://doi.org/10.1128/AEM.00148-11

Kortekamp, A. 2011. *Unexpected side effects of herbicides: modulation of plant-pathogen interactions*. INTECH Open Access Publisher.

Krauthausen, H. J.; Laun, N.; Wohanka, W. 2011. Methods to reduce the spread of the black rot pathogen, Xanthomonas campestris pv. campestris, in brassica transplants, *Journal of Plant Diseases and Protection* 118: 7–16. https://doi.org/10.1007/BF03356375

Lecroy, C.; Masiello, C. A. Rudgers, J. A.; Hockaday, W. C.; Silberg, J. J. 2013. Nitrogen, biochar, and mycorrhizae: alteration of the symbiosis and oxidation of the char surface, *Soil Biology & Biochemistry* 58: 248–254. https://doi.org/10.1016/j.soilbio.2012.11.023

Lu, Y.; Rao, S.; Huang, F.; Cai, Y.; Wang, G.; Cai, K. 2016. Effects of biochar amendment on tomato bacterial wilt resistance and soil microbial amount and activity, *International Journal of Agronomy*. Article ID 2938282. https://doi.org/10.1155/2016/2938282

Matsubara, Y.; Hasegawa, N.; Fukui, H. 2002. Incidence of Fusarium root rot in asparagus seedlings infected with arbuscular mycorrhizal fungus as affected by several soil amendments, *Journal of the Japanese Society for Horticultural Science* 71: 370–374. https://doi.org/10.2503/jjshs.71.370

Mehari, Z. H.; Elad, Y.; Rav-David, D.; Graber, E. R.; Harel, Y. M. 2015. Induced systemic resistance in tomato (Solanum lycopersicum) against *Botrytis cinerea* by biochar amendment involves jasmonic acid signaling, *Plant and Soil* 395: 31–44. https://doi.org/10.1007/s11104-015-2445-1

Meller Harel, Y.; Elad, Y.; Rav-David, D.; Borenstein, M.; Shulchani, R.; Lew, B.; Graber, E. R. 2012. Biochar mediates systemic response of strawberry to foliar fungal pathogens, *Plant and Soil* 357: 245–257. https://doi.org/10.1007/s11104-012-1129-3

Mendez, A.; Paz-Ferreiro, J.; Gil, E.; Gasco, G. 2015. The effect of paper sludge and biochar addition on brown peat and coir based growing media properties, *Scientia Horticulturae* 193: 225–230. https://doi.org/10.1016/j.scienta.2015.07.032

Nerome, M.; Toyota, K.; Islam, T.; Nishijima, T.; Matsuoka, T.; Sato, K.; Yamaguchi, Y. 2005. Suppression of bacterial wilt of tomato by incorporation of municipal biowaste charcoal into soil, *Soil Microorganisms* 59: 9–14.

Nieto, A.; Gasco, G.; Paz-Ferreiro, J.; Fernandez, J. M.; Plaza, C.; Mendez, A. 2016. The effect of pruning waste and biochar addition on brown peat based growing media properties, *Scientia Horticulturae* 199: 142–148. https://doi.org/10.1016/j.scienta.2015.12.012

Obia, A.; Cornelissen, G.; Mulder, J.; Dorsch, P. 2015. Effect of soil pH increase by biochar on NO, N2O and N-2 production during denitrification in acid soils, *Plos One* 10 (9): e0138781. https://doi.org/10.1371/journal.pone.0138781

Postma, J.; Clematis, F.; Nijhuis, E. H.; Someus, E. 2013. Efficacy of four phosphate-mobilizing bacteria applied with an animal bone charcoal formulation in controlling *Pythium aphanidermatum* and *Fusarium oxysporum f.sp radicis lycopersici* in tomato, *Biological Control* 67: 284–291. https://doi.org/10.1016/j.biocontrol.2013.07.002

Rajkovich, S.; Enders, A.; Hanley, K.; Hyland, C.; Zimmerman, A. R.; Lehmann, J. 2012. Corn growth and nitrogen nutrition after additions of biochars with varying properties to a temperate soil, *Biology and Fertility of Soils* 48: 271–284. https://doi.org/10.1007/s00374-011-0624-7

Rondon, M. A.; Lehmann, J.; Ramirez, J.; Hurtado, M. 2007. Biological nitrogen fixation by common beans (*Phaseolus vulgaris L.*) increases with bio-char additions, *Biology and Fertility of Soils* 43: 669–708. https://doi.org/10.1007/s00374-006-0152-z

Ruysschaert, G.; Nelissen, V.; Postma, R.; Bruun, E.; O'toole, A.; Hammond, J.; Rödger, J. M.; Hylander, L.; Kihlberg, T.; Zwart, K.; Hauggaard-Nielsen, H.; Shackley, S. 2016. Field applications of pure biochar in the North Sea region and across Europe. *Biochar in European Soils and Agriculture.* Routledge, 99–135.

Saxena, J.; Rana, G.; Pandey, M. 2013. Impact of addition of biochar along with Bacillus sp. on growth and yield of French beans, *Scientia Horticulturae* 162: 351–356. https://doi.org/10.1016/j.scienta.2013.08.002

Silber, A.; Levkovitch, I.; Graber, E. R. 2010. pH-dependent mineral release and surface properties of cornstraw biochar: agronomic implications, *Environmental Science and Technology* 44: 9318–9323. https://doi.org/10.1021/es101283d

Spokas, K. A.; Baker, J. M.; Reicosky, D. C. 2010. Ethylene: potential key for biochar amendment impacts, *Plant and Soil* 333: 443–452. https://doi.org/10.1007/s11104-010-0359-5

Spokas, K. A.; Cantrell, K. B.; Novak, J. M.; Archer, D. W.; Ippolito, J. A.; Collins, H. P.; Boateng, A. A.; Lima, I. M.; Lamb, M. C.; Mcaloon, A. J.; Lentz, R. D.; Nichols, K. A. 2012. Biochar: a synthesis of its agronomic impact beyond carbon sequestration, *Journal of Environmental Quality* 41: 973–989. https://doi.org/10.2134/jeq2011.0069

Steiner, C.; Harttung, T. 2014. Biochar as a growing media additive and peat substitute, *Solid Earth* 5: 995–999. https://doi.org/10.5194/se-5-995-2014

Tian, Y.; Sun, X. Y.; Li, S. Y.; Wang, H. Y.; Wang, L. Z.; Cao, J. X.; Zhang, L. 2012. Biochar made from green waste as peat substitute in growth media for *Calathea rotundifola cv. Fasciata*, *Scientia Horticulturae* 143: 15–18. https://doi.org/10.1016/j.scienta.2012.05.018

Vaughn, S. F.; Kenar, J. A.; Thompson, A. R.; Peterson, S. C. 2013. Comparison of biochars derived from wood pellets and pelletized wheat straw as replacements for peat in potting substrates, *Industrial Crops and Products* 51: 437–443. https://doi.org/10.1016/j.indcrop.2013.10.010

Viger, M.; Hancock, R. D.; Miglietta, F.; Taylor, G. 2014. More plant growth but less plant defence? First global gene expression data for plants grown in soil amended with biochar, *GCB Bioenergy* 7: 658–672.

Yamato, M.; Okimori, Y.; Wibowo, I. F.; Anshori, S.; Ogawa, M. 2006. Effects of the application of charred bark of Acacia mangium on the yield of maize, cowpea and peanut, and soil chemical properties in South Sumatra, Indonesia, *Soil Science and Plant Nutrition* 52: 489–495. https://doi.org/10.1111/j.1747-0765.2006.00065.x

Zhang, L.; Sun, X. Y.; Tian, Y.; Gong, X. Q. 2014. Biochar and humic acid amendments improve the quality of composted green waste as a growth medium for the ornamental plant Calathea insignis, *Scientia Horticulturae* 176: 70–78. https://doi.org/10.1016/j.scienta.2014.06.021

Zwart, D. C.; Kim, S. H. 2012. Biochar amendment increases resistance to stem lesions caused by Phytophthora spp. in tree seedlings, *Hortscience* 47: 1736–1740.

Omer FRENKEL, PhD. Research Scientist in the Institute of Plant Protection, The Volcani Center, Agricultural Research Organization, Bet Dagan 50250 Israel. 25 publications in international peer-reviewed journals, 724 citations, H-index 11 (Google Scholar).

Amit K. JAISWAL, MSc. PhD student in the Department of Plant Pathology, The Volcani Center, Agricultural Research Organization, Bet Dagan 50250 Israel and Faculty of Agriculture, Hebrew University, Rehovot 76100 Israel. Five publications in international peer-reviewed journals, 89 citations, H-index 3 (Google Scholar).

Yigal ELAD, PhD. Senior Principal Research Scientist in the Institute of Plant Protection, The Volcani Center, Agricultural Research Organization, Bet Dagan 50250 Israel. 188 publications in international peer-reviewed journals, 42 book chapters, 15285 citations, H-index 68 (Google Scholar).

Beni LEW, PhD. Research Scientist in the Institute of Agricultural Engineering, The Volcani Center, Agricultural Research Organization, HaMaccabim Road. P.O.Box 15159, Rishon LeZion, 7528809 Israel. 31 publications in international peer-reviewed journals, 570 citations, H-index 12 (Google Scholar).

Claudia KAMMANN, PhD, Professor for Climate Change Research for Special Crops, Institute for Soil Science and Plant Nutrition, Hochschule Geisenheim University, Geisenheim, Germany, 51 peer-reviewed articles in scientific journals, 5 book chapters, 2842 citations, H-index 27 (Google Scholar).

Ellen R. GRABER, PhD. Senior Principal Research Scientist in the Institute of Soil, Water and Environmental Sciences, The Volcani Center, Agricultural Research Organization, Bet Dagan 50250 Israel. 87 publications in international peer-reviewed journals, 8 book chapters, 4427 citations, H-index 36 (Google Scholar).

WOOD PROCESSING BY-PRODUCTS TREATED WITH THE LIGNIN-BASED CONDITIONER AS MULCH FOR SOIL PROTECTION

Galia SHULGA[a], Talrits BETKERS[b], Sanita VITOLINA[c], Brigita NEIBERTE[d], Anrijs VEROVKINS[e], Olga ANNE[f], Dalia AMBRAZAITIENE[g], Audrone ZUKAUSKAITE[h]

[a, c, d, e]*Laboratory of Lignin Chemistry, Latvian State Institute of Wood Chemistry, Dzerbenes iela 27, LV-1006 Riga, Latvia*

[b]*Faculty of Building and Civil Engineering, Riga Technical University, Azenes iela 16/20, LV-1048 Riga, Latvia*

[f, g, h]*Faculty of Natural Sciences and Mathematics, Klaipėda University, H. Manto g. 84, LT-5808 Klaipėda, Lithuania*

[h]*Faculty of Marine Engineering, Klaipėda University, Bijūnų g. 17, LT-91225 Klaipėda, Lithuania*

Abstract. Wood processing by-products such as bark of different wood species and hydrolysis lignin were applied as soil mulch. The lignin-based soil conditioner (LSC) representing a lignin-based polyelectrolyte complex with the different composition (a lignin/polymer mass ratio) was obtained under laboratory conditions and was intended to protect sandy soil from erosion by simultaneous application of a soil conditioner as an adhesive, mulch and plant seeds by hydroseeding. The study revealed the pronounced dependence of the properties of the treated wood-originated mulch on the wood species as well as on the composition, the applied concentration and the application rate of LSC. A comparison of the obtained results showed that the treated hydrolysis lignin was characterised by higher compressive strength, higher water resistance and lower moisture losses from sandy soil for a given LSC composition and application rate than the bark-based mulch. The conditioner concentrations don't have negative impact on the germination of seeds plants of the coastal dune zone.

Keywords: bark, hydrolysis lignin, lignin-based soil conditioner, mulch, soil erosion, seed germination.

Introduction

The peculiarities of the climatic conditions of the coastal zone of Latvia and Lithuania and non-coherent, structureless dune sand, on which the most popular species of conifers such as Pinus sylvestris, Picea excelsa and Pinus monticota grow, are the main reasons for riparian forest soil erosion. Unfortunately, methods of erosion control are labour intensive and sometimes are not very effective. To protect land by reducing soil erosion, afforestation and erection of hedges have been historically applied (Dymond *et al.* 2008; Shao *et al.* 2011).

The known management practice for mitigating soil erosion on non-agricultural land is based on the application of various mulching materials (Kim *et al.* 2008; Gimenez, Govers 2008; Foltz, Copeland 2009; Zhang *et al.* 2013). The application of fibre mulch for protection of erodable soil has been demonstrated for straw, wood chips, sawdust and synthetic products, including erosion control geonets, geotextiles, blankets and geomattings (Curnoe *et al.* 2006; Pack 2007; Calatrava, Franko 2011; Lee, Douglas 2012). Owing to the notable wood processing industry in the Baltic countries, a wide-scale wood waste such as sawdust and bark as well as hydrolysis lignin, a waste of the post-hydrolysis industry in Lithuania, in the form of organic mulch are economically available and promising erosion control products.

At the same time, the application of unanchored mulch may lead to its washout and blowing off and destroying of the soil mulched layers. An attractive solution to the erosion mitigation and control in the dune zone may be the biological reinforcement of the sandy soil surface

Corresponding author: Galia Shulga
E-mail: shulga@junik.lv

by hydroseeding, combining simultaneous application of mulch, grass seeds, fertilisers and a soil polymer conditioner. A polymer conditioner attaches the mulch and grass seeds to the sandy surface, glues them by forming upper stabilised layers, and thereby prevents the washout and blowing off of mulch and reduces water runoff and soil loss (Sojka *et al.* 2007; Li *et al.* 2011).

The use of a natural polymer such as lignin is of great scientific and practical interest as a soil conditioner. Lignin is the world's second most abundant natural substance performing the function of polymeric glue in the wood matrix. Furthermore, lignin is a precursor of humus substances in soil. Commercial technical lignins are wide-scale and low-cost by-products of pulp and paper mills, and their application for the production of soil improving agents is well known (Dong *et al.* 2008; Shulga *et al.* 2008a; Aimi *et al.* 2009; Passauer *et al.* 2012). It has been found that the interaction of the lignin-based soil conditioners with the mineral surface of soil particles has a complicated character and is realised via various types of binding, including the Van der Waals forces of attraction as well as electrostatic and electron-donor-acceptor interactions (Kullmann 1982). The commercial lignin-based conditioners include the following: Sandstop, Dustex, Lima, Lia, Copolima, Copoliba, RoadBond, RP Ultra Plus™, BorrePlex CA, etc. In comparison to synthetic polymeric soil conditioners, the lignin-based conditioners biodecay in soil, increasing its fertility (Popa *et al.* 2010).

Various non-expensive lignin-based soil conditioners (LSCs) have been synthesised under laboratory conditions by the chemical modification of technical lignins, low-cost wide-scale by-products of pulp and paper mills, using various environmentally friendly polymeric or oligomeric modifiers, containing soil nutrient elements (Shulga *et al.* 2001). From the chemical point of view, the developed LSCs represent lignin-based polycomplexes, in which the macromolecules of lignin and the modifier are linked together with physicochemical bonds.

In the previous works (Shulga *et al.* 2007, 2008b), the developed lignin-based conditioner was applied for gluing mulch from wood sawdust and structuring sandy soil. It was found (Shulga *et al.* 2008b) that, after applying the water LSC solutions to the top of the soil mulched with wood sawdust, the layer soil-mulch composites were formed upon drying. They consisted of a thick upper glued layer (20–50 mm) of the sawdust and a thin lower glued layer (1–3 mm) of the soil.

It is quite important to emphasize, that lignin based conditioners and mulch are bind (fix) soil particles together and at the same time should be harmless for dune sand and for plants growing under the trees, especially for seeds germination. The sensitivity of Coastal zone "obligates" any technologies to be friendly for environment and don't have any negative impact to ecosystems. Only comprehensive research of proposed technology could contribute to successful results of its application.

The aim of this work was to study the comparison effect of the developed LSC on the properties of sandy soil-mulch composites obtained from wood-originated by-products, namely, wood bark and hydrolysis lignin.

2. Objects and methods

2.1. Lignin-based soil conditioner

Commercial lignosulphonate (LS), a by-product of softwood delignification, with an average molecular weight of 28 KDa was used to obtain the new soil conditioner. The LS elementary and functional analysis gave the following formula: $C_9 H_{6.89} O_{2.57} (OCH_3)_{0.71} (SO_3)_{0.35} (OHph)_{0.68} (CO)_{0.36}$. A lignin-based polyelectrolyte complex having different composition Z (from 0.02 to 1.00), which corresponded to the polymer/lignin mass ratio, was obtained via the chemical modification of LS with the acrylic polymer modifier according to the procedure (Shulga *et al.* 2007). It differed from the conditioners developed earlier by the enhanced colloidal stability at different pH values in water solutions.

2.2. Soil

Soil was taken from the coastal dune zone of the Curonian Spit of the Baltic Sea near Klaipeda (N-55, 67725, E-21, 10336, Lithuania). It was a very poor light sandy soil with the humus content below 0.1 wt % and a pH of 6.6. The sand was rich in phosphorus (185 mg/kg), but poor in potassium (54 mg/kg), which is typical for a dune sandy soil. The soil sample was obtained over a depth of 0–20 cm and had the following mechanical composition: 3.1% of particles ranging from 1.0 to 0.25 mm, 93.9% of particles ranging from 0.25 to 0.05 mm, and 3.0% of particles less than 0.01 mm in size.

2.3. Mulch

The bark of the following wood species was used: aspen, birch, spruce and pine, which are readily available from Latvia's woodworking enterprises. Hydrolysis lignin was obtained from the manufacture of fodder yeast, the remaining wide-scale waste of the former Lithuanian hydrolysis industry.

The chemical composition of the applied bark (cellulose, lignin and extractives) is presented in Table 1. Before the study, the hydrolysis lignin was washed with hot distilled water for removing organic and mineral acid residues. The chemical composition of the hydrolysis lignin was the following: pH 5.1; organic carbon 30.91 wt %; total nitrogen 0.34 wt %; P_2O_5 404 mg/kg; K_2O_5 831 mg/kg; moisture content 7.1 wt %. Besides lignin substances, this organic waste also contained non-hydrolysed polysaccharides

(25.8%) as well as small quantities of sugars (1.3%) and cations of polyvalent metals (5.47%).

Table 1. Chemical composition of bark

Wood species	Moisture, %	Ash, %	OCH_3, %	Cellulose, %	Lignin, %	Extractives, %, soluble in	
						Hot water	Organic mixture[x]
Pine	6.9	2.4	2.4	18.1	67.4	1.7	19.7
Spruce	6.9	2.1	2.5	21.8	44.8	2.9	12.7
Birch	6.3	4.3	5.1	25.2	40.0	9.5	3.9
Aspen	5.9	4.2	3.6	22.1	26.6	9.6	8.0

Note: [x] ethyl alcohol : benzene ratio is close to 1:2.

After mechanical grinding, fractions with the size ranging from 1 to 2 mm and less than 0.5 mm prevailed in the mechanical composition of the used mulches. Two layer thicknesses of mulch, namely, 2 and 5 cm were used to study the effect of the treated mulch on the soil chemical composition.

2.4. Soil-mulch composites

Because the composites formed after treating the mulched soil with a LSC solution were composed of an upper glued mulch layer and a lower glued soil one, it was expedient to investigate the effect of LSC on the properties of each layer separately.

2.4.1. Glued soil layers

For obtaining a glued soil surface layer, the set of the following procedures was carried out: (a) placing the slightly moistened sandy soil into aluminium containers 0.4 m square and 0.1 m in height; (b) treating the surface of the samples by spraying aqueous solutions of LSC with a defined concentration of 50–200 g/*l* at an application rate of 1.0–2.5 l/m^2; (c) drying the treated soil samples at room temperature.

The penetration resistance and thickness of the glued upper sandy layers were measured using a manual laboratory coner penetrometer (angle 30°) of original construction and a calliper square, described in detail earlier (Shulga *et al.* 2001). The wetting-drying test to investigate the water resistance of the formed glued layers was carried out according to the procedure (Shulga *et al.* 2001). Penetration resistance and thickness were measured at 8 locations on each coating, were averaged, and their standard deviation values were calculated.

To study the effect of climatic factors on the change in the penetration resistance of the glued surface soil layers, a climatic chamber Binder KBF LQC 240 (Binder, Germany) was used. The parameters of the test were chosen, taking into account the climatic conditions of the Baltic coastal zone – the high humidity and moderately high temperature of the soil upper layers at sunlight. The air-dried glued sandy samples were put into the climatic chamber having a temperature of 60 °C, a humidity of 75% and the UV – radiation of 1.7 Wm^{-2} (in the range of 320–400 nm) and then exposed to the action of the climatic parameters during 5, 15 and 30 days. The number of replicates was three, and the presented results were average arithmetic values.

2.4.2. Mulch-based composite materials and their properties

To investigate the properties of the upper glued layer of mulch, the composite materials were made in accordance with the following set of procedures: (a) mixing the slightly moistened mulch with LSC water solutions with the defined concentration; (b) compacting the obtained raw mixtures in the form of cylinders (2×3 cm) at a pressure of 0.1 MPa; (c) drying the cylinders at room temperature. The content of the mulch in the obtained composites was varied from 95 wt % to 75 wt % on the dry composite weight.

The ability of the conditioner to glue mulch particles was evaluated by the value of the destructive compressive strength of the obtained composite cylinders. For this purpose, the cylinders were placed between the platens of a universal testing machine (Instron, USA) and then compressed at a constant rate. Compressive strength was presented as an average value with a standard deviation for seven samples of each applied composition of LSC.

Water resistance of the obtained composites was evaluated visually by the time of the initiation of the disintegration of the cylinder fully immersed in cold distilled water (287 K). The time of the beginning of the sample decomposition in water was presented as an average value with a standard deviation measured for seven samples of each applied composition of LSC.

To study the effect of the mulch treated with a LSC solution on the moisture evaporation from sandy soil, procedures for preparing the samples were as follows: (a) covering the slightly moist sandy soil, placed in round plastic baths 0.12 m in height and a surface square of 0.016 m^2, with a 2 cm layer of the slightly moist mulch; (b) treating the mulch surface with a water solution of LSC with a concentration of 50–200 g/l at an application rate of 2–6 $l\ m^{-2}$; (c) drying the obtained soil samples to an air-dry state at room temperature. Then equal amounts of water were applied to the surface of each sample so that the total content of moisture in all samples would be equal and would make up 12.5–12.8 wt %. The water evaporation from the soil surface was evaluated from the weight loss of the bath using the electron balance Kern 572 (Kern & Sohn GmbH,

Germany) with the readout 0.01 g upon its standing in the open air at the temperature 293±2K and an air humidity of 60±2% every 12 hours. The soil sample was considered air-dried, if the weight of the bath stopped varying with increasing storage time in the given experimental conditions. The average weight loss with a standard deviation for seven samples with the same LSC application concentration and rate was calculated.

2.5. Plants

After evaluating a variety of coastal zone plants and dune flora, the following sorts of plants were chosen: *Festuca rubra L., Gypsophila paniculata L., Thymus serpyllum L.* These greeneries, according to their morphological and ecological properties, are most suitable for prevention of soil erosion in the dune zone of the Baltic Sea. The viability and germination energy of plant seeds were estimated at the Lithuanian State Seeds Laboratory in accordance with international research methods (International Seed Testing Association 1996). Viability and germination energy of the seeds were analysed on Petri plates in five replicates, using distilled water and nutrient solutions wetting the filter paper. Dishes with the seeds were incubated in a biological incubator using: an incubation temperature of 20 °C for *Gypsophila paniculata L.*, evaluation of seeds viability after 4, 10 and 14 days; an incubation temperature of 20–30 °C for *Thymus serpyllum L.* and *Festuca rubra L.*, evaluation of seeds viability after 7 and 21 days. Germinated, non-germinated and abnormal seeds were counted every time. Statistical analysis was made using *Statgraphics plus software* ±x –mean square deviation. The significance levels of the results are given for seed germination characteristics (ANOVA): $S_{x\%}$ – standard deviation; LSD_{05} – least significant difference.

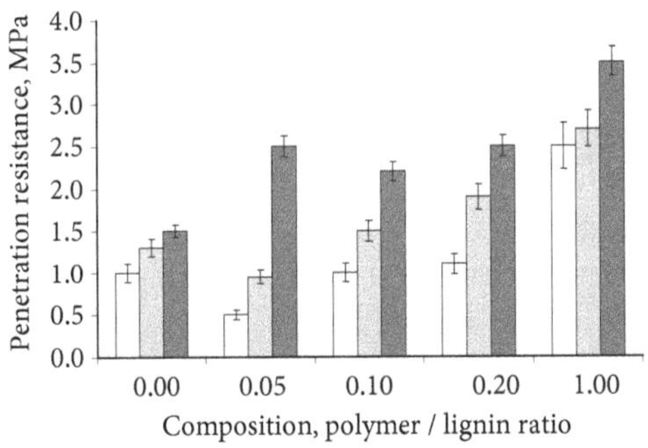

Fig. 1. Penetration resistance of reinforced soil layers versus the LSC composition; concentration of the applied solutions: ☐, 50 g/*l*; ▒, 100 g/*l*; ■, 200 g/*l*; application rate 1.5 l/m²

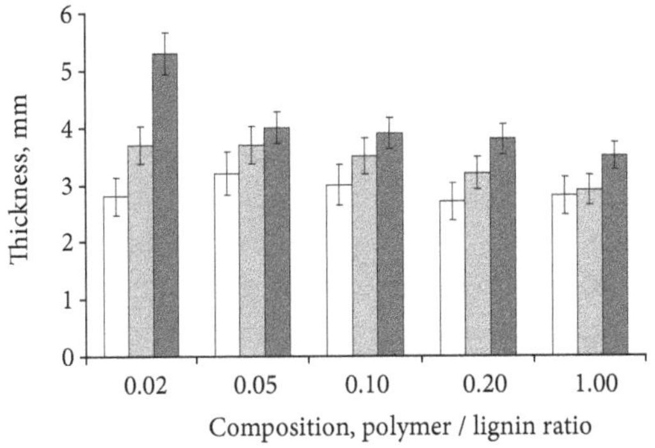

Fig. 2. Thickness of reinforced soil layers versus the LSC composition; concentration of the applied solutions: ☐ 50 g/*l*; ▒ 100 g/*l*; ■ 200 g/*l*; application rate 1.5 l/m²

3. Results and discussion

3.1. Effect of the lignin-based soil conditioner on the properties of sandy soil layers

According to our findings, the penetration resistance and thickness of the glued sandy layers formed depend on the concentration of the applied water solutions and the application rate, as well as on the LSC composition. The penetration resistance and thickness of the glued upper soil layers increase from 0.05 to 3.50 MPa and from 0.5 to 6.1 mm, respectively, with increasing application rate of the LSC solutions from 1.0 to 2.5 *l*/m², the concentration from 10 to 200 g/*l* and Z from 0.02 to 0.2. Figures 1 and 2 demonstrate the dependences of the penetration strength and thickness of the glued soil layers for the different concentrations of the applied solutions on the LSC composition, at an application rate of 1.5 *l*/m². Reasoning from the obtained data and taking into account the results of our previous work (Shulga *et al.* 2001), this application rate has to be sufficient to impart the optimum value of mechanical strength and thickness to the sandy layer. With increasing the composition of the LSC from 0.05 to 1.00 and its concentration in a water solution from 50 g/*l* to 200 g/*l* at the same application rate, the penetration strength of the layers increases 1.8–2.5 times.

The observed gain in the mechanical properties may be governed by increasing the LSC flocculation ability relative to the soil particles as a result of the enhancement in the molecular weight of the lignin/polymer complex, which was testified by an increase in the maximum dynamic viscosity of the LSC solutions with an increase in Z (Shulga *et al.* 2008b). At the same time, the thickness of the glued sandy layers decreases with the growth in the composition (Z ≥ 0.1) and concentration (>100 g/*l*) of the LSC solutions (Fig. 2) that may be caused by their lower infiltration into the soil due to the growth of dynamic viscosity. The higher values of thickness for the glued soil layers are provided by the application of the solutions of the LSC with Z = 0.02–0.05, having lower values of dynamic viscosity.

It was found that the changes in the penetration resistance of the glued soil layers during a test on wetting/drying cycles depended non-linearly on the LSC composition. The sandy layers treated with the solutions of the LSC with 0.1 ≤ Z ≤ 0.2 and the concentration 50–200 g/*l* had the highest resistance to the action of the wetting/drying cycles. Obviously, the ability of the glued soil layers to resist the destructive action of the wetting/drying cycle is defined by the cohesion strength of the LSC polymer network formed, which is directly connected with the reduced viscosity of the lignin/polymer solutions. The obtained correlations are similar to those that have been earlier found for the water resistance of the artificial sandy aggregates (Shulga *et al.* 2008b).

It is known that the combined prolonged effect of climatic factors such as temperature, humidity and solar radiation could have a negative action on the gluing properties of a polymer conditioner due to irreversible chemical transformations in its structure, which can be governed by both the hydrolysis of its chemical bonds owing to the presence of elevated humidity and destruction reactions as a result of the integrated action of elevated temperature and UV-radiation. The revealed changes in the properties of the glued soil layers after their exposure in a climatic chamber allows evaluating weather resistance of the developed soil conditioner itself. Table 2 shows the values of the penetration resistance of the glued soil layers subjected to the simultaneous action of the elevated temperature, the humidity and the UV-radiation in different periods of time. It can be seen that, with increasing the exposure time, the penetration resistance of the glued soil layers decreases. The values of the decrease in the penetration resistance depend on the composition of LSC.

After 30 days of the exposure in the chamber, the decrease in the penetration resistance for the lowest LSC composition Z = 0.02 is equal to 26%, but, with increasing Z to 0.1–0.2, the changes of the resistance decrease and vary in the range of 12–17%. Taking into account the moderate changes of the mechanical strength of the glued soil layers, it may be concluded that the weather resistance of the developed LSC is sufficient for retaining its binding properties during the period of plant development and the formation of turf on mulched soil.

3.2. Effect of the lignin-based soil conditioner composition on the properties of mulch-based composites

The binding properties of the soil conditioner relatively to the mulch were investigated using several series of the composite cylinders, in which bark and hydrolysis lignin were used as a filler and LSC was applied as a binder. It was found that, with growing content of LSC with Z = 0.02–0.2 from 5 to 25 wt % in the composites, their compressive strength increased 2.5–4.5 times, while the density of the bark- and hydrolysis lignin-based composites increased from 521 to 598 kg/m^3 and from 557 to 612 kg/m^3, respectively. The effect of the LSC composition on the compressive strength of the composites from the bark of the different wood species and the hydrolysis lignin at the maximum content of the soil conditioner is demonstrated in Figure 3. Evidently, the presence of active reaction centres in the hydrolysis lignin in the form of polyvalent metal ions increases its specific surface, which, together with the high-developed porous structure, reinforces the adhesion interaction at the LSC/filler interface.

Table 2. Penetration resistance of air-dried glued sandy soil layers[xx] after exposure in the climatic chamber "Binder" during different time

Composition, Z	Penetration resistance, MPa/Time, days			
	0	5	15	30
0.02	1.30±0.12	1.23±0.14	1.09±0.13	0.96±0.16
0.1	1.51±0.17	1.48±0.18	1.39±0.14	1.26±0.19
0.2	1.92±0.21	1.90±0.20	1.82±0.22	1.70±0.24

Note: [xx] concentration of LSC solutions – 100 g/*l*; application – 1.51 m^2.

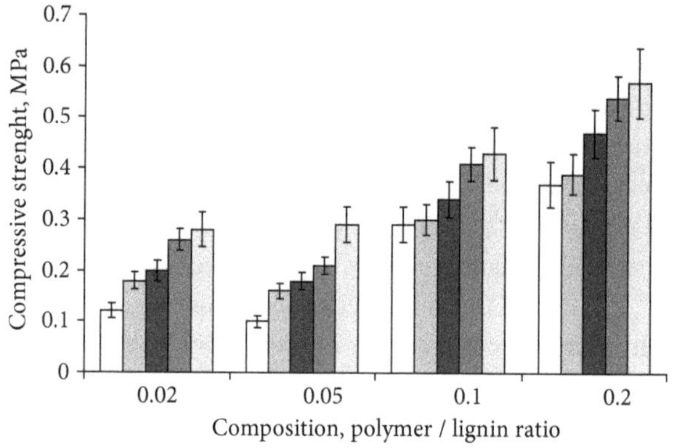

Fig. 3. Compressive strength of composites made from □ pine bark; □, spruce bark; ■, birch bark; ■, aspen bark; □, hydrolysis lignin; LCS content 25 wt % on the dry matter of the composite

It can be seen (Fig. 3) that the compressive strength of the obtained composites increases with growing the Z values and reaches a maximum at Z = 0.2. The mulch-based composites obtained with Z = 0.02–0.05 are characterised by weak mechanical properties. The increase in the composition of LSC from 0.05 to 0.2 results in a 1.5–3.8-fold growth in compressive strength, depending on the type of the wood filler. It should be noted that the composites filled with the bark of deciduous wood species have higher values of compressive strength than those filled with the bark of coniferous species. The found dependence of the gluing ability of LSC on the wood species of bark may be governed by the peculiarities of the chemical composition of bark of deciduous and coniferous wood species, which is confirmed by the data of extraction with an alcohol-benzene mixture of the latter (Table 1). A considerable amount of resin, extracted from the bark of conifers in the organic mixture, can weaken the adhesive interaction between the soil conditioner and the surface of the bark particles. At the same time, the hydrolysis lignin-based composites are characterised by the highest compressive strength for a given Z (Fig. 3). Taking into account these results, it may be supposed that the deformation of the obtained mulch-based composites subjected to the shear forces of wind and water flow during their field application will not be dramatic.

With increasing the content of LSC from 5 wt % to 25 wt % in the range of Z = 0.02–0.2, the time of the onset of the disintegration of the composite cylinders immersed in cold water increases 5–15 times, depending on the kind of the wood filler. Figure 4 shows the values of time for the composites filled with the bark of the different wood species and the hydrolysis lignin, depending on the composition of LSC at its content 25 wt %. According to the obtained results, the composites filled with the bark of deciduous wood species are characterised by a higher resistance to destruction under the action of water than the composites based on the bark of coniferous wood species. It can be seen that, in the range of Z = 0.02–0.05, the water resistance of the obtained composites is low and practically does not depend on the LSC composition. With increasing Z from 0.05 to 0.20, the time of the onset of disintegration for the composites made from the bark of coniferous wood species grows twice, while, for the composites filled with the bark of deciduous wood and the hydrolysis lignin, the time increases only 1.3–1.5 times.

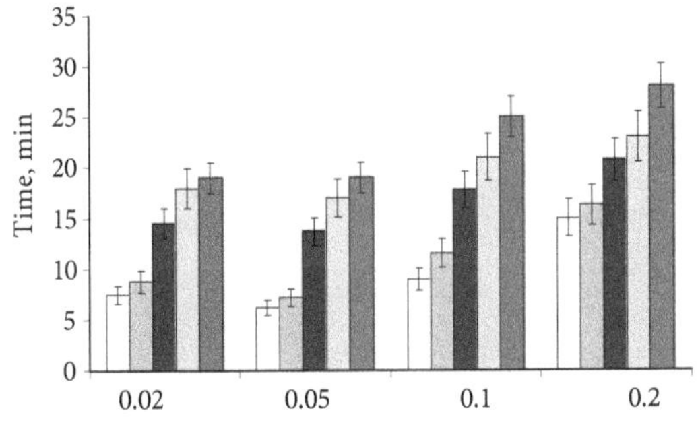

Fig. 4. Time of the onset of disintegration in cold water for composites made from □, pine bark; ▒, spruce bark; ■, birch bark; ░, aspen bark; ▓, hydrolysis lignin; LCS content 25 wt % on the dry matter of the composite

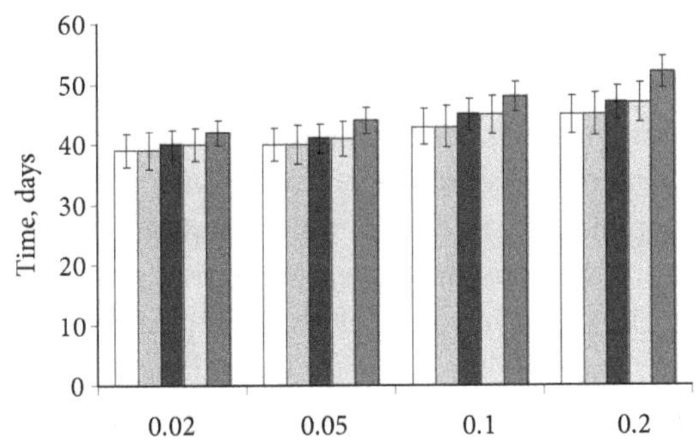

Fig. 5. Effect of moisture evaporation for sandy soil mulched with composite materials formed from □, pine bark; ▒, spruce bark; ■, birch bark; ░, aspen bark (4), ▓, hydrolysis lignin (5) on the composition of LSC; concentration of the solutions 100 g/*l*, the application rate 4.0 *l*/m^2

At the same time, the more pronounced water resistance of the hydrolysis lignin-based composite (Fig. 4) may be connected with the more developed system of the interfacial physico-chemical bonds between LSC and the filler, which has a favourable effect on the hydrophobicity of the composite materials.

3.3. Effect of mulch-based composites on the evaporation of soil moisture

The results of the performed tests have shown that the uncovered soil sample dry out in 22 days, while the samples covered with the bark and the hydrolysis lignin not glued with the soil conditioner dry out in 33 and 34 days, respectively. The ability to retain more soil moisture for hydrolysis lignin in comparison with the case of bark as mulch may be governed by its developed specific surface. The soil samples treated with the LSC solution dry out in 39–52 days, depending on the composition of the soil conditioner, the concentration of its solution, the application rate as well as the type of mulch. These data exceed by 77–163% and 18–57% the drying time of the uncovered soil and the untreated mulch samples, respectively. With higher Z values, at the same LSC concentration and application rate, the mulching effect tends to grow (Fig. 5). Higher LSC composition results in longer drying periods. This tendency manifests itself most dramatically in Z values in the range of 0.05–0.2. The lowest rate of the moisture evaporation is achieved with the deciduous wood bark and the hydrolysis lignin treated with the LSC solutions having Z = 0.1–0.2 and a concentration of 100–150 g/*l*.

It has been found that the effect of the studied concentration of the applied solutions and the application rate in the range 2–6 *l*/m^2 on the moisture losses from the mulched soil samples has a more complicated character. At the solution concentration >100 g/*l* and the consumption 2 *l*/m^2, the soil conditioner, passing through the mulch, is mainly adsorbed on its particles, gluing them together in the air-dry sate. In this case, the adherence of the mulch to the soil as well as the mechanical strength of the upper soil layers is relatively low due to the shortage of the LSC solution. An increase in the application rate of the conditioner solutions up to 4 *l*/m^2 in the same concentration range enhances the adhesion interaction at the mulch/soil interface and forms a glued soil layer under the mulch. In this case, the degree of binding the mulch practically does not change. At the application rate of the solution 6 *l*/m^2, the excess of the LSC solution, concentrating mainly at the mulch/soil interface, enhances significantly the penetration resistance of the glued soil layers. However, owing to the features of the LSC polymer structure and the porous surface of the formed glued sandy soil layers, no considerable decrease in the moisture evaporation from the mulched soil has been achieved. On the other hand, the decrease in the concentration of the solutions of LSC <100 g/*l* within the application rates under this study results in the worsening of gluing both the mulch and the soil particles, which leads to a faster evaporation of moisture from the mulched soil. A comparative analysis of the obtained results suggests that the application of 100–150 g/*l* solutions of LSC with Z = 0.1–0.2 at the rate of 4 *l*/m^2 may be the optimum one for treating a 2 cm mulched layer with the aim to diminish the evaporation of moisture from sandy soil.

3.4. Effect of the lignin-based soil conditioner composition on seeds' viability and germination

To assess the effect of the developed soil conditioner on the seeds' viability and germination energy, the water solutions of LSC, having Z = 0.02, 0.1 and 0.2, were used. The conducted study has shown that the composition and concentration of LSC in the solution are crucial factors, which influence variously the Gypsophila paniculata, Thymus serpyllum and Festuca rubra seeds' germination and viability. According to Figure 6 (a, b, c), the 10–100 g/*l* and 10–50 g/*l* LSC solutions, irrespective of the Z values, have a favourable action on the viability and germination energy of all the plant seeds.

At these LSC solution concentrations, the germinative energy for *Gypsophila paniculata L.* seeds after 4 days and for *Thymus serpyllum* and *Festuca rubra* seeds after 7 days is increased compared to the case of the control.

After 14 and 21 days, the quantity of normally developed seeds for *Gypsophila paniculata* as well as *Thymus serpyllum* and *Festuca rubra,* respectively, on the average, is equal to the control.

For the higher concentrations (>100 g/*l*) of LSC with Z ≥ 0.2, the seed germinative energy and viability drop. According to Figure 6 (b, c), the effect on the seeds' germination for *Thymus serpyllum* and *Festuca rubra* is dramatic. The LSC solutions with Z = 0.02 and Z = 0.1 are more favourable for the seed germination of these kinds of plants. It should be noted that the high concentrations of LSC are not harmful to all the plants, since most parts of the non-germinated seeds remain undamaged and survive in the Petri dishes. Using an optical microscope, we have found that the high LSC concentrations, characterised by high values of dynamic viscosity, form thick polymeric films at the seeds' surfaces, which can hinder the access of water to plant seeds. It is known that, during the germination period, seeds should get 35–60% of the total water quantity needed, and water shortage dramatically reduces the seed germination.

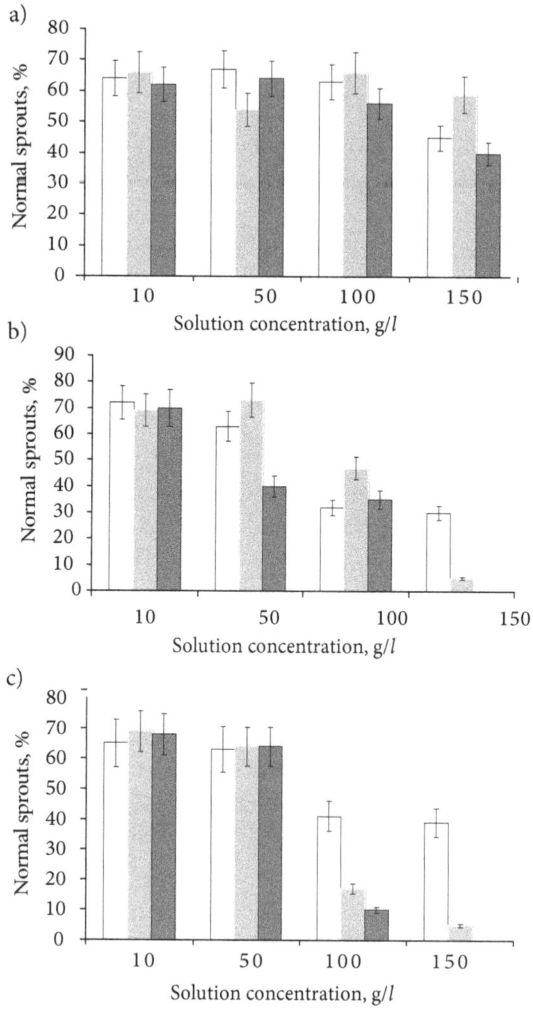

Fig. 6. Seed germination of *Gipsophila paniculata L.* (a) on 14th day after the onset of the experiment (LSD$_{05}$ 6.10, $S_{X\%}$ = 4.04); *Thymus serpyllum L.* (LSD$_{05}$ = 6.51, $S_{X\%}$ = 2.29) (b) and *Festuca rubra L.* (LSD$_{05}$ = 4.98, $S_{X\%}$ = 5.01) (c) on 21st day after the onset of the experiment versus the concentration of the applied solutions of LSC with different composition: □, 0.02; ▒, 0.10; ■, 0.20

Conclusions

1. Due to a pronounced gluing ability, the developed lignin-based conditioner is capable of reinforcing the surface layers of sandy soil with a thickness of 0.5–6.1 mm and a penetration resistance of 0.5–3.5 MPa with increasing an application rate of its water solution from 1.0 to 2.5 *l*/m^{-2} and a solution concentration from 10 to 200 g/*l*.

2. The two layered soil-mulch composite coverings, formed after applying the water conditioner solution to the top of the sandy soil mulched with the wood processing by-products, more efficiently suppressed the moisture evaporation from the soil in comparison with the untreated mulch covering. Evidently, the developed system of the interfacial physico-chemical bonds between the conditioner and the mulch has a favourable effect on the hydrophobicity of the formed soil-mulch composite. The

gluing efficiency of the soil conditioner towards the mulch depends on its composition, the applied solution concentration and the application rate.

3. Owing to the presence of a considerable amount of resin substances in the bark of conifers that can weaken the adhesive interaction between the conditioner and the mulch particles, the composite covering from the conifer wood bark has worse mulching properties than the mulch covering from the bark of deciduous wood.

4. The treated hydrolysis lignin-based mulch was characterised by higher compressive strength, higher water resistance and lower moisture losses from sandy soil for a given conditioner composition and application rate than the treated bark-based mulch due to the enhanced specific surface of the hydrolysis lignin as well as its high-developed porous structure, promoting the adhesion interaction between the conditioner and the mulch.

5. The 10–100 g/*l* LSC solutions, irrespective of the conditioner composition, have a favourable action on the viability and germination energy of the plant seeds from typical plants of the coastal dune zone. More concentrated solutions of LSC form the thick polymeric films at the seeds' surfaces, which can hinder the access of water to plant seeds.

Acknowledgements

The authors are grateful to the Latvian and Lithuanian Councils of Science for the financial support of the Eureka project E!2622 "Prevention and Control of Forest Soil Erosion on the Coastal Strip of the Baltic Sea by Using the Lignin-Based Polymers".

References

Aimi, H.; Ohmura, S.; Uetake, M.; Shimizu, K. 2009. Development of acid soil conditioning agent from lignin by ozone treatment II, *Journal of Wood Science* 55: 121–125. http://dx.doi.org/10.1007/s10086-008-1010-4

Calatrava, J.; Franco, J. A. 2011. Using pruning residues as mulch: analysis of its adoption and process of diffusion in Southern Spain olive orchards, *Journal of Environmental Management* 92: 620–629. http://dx.doi.org/10.1016/j.jenvman.2010.09.023

Curnoe, W. E.; Irving, D. C.; Dow, C. B.; Velema, G.; Unc, A. 2006. Effect of spring application of a paper mill soil conditioner on corn yield, *Agronomy Journal* 98: 423–429. http://dx.doi.org/10.2134/agronj2005.0041

Dymond, J. R.; Ausseil, A. G. E.; Overton, J. M. 2008. A landscape approach for estimating the conservation value of sites and site-based projects, with examples from New Zealand, *Ecological Economics* 66: 275–281. http://dx.doi.org/10.1016/j.ecolecon.2008.03.008

Dong, Z.; Wang, L.; Zhao, S. 2008. A potential compound for sand fixation synthesized from the effluent of pulp and paper mills, *Journal of Arid Environments* 72: 1388–1393. http://dx.doi.org/10.1016/j.jaridenv.2008.02.008

Foltz, R. B.; Copeland, N. S. 2009. Evaluating the efficacy of wood shreds for mitigating erosion, *Journal of Environmental Management* 90: 779–785. http://dx.doi.org/10.1016/j.jenvman.2008.01.006

Gimenez, R.; Govers, G. 2008. Effects of freshly incorporated straw residue on rill erosion and hydraulics, *Cattena* 72: 214–223. http://dx.doi.org/10.1016/j.catena.2007.05.004

International Seed Testing Association. 1996. *International Rules for Seed Testing*. International Society for Horticultural Science.

Kim, C.-G.; Kwangil, S.; Joo, K. J.; Lee, K. S.; Shin, S. S.; Yeonsook, C. 2008. Effects of soil conservation measures in a partially vegetated area after forest fires, *Science of the Total Environment* 399: 158–164. http://dx.doi.org/10.1016/j.scitotenv.2008.03.034

Kullmann, A. 1982. *Synthetic soil conditioners*. Moscow: Kolos, 84–86 (in Russian).

Lee, E. C.; Douglas, R. S. 2012. Geotextile tubes as submerged dykes for shoreline management in Malaysia, *Geotextiles and Geomembranes* 30: 8–15. http://dx.doi.org/10.1016/j.geotexmem.2011.01.003

Li, Y.; Shao, M.; Horton, R. 2011. Effect of polyacrylamide applications on soil hydraulic characteristics and sediment yield of sloping land, *Procedia Environmental Sciences* 11(B): 63–73. http://dx.doi.org/10.1016/j.proenv.2011.12.118

Pack, J. 2007. Hydromulch passes test for steep slope stabilization with the North Carolina DOT, *Land and Water* 51: 15–17.

Passauer, L.; Struch, M.; Schuldt, S.; Appelt, J.; Schneider, Y.; Jaros, D.; Rohm, H. 2012. Dynamic moisture sorption characteristics of xerogels from water-swellable oligo (oxyethylene) lignin derivatives, *ACS Applied Materials & Interfaces* 4: 5852–5862. http://dx.doi.org/10.1021/am3015179

Popa, V . I.; Stingu, A. P.; Volf, I. 2010. Lignin and Polyphenols in Bioremediation, in *Bioremediation Technology: Recent Advances*, M. H. Fulekar (Ed.). Capital Publishing Company, Springler, Heidelberg. http://dx.doi.org/10.1007/978-90-481-3678-0_4

Shao, G.; Dai, L.; Dukes, J. S.; Jackson, R. B.; Tang, L.; Zhao, J. 2011. Increasing forest carbon sequestration through cooperation and shared strategies between China and the United States, *Environmental Science Technology* 45: 2033–2034. http://dx.doi.org/10.1021/es200147k

Shulga, G.; Rekners, F.; Varslavans, J. 2001. Lignin-based interpolymer complexes as a novel adhesive for protection against erosion of sandy soil, *Journal of Agricultural Engineering Research* 78: 309–316. http://dx.doi.org/10.1006/jaer.2000.0599

Shulga, G.; Betkers, T.; Shakels, V.; Neiberte, B.; Verovkins, A.; Brovkina, J.; Belous, O.; Ambrazaitene, D.; Žukauskaite, A. 2007. Effect of the modification of lignocellulosic materials with a lignin-polymer complex on their mulching properties, *BioResources* 2: 572–582.

Shulga, G.; Betkers, T.; Brovkina, J.; Neiberte, B.; Verovkins, A.; Belous, O.; Ambrazaitene, D.; Žukauskaite, A. 2008a. New lignin-based polymers for ecological rehabilitation, *Molecular Crystals and Liquid Crystals* 483: 1333–1347.

Shulga, G.; Betkers, T.; Brovkina, J.; Aniskevicha, O.; Ozolinš, J. 2008b. Relationship between composition of the lignin-based interpolymer complex and its structuring ability, *Environmental Engineering and Management Journal* 7(4): 397–400.

Sojka, R. E.; Bjorneberg, D. L.; Entry, J. A.; Lentz, R. D.; Orts, W. J. 2007. Polyacrylamide in agriculture and environmental land management, *Advances in Agronomy* 92: 75–162. http://dx.doi.org/10.1016/S0065-2113(04)92002-0

Zhang, G.; Zhang, X.; Hu, X. 2013. Runoff and soil erosion as affected by plastic mulch patterns in vegetable field at Dianchi lake's catchment, China, *Agricultural Water Management* 122: 20–27. http://dx.doi.org/10.1016/j.agwat.2013.02.004

Galia SHULGA. Doctor of Polymer Chemistry (1982); Doctor Habilitatus of Wood Chemistry (1998). Leading researcher, Department of Lignin Chemistry, Latvian State Institute of Wood Chemistry (LSIWC). Publications: author/co-author over 200 scientific papers, 17 patents. Research interests: wood chemistry and technology, soil melioration, environmental technologies.

Talrits BETKERS. Master of Engineering Sciences (2007). Researcher, Faculty of Building and Civil Engineering, Riga Technical University (RTU). Publications: author/co-author over 50 scientific papers. Research interests: engineering sciences, environmental technologies.

Sanita VITOLINA. Master of Chemistry (2011). Researcher, Department of Lignin Chemistry, Latvian State Institute of Wood Chemistry (LSIWC). Publications: author/co-author over 30 scientific papers. Research interests: lignin chemistry and modification, environmental technologies.

Brigita NEIBERTE. Master of Chemistry (2006). Researcher, Department of Lignin Chemistry, Latvian State Institute of Wood Chemistry (LSIWC). Publications: author/co-author over 90 scientific papers, 3 patents. Research interests: lignin and lignocellulosic material modification, analytical chemistry of lignin, wood products.

Anrijs VEROVKINS. Master of Chemistry (2007); Doctor of Wood Chemistry (2013). Researcher, Department of Lignin Chemistry, Latvian State Institute of Wood Chemistry (LSIWC). Publications: author/co-author over 40 scientific papers. Research interests: lignin and bark modification, wood products, engineering sciences.

Olga ANNE. Doctor of Technological Sciences (1993). Professor of Ecology Department of Klaipėda University. Main areas of activity: Environmental Management, Environmental Impact Assessment, Renewable Energy. Publications: author/co-author over 50 scientific papers.

Dalia AMBRAZAITIENE. Doctor of Natural Sciences (1987), HP in Biomedical sciences; Professor of Ecology Department of Klaipėda University. Research interests: technological and environmental aspects related to use of nitrogen fixing bacteria preparations, biodegradation of organic materials and oil products in the soil, estimation of soil microbiological properties.

Audrone ZUKAUSKAITE. Doctor of Natural Sciences (1993). Assoc. Prof. Dr, Department of Technological Processes, Klaipėda University. Publications: author of more than 50 research papers. Research interests: waste treatment technologies, biodegradation of oil products.

DISSOLVED ORGANIC CARBON CONTENT AND LEACHABILITY OF BIOMASS WASTE BIOCHAR FOR TRACE METAL (Cd, Cu AND Pb) SPECIATION MODELLING

Enrico MANCINELLI[a], Edita BALTRĖNAITĖ[b], Pranas BALTRĖNAS[b], Eglė MARČIULAITIENĖ[b], Giorgio PASSERINI[a]

[a]*Department of Industrial Engineering and Mathematical Sciences, Università Politecnica delle Marche, via Brecce Bianche 12, 60131 Ancona, Italy*

[b]*Vilnius Gediminas Technical University, Saulėtekio al. 11, LT-10223 Vilnius, Lithuania*

Abstract. Dissolved organic carbon (DOC) interacts with dissolved trace metal affecting their mobility and bioavailability through the formation of DOC–metal complexes. Several types of biochar (BC) produced from slow pyrolysis of wood chips (WC), lignin (LG), and digested sewage sludge at 450 and 700 °C were tested for DOC leaching via batch and up-flow percolation test methods. Trace metal (Cd, Cu, and Pb) speciation modelling in BC eluates was carried out combining measured data (i.e., DOC, pH, temperature, and dissolved trace metal concentrations) with data reported in the literature regarding fractions of DOC that are inert or active (i.e., fulvic acids (FA) and humic acids (HA)) in metal binding. BC from LG (BCLG) and WC (BCWC) at 700 °C released lower cumulative amounts of DOC compared with BC at 450 °C in the range 0.02–0.07% and 0.06–0.09% of total carbon content, respectively. For both pyrolysis temperatures, BCWC exhibited a higher tendency to release DOC compared to BCLG. Speciation modelling results showed the predominance of FA and HA complexes of Cd, Cu, and Pb in all the eluates from BCWC and BCLG irrespective of the inert fractions of DOC or the different fractions of active FA and HA considered.

Keywords: biochar, dissolved organic carbon, trace metal speciation, leaching, waste management technologies.

Introduction

In water systems, dissolved organic carbon (DOC) interacts with dissolved trace metal affecting their mobility and bioavailability through the formation of DOC–metal complexes.

Biochar-borne DOC may represent a source of impairment of aquatic environment containing organic species producing an inhibitory effect on the growth of aquatic micro-organisms (Smith *et al.* 2012) or increasing metal mobility through complexation.

Several studies have reported a decrease in metal availability and leachability from biochar (BC) as a result of physical and chemical alterations occurring during pyrolysis of waste biomass (Lu *et al.* 2011; Lin *et al.* 2012; Devi, Saroha 2014).

Feedstock selection and pyrolysis conditions (i.e., heating rate, residence time and temperature) have been reported to influence the content, composition, and extent to which DOC is released from BCs (Mukherjee, Zimmerman 2013; Alburquerque *et al.* 2014; Jamieson *et al.* 2014).

As for the influence of feedstock type on DOC content as well as the extent to which it is prone to leaching from BC, several authors have reported that BCs from wood and woody biomass generally release less DOC compared to non-wood biomass (e.g., agricultural residue and animal wastes) (Mukherjee, Zimmerman 2013; Graber *et al.* 2014). Alburquerque *et al.* (2014) reported DOC content highest in BC from woody biomass (0.8–6.7 g/kg) consisting of leaves and thin branches (i.e., olive tree pruning), followed by BC from agricultural residue, such as almond shell (0.6–2.9 g/kg) and olive stone (0.3–1.5 g/kg), and wood biomass, such as pine wood (0.3–1.4 g/kg). In the work of Graber *et al.* (2014), DOC was detected at 1.5 to 9.7 times higher concentration in the extracts of BCs from non-wood biomass (i.e., green house waste and olive pomace) than in the extracts of BC from wood biomass (i.e., wood chips). Furthermore, these authors observed

Corresponding author: Enrico Mancinelli
E-mail: e.mancinelli@pm.univpm.it

that the influence of feedstock selection for BC production on the measured DOC concentrations was highest for BC at lower temperatures (i.e., 350 and 450 °C) compared with higher temperature (i.e., 600 °C).

Several authors have noted a decrease in DOC content with an increase in pyrolysis temperature (Gaskin *et al.* 2008; Lin *et al.* 2012; Mukherjee, Zimmerman 2013; Zhang *et al.* 2015; Alburquerque *et al.* 2014). In the work by Smith *et al.* (2016), each increase in pyrolysis temperature of 100 °C led to a decrease in DOC content in the range 41–44% in BC produced from lignin in the temperature range 300–500 °C. Al-Wabel *et al.* (2013) reported the largest decrease (from a DOC content of about 1.8 g/kg to <0.6 g/kg) for BC produced in the range 200–400 °C, probably due to the low carbonisation reached at 200 °C and the resulting higher content of organic materials suggested by the highest H/C ratio.

Conversely, some studies reported an increase in the DOC content of BC with increases in pyrolysis temperature from 350 to 400 °C (Luo *et al.* 2015) and from 350 to 450 °C (Graber *et al.* 2014).

Jamieson *et al.* (2014) suggested that pyrolysis conditions such as fast heating rates and low residence times may lead to produce BCs that are more prone to leaching DOC. These authors observed an increase of about 90% in DOC concentration in the leachate from fast pyrolysis BC at 600 °C compared with the slow pyrolysis BC at 380 °C.

DOC composition of BC is affected by the processing conditions (e.g., pyrolysis temperature) selected for producing BC (Trompowsky *et al.* 2005; Lin *et al.* 2012; Uchimiya *et al.* 2013). Trompowsky *et al.* (2005) reported a temperature trend for the yield of fulvic acids (FA) and humic acids (HA) extracted from wood BC pyrolysed in the range 300–550 °C. At the lowest pyrolysis temperature (300–350 °C), the yield of FA was highest whereas the yield of HA was the lowest. With each temperature increase of 50 °C from 350 to 550 °C, the yield of FA exhibited a steady decrease, whereas HA reached its maximum yield at 450 °C and subsequently decreased. Uchimiya *et al.* (2013) observed that DOC released by biochars in the temperature ranging 350–500 °C was enriched with carboxyl and poly(phenolic) functionalities that are able to affect metal mobilisation via complexation with metal ions.

As for the mechanism that leads to the release of DOC from BC, Jones *et al.* (2011) suggested that DOC is released at first from the surface and later from the BC matrix following a diffusion-limited mechanism. Observing a good fit between the curves showing the cumulative released quantity of DOC determined via leaching and batch methods, Mukherjee and Zimmerman (2013) suggested that the amount of DOC released from biochars depends on the ratio of the solvent to solute rather than parameters, such as contact time and energy of mixing. In column leaching tests performed by Beesley and Marmiroli (2011) and Iqbal *et al.* (2015), DOC was observed to be rapidly released from biochar with no substantial amount of DOC recorded after a few eluate samples.

As for trace metal speciation in water, dissolved organic matter (DOM) can be subdivided into an active fraction comprising humic substances able to complex metals (i.e., fraction of DOM active for metal binding) and other biogenic molecules (e.g., amino acids, proteins and carbohydrates) not affecting metal speciation (i.e., fraction of DOM inert for metal binding) (Mueller *et al.* 2012). Based on an extensive literature review, Groenenberg and Lofts (2014) reported active fractions of DOM in the range 31–100%. DOM is generally quantified as DOC (Wood *et al.* 2011) (e.g., 2 times the DOC values measured in the sample analysis) (Ahmed *et al.* 2013).

The capacity of DOC to form metal complexes depends on the presence and availability of carboxylic and phenolic functional groups. Through deprotonation, these functional groups provide DOC with negatively charged sites that are able to bind metal cations (Dee 2016). Furthermore, as these functional groups exhibit an acid-base behaviour, the metal binding ability of DOC depends on the pH of the system (Christensen, J., Christensen, T. 2000).

The chemical speciation program WHAM7 (Tipping *et al.* 2011) is based on WHAM (Tipping 1994), a mechanistic-based equilibrium model simulating the interactions of protons and metal cations with humic substances. As input data, WHAM model requires the fraction of DOM that is active for proton/metal binding and the sub-fractions that act as isolated purified humic substances (i.e., active fractions of HA or FA). The concentration of humic substances to be input in the model is often based on the measured DOC concentration and on some assumptions about DOC composition and reactivity (i.e., HA or FA) (Ahmed *et al.* 2013). For a comprehensive review about the most advanced ion-binding models (including WHAM) applied so far for predicting trace element speciation in water systems, we recommend the work by Groenenberg and Lofts (2014).

In order to avoid potential drawbacks in BC water system application (e.g., filtering media) and to ensure the durability of BC in biofiltration systems (Baltrėnas *et al.* 2016) or its application in soil (Baltrėnaitė *et al.* 2016) resulting from leaching of DOC, it is fundamental to determine the rate of carbon content that is prone to leaching as DOC and its effect on trace metal speciation and mobility.

The goals of the present research were:

1. To evaluate and compare changes in DOC leachability from BCs produced from different types of biomass feedstock (i.e., wood chips, lignin, and digested sewage sludge) at different temperatures.

2. To model trace metal (Cd, Cu, and Pb) speciation in the BC eluates.

To the best of the authors' knowledge, few studies have assessed DOC content and composition of BC from lignin (Smith *et al.* 2016) or trace metal concentration, leachability and bioavailability in BC from lignin (Mancinelli *et al.* 2016). Hence, the need for investigating DOC content and leachability of BC and for modelling trace metal speciation in the BC eluates.

1. Materials and methods

1.1. Biochar production

Wood chips (WC) from a sawmill in Kaliningrad (Russian Federation), lignin (LG) from a sugar refining industry located in Lithuania, and digested sewage sludge (DSS) from an urban waste water treatment plant in Klaipeda (Lithuania) were used as raw materials for BC production.

The pyrolysis of the raw materials was carried out according to the procedure described elsewhere (Komkiene, Baltrenaite 2016). Briefly, the slow pyrolysis was performed for 120 min with a heating rate of approximately 10 °C/min until the two desired pyrolysis temperatures of 450 and 700±5 °C were reached.

The obtained BCs were assigned the following codes:

- BCWC450 and BCWC700 for BC from wood chips (BCWC) at the pyrolysis temperatures of 450 and 700 °C, respectively,
- BCLG450 and BCLG700 for BC from lignin (BCLG) at the pyrolysis temperatures of 450 and 700 °C, respectively,
- BCDSS450 and BCDSS700 for BC from digested sewage sludge (BCDSS) at the pyrolysis temperatures of 450 and 700 °C, respectively.

1.2. Up-flow percolation test

An up-flow percolation test was performed in order to determine the leaching behaviour of BCs, as fully described in Peckyte and Baltrenaite (2015). Columns were packed with 137.5 g BCWC450, 130.8 g BCWC700, 210.5 g BCLG450, and 316.7 g BCLG700. BCs were soaked with a leachant (i.e., deionised water) for three days, and then hydrostatic pressure was applied resulting in a flow rate of approximately 12 ml/h. The first volume of eluate equal to a cumulative liquid to solid (L/S) ratio of 1 l/kg dry weight (DW) (i.e., 0.137, 0.131, 0.210, 0.317 litres of eluate from BCWC450, BCWC700, BCLG450, and BCLG700, respectively) was not analysed for DOC. At the beginning of the column leaching test (i.e., until L/S ratios of 0.1 or 0.5 l/kg DW), labile species may be rapidly released by the tested material with the resulting elution curve showing a first flush behaviour (Van der Sloot, Dijkstra 2004). In the present research, DOC measurements were taken for the eluate collected at L/S ratios equal to 1, 3, and 5 l/kg DW, thus not considering the fraction of DOC in BC that is subject to rapid washout.

1.3. Chemical analysis

The eluate collected during the up-flow percolation test of BCWC and BCLG at L/S ratios equal to 1, 3, and 5 l/kg DW was transferred to the laboratory for electrical conductivity (EC), pH and DOC measurements and trace metal analysis.

EC and pH were measured immediately after sampling. EC was measured with a conductivity meter (inoLab Cond 740 WTW) in accordance with the methodology of *ISO 7888: Water quality – Determination of electrical conductivity*. pH was measured with an electronic glass electrode (SevenMulti ion/pH/ORP module Mettler Toledo) after calibrating the pH-meter with buffer solutions at pH values of 4, 7, and 10. pH measurements were carried out in accordance with the methodology described in *ISO 10523: Water quality – Determination of pH*. For DOC analysis, samples of about 40 ml volume were taken from the eluate collected during the up-flow percolation test. After filtering the samples through a syringe (pore size 0.45 μm, Pall Corporation), DOC was determined by employing a total organic carbon (TOC) analyser (Shimadzu-V CSN).

DOC concentrations in BCs were also determined via a batch test method. Water extracts were prepared using a 1:20 (w:v) suspension after shaking (RS12 Rotoshake, Gerhardt) for 180 min according to a method described by Dias *et al.* (2010). Samples were filtered through a syringe (pore size 0.45 μm, Pall Corporation), and DOC was determined by a TOC analyser (Shimadzu-V CSN).

DOC concentrations were corrected by subtracting the mean concentration (20.4 mg/l), as measured in the blank samples. Samples were stored at 4 °C in the dark until DOC analysis was accomplished.

As for the batch test and up-flow percolation test, the dissolved organic carbon (DOC; g/kg) contents of the BCs were calculated using the following equation, adapted from Jamieson *et al.* (2014):

$$\text{DOC} = \frac{V \times C}{M}, \tag{1}$$

where V – volume of deionised water utilised as eluate or extractant, l; C – measured concentration of DOC in the sample, mg/l; M – dry mass of biochar, g.

As far as the up-flow percolation test is concerned, the cumulative released quantity $\left(\sum_i \text{DOC}_i\right)$ of DOC, g/kg, was calculated for analysis in the eluate fraction as described below:

$$\sum_i \text{DOC}_i = \sum_i \frac{V_i \times C_i}{M}, \tag{2}$$

where M – dry mass of biochar, g; i – index of the eluate fraction from 1 to 3; C_i – measured concentration of DOC in the sample, mg/l; V_i – volume of eluate fraction varying with the L/S ratios set in the up-flow percolation test, l.

The cumulative released quantity $\left(\sum_i \mathrm{DOC}_i\right)$ of DOC expressed as a percentage of total carbon content of the biochar was calculated as follows:

$$\sum_i \frac{\mathrm{DOC}_i}{\mathrm{TC}} \times 100 = \sum_i \frac{V_i \times C_i}{M} \times \frac{1}{\mathrm{TC}} \times 100, \quad (3)$$

where DOC_i – dissolved organic carbon calculated as in equation (2), g/kg; TC – total carbon content of biochar, g/kg; i – index of the eluate fraction; V_i – volume of deionised water utilised as eluate, l; C_i – measured concentration of DOC in the sample, mg/l; M – dry mass of biochar, g.

For trace metals (Cd, Cu, and Pb) analysis, 50 ml of the eluate from the collected sample was filtered through sterile Acrodisc syringe filters with a Supor (hydrophilic polyethersulfone) membrane and pore size 0.45 μm (Pall Corporation) (Mancinelli *et al.* 2015). Measurements were taken by flame atomic absorption spectrophotometry (FAAS) or graphite furnace (GFAAS) analysis according to the procedure described by Baltrėnaitė *et al.* (2016).

1.4. Quality assurance

For quality assurance, measurements were repeated three times. All samples were prepared in duplicate. For DOC determination in BCs, two blank samples were prepared following the entire sequence of steps, as described for DOC analysis.

1.5. Calculation and statistical analysis of the experimental data

Statistics were accomplished using Excel (Microsoft). Analysis of Variance (ANOVA) was used to test differences between the data using Excel (Microsoft). Assuming the truth of the null hypothesis (i.e., the population means are all equal), the p-value is the probability of observing an event at least as extreme as the one in the tested sample data. In the present study, the differences were significant with $p \leq 0.05$ (i.e., 5% risk of rejecting the null hypothesis when it is true).

Comparisons between the DOC concentrations in the water extracts of BC (i.e., measurement value) determined via the batch test method, pyrolysis temperature and the feedstock (i.e., the nominal variables chosen to organise the measurements into groups) were carried out using the two-way ANOVA with replications. The two-way ANOVA with replications was calculated within each feedstock type in order to determine any statistical differences between the pH or EC values of the eluates collected from the BC up-flow percolation test (i.e., the measurement value), L/S ratios, and the pyrolysis temperatures (i.e., the nominal variables).

Comparisons between the DOC concentrations of the eluates collected from the BC up-flow percolation test (i.e., measurement value), pyrolysis temperatures, and the *L/S* ratios (i.e., the nominal variables chosen to organise the measurements into groups) were carried out using the two-way ANOVA with replications.

1.6. WHAM modelling

Trace metal speciation modelling was done using WHAM/ Model 7 (Tipping *et al.* 2011) for waters. The WHAM model determines metal binding to DOM, such as metal binding to HA and FA. In order to define the binding parameters that represent the nature and number of sites available for metal binding with organic ligands, input data WHAM model requires the fraction of DOM that is active for proton/metal binding and sub-fractions that act as HA or FA. In the present study, DOM was considered colloidal, and it was assumed to consist of 50% *DOC*, which is 2 times the DOC values measured in the sample analysis (Ahmed *et al.* 2013). Since DOC composition and reactivity (i.e., active fractions of HA and FA) were not determined in the present study, a range of fractions of DOC acting as active HA and FA according to values reported in the literature (Table 1) was assumed. We also assumed that the active fraction of DOC in BC has the binding properties of FA and HA, as these fractions were detected in the BC extract in a study by Trompowsky *et al.* (2005).

Table 1. Summary of the input values for trace metal speciation modelling: fractions of dissolved organic carbon (DOC) considered inert and active (i.e., colloidal fulvic acid (FA) and humic acid (HA)) to metal complexation

	Inert, %	Active FA, %	Active HA, %	References
DOC	0	90	10	Unsworth *et al.* 2006
	35	65	0	Ahmed *et al.* 2013
	35	55	10	

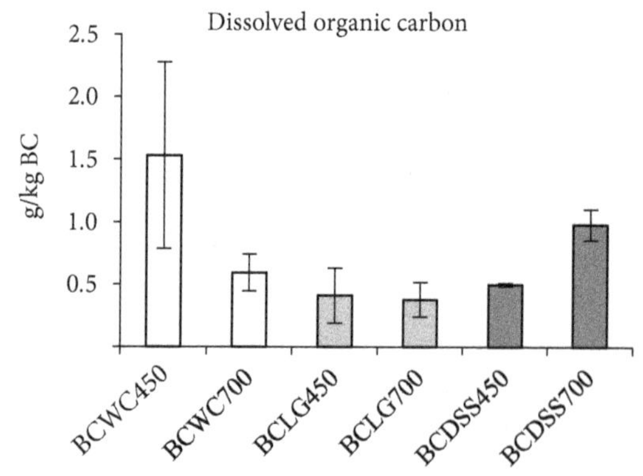

Fig. 1. Dissolved organic carbon content in biochars (BCs) determined via batch method. Values are shown as the mean of two observations (n = 2) ± standard deviation. BCs at 450 and 700 °C from wood chips, lignin and digested sewage sludge are BCWC450, BCWC700, BCLG450, BCLG700, BCDSS450, and BCDSS700, respectively

The values of Cd, Cu, and Pb concentration were input as measured in the BC eluates (Figs S1–3 in supplemental material). Regarding the solute type adopted for the modelling, the concentration in BC eluates of trace metals in soluble form (syringe filtered water, 0.45 μm pore size) was identified as total dissolved concentration.

2. Results and discussion

2.1. Dissolved organic carbon in a water extract of biochars via batch test method

Figure 1 shows the results of the DOC analysis in the water extract of BCs via the batch test method. Several studies have observed that wood and woody biomass feedstock tend to release less DOC than non-wood biomass feedstock (Mukherjee, Zimmerman 2013; Alburquerque *et al.* 2014; Graber *et al.* 2014).

At the pyrolysis temperature of 450 °C, BCWC450 showed the highest DOC content which was about 3 and 4 times higher than the DOC contents of both BCLG450 and BCDSS450. At the pyrolysis temperature of 700 °C, BCDSS700 exhibited the highest DOC content, which was about 2 and 3 times higher than the DOC contents of both BCWC700 and BCLG700.

With regard to the observed differences in DOC content of BC from different types of feedstock, transformations of organic structure into recalcitrant forms (Wu *et al.* 2011) as well as secondary reactions (Lin *et al.* 2012) occurring during pyrolysis process may have resulted in lowering the content of DOC in the batch extracts of BCWC and BCLG.

During the pyrolysis process, physical alterations (e.g., formation of larger pores and increases in surface area) occurring in the char matrix (Liaw, Wu 2015) or volatilisation of organic volatiles (Zhang *et al.* 2015) could lead to opening blocked pores, resulting in a higher concentration of DOC in the batch extracts of BCDSS700.

The DOC temperature trend varied significantly ($p < 0.05$) depending on the feedstock. With the increase in pyrolysis temperature, BCWC and BCLG showed a significant ($p < 0.05$) decrease in DOC concentration in the water extract by about 61 and 7%, respectively. A similar trend in a decrease in the amounts of DOC with an increase in pyrolysis temperatures has been reported by several authors over a wide range of biomass feedstock and pyrolysis temperatures (Gaskin *et al.* 2008; Mukherjee, Zimmerman 2013; Graber *et al.* 2014; Smith *et al.* 2016). Conversely, the amount of DOC in BCDSS increased by 96% with the increase in pyrolysis temperature. The observed trend was inconsistent with the results reported by Zhang *et al.* (2015), who noted a decrease in DOC content associated with each increase in temperature of 100 °C in BC from sewage sludge pyrolysed in the range 300–900 °C.

The observed differences in the DOC temperature trend may be related to differences in molecular weight distribution of DOC between the different types of BC. With increases in pyrolysis temperatures the DOC content of BC decreases because of the decomposition of DOC with a lower molecular weight (Luo *et al.* 2015), resulting in a lower concentration of DOC in the batch extracts of BC from wood chips and lignin at 700 °C. The observed increase in DOC content of BC from digested sewage sludge with an increase in pyrolysis temperature may result from the thermo-stable behaviour of DOC with higher molecular weight and the concurrent decrease in BC yield (Luo *et al.* 2015).

2.2. Results of the up-flow percolation test method

Due to time and resource constraints, DOC measurements were taken only in the eluates collected from the up-flow percolation test performed on BCWC450, BCWC700, BCLG450, and BCLG700. These BCs were selected for

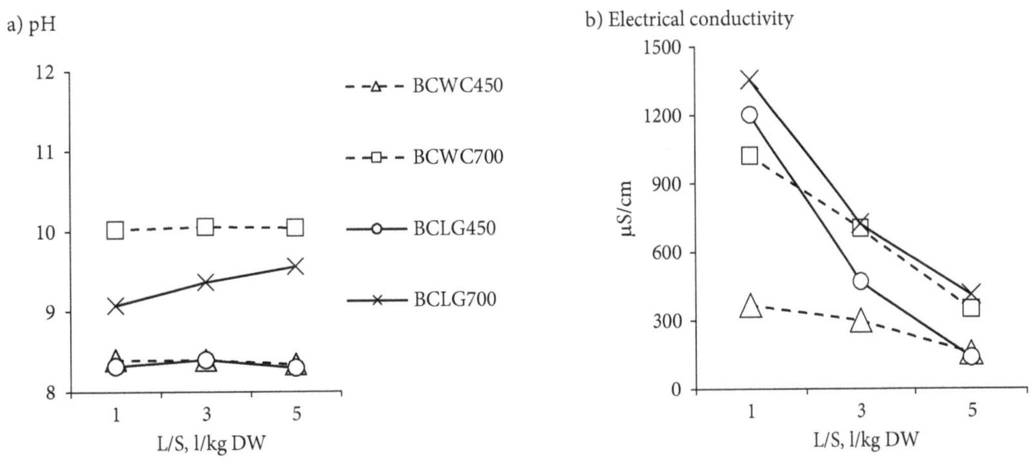

Fig. 2. Mean values of (a) pH and (b) electrical conductivity of biochar (BC) eluates collected at various liquid to solid (L/S) ratios. Values are shown as the mean of two observations (n = 2). Error bars of each data point would be smaller than size of symbol. DW – dry weight. BCs at 450 and 700 °C from wood chips and lignin are BCWC450, BCWC700, BCLG450, and BCLG700, respectively

testing DOC leaching behaviour because of the TC analysis results reported in a previous study by Mancinelli *et al.* (2016). These authors measured the highest TC content in BCWC and BCLG for both the applied pyrolysis temperatures of 450 and 700 °C which were between 54 and 68%, respectively, higher than the content of TC in BCDSS.

The pH and EC of the eluates collected from BCWC and BCLG up-flow percolation test increased significantly ($p < 0.05$) with increases in pyrolysis temperatures (Fig. 2). For each feedstock type, EC values of the eluates significantly ($p < 0.05$) decreased with increases in liquid to solid ratios. EC provides information about the presence of total soluble salts in the eluate. For all the tested BCs, with increasing volume leached through the columns, BCs released decreasing quantities of soluble salts.

For each feedstock type, pH values of the eluates showed significant ($p < 0.05$) differences for liquid to solid ratios and pyrolysis temperatures. BCWC showed the highest rise in alkalinity, with the pH of eluates from BCWC700 being between 1.6 and 1.7 units higher with each increase in liquid to solid ratio compared to the respective pH levels from BCWC450. The pH levels of the eluate samples from BCLG700 were between 0.8 and 1.3 units higher than the respective levels from BCLG450. For all the liquid to solid ratios, BCs at 450 °C from wood chips and lignin showed a similar alkaline pH in the range 8.3–8.4.

Figure 3 shows the DOC concentrations determined in the eluates collected at various liquid to solid ratios in the up-flow percolation test of BCWC and BCLG. Irrespective of the temperature adopted for pyrolysis treatment, BCWC and BCLG showed a similar release trend characterised by a significant ($p < 0.05$) decrease in DOC concentrations of the eluates collected at increasing liquid to solid ratios.

An increase in L/S ratio from 1 to 3 l/kg DW produced the highest decreases in the DOC concentrations of the eluates collected from BCLG700 (50%) and BCWC450 (49%), followed by BCLG450 (30%) and BCWC700 (21%). A further increase in L/S ratio from 3 to 5 l/kg DW showed a substantial decrease (71%) in the DOC concentration of the eluate collected from BCLG450, followed by BCWC700 (49%), BCWC450 (18%), and BCLG700 with the slightest decrease (1%). In the works by Beesley and Marmiroli (2011) and Iqbal *et al.* (2015), DOC was observed to be rapidly released from BC with no substantial amount of DOC recorded after a few eluate samples.

Low temperature (450 °C) BCLG and BCWC significantly ($p < 0.05$) released greater amounts of DOC compared with the respective high temperature (700 °C) BC. A decrease in DOC content with an increase in pyrolysis temperature was reported in a previous study on BC leaching trends by Mukherjee and Zimmerman (2013). A decrease in water leachability of organic matter (quantified as total organic carbon as percentage of TC) in BC is related to the transformation of organic structure into recalcitrant forms occurring during pyrolysis (Wu *et al.* 2011).

In order to analyse the fraction of carbon that is released from BC, the DOC content of BCs was expressed as a percentage of the respective TC content (Table 2).

With regard to the comparison between the DOC results of the batch test and the TC content of biochars, our results were comparable with those reported by Jones *et al.* (2011), who calculated DOC in water extracts of BC from mixed hardwood at 450 °C, which was 0.04% of the TC content of BC. We calculated the highest percentage of DOC for BCWC450 (0.24% TC), followed by BCDSS450 (0.17% TC), and BCLG450 (0.05% TC).

Regarding the up-flow percolation test, the calculated cumulative released quantity of DOC at the cumulative liquid to solid ratio of 9 l/kg DW was highest for BCWC450 (0.09% of the TC content), followed by BCWC700 (0.07% of the TC content), BCLG450 (0.06% of the TC content), and BCLG700 (0.02% of the TC content) (Table 2).

For both pyrolysis temperatures, BCWC exhibited a higher tendency to release DOC compared with BCLG. The cumulative released quantity of DOC of BCWC expressed as a percentage of TC content was 1.5 times higher at 450 °C and 3.5 times higher at 700 °C than the respective quantity of DOC calculated for BCLG. In the work by Li *et al.* (2014), BC from wood and lignin pyrolysed at 400 and 600 °C had different surface properties, with BC from wood exhibiting higher micropore structure (specific surface area, micropore area and volume) compared with BC from lignin at both pyrolysis temperatures. Closed or blocked pores that were formed in BC during pyrolysis open due to physical rearrangement occurring during thermochemical treatment (e.g., partial steam gasification)

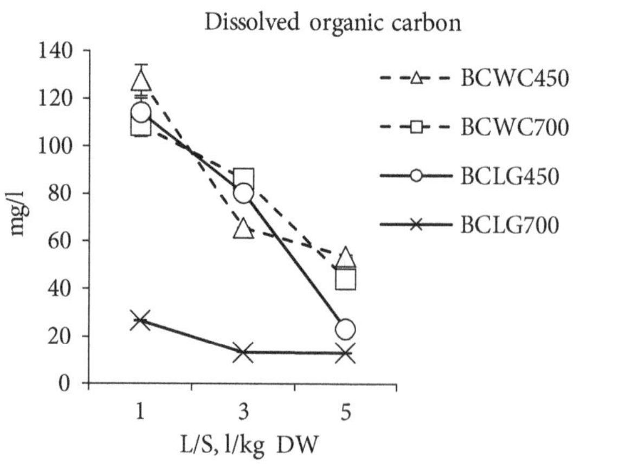

Fig. 3. Mean values of dissolved organic carbon (DOC) of biochar (BC) eluates collected at various liquid to solid (L/S) ratio. Values are shown as the mean of two observations ($n = 2$) ± standard deviation. DW – dry weight. BCs at 450 and 700 °C from wood chips and lignin are BCWC450, BCWC700, BCLG450, and BCLG700, respectively

Table 2. Dissolved organic carbon (DOC) content in biochars (BCs) determined via batch test and up-flow percolation test expressed as a percentage of the respective total carbon (TC) content

	Batch test		Up-flow percolation test			
	TC[1], g/kg BC	Amount of DOC as a % of TC	Amount of DOC as a % of TC			Cumulative amount of DOC as a % of TC
		(1:20) w:v	L_1/S	L_2/S	L_3/S	$\Sigma L_i/S$
BCWC450	629.2 ±44.1	0.24 ±0.14	0.02 ±0.0	0.03 ±0.0	0.04 ±0.0	0.09
BCWC700	876.9 ±67.2	0.07 ±0.02	0.01 ±0.0	0.03 ±0.0	0.03 ±0.0	0.07
BCLG450	839.2 ±19.0	0.049 ±0.03	0.01 ±0.0	0.03 ±0.0	0.01 ±0.0	0.06
BCLG700	717.4 ±49.2	0.053 ±0.02	0.00 ±0.0	0.01 ±0.0	0.01 ±0.0	0.02
BCDSS450	290.4 ±2.2	0.17 ±0.00	N.D.	N.D.	N.D.	N.D.
BCDSS700	279.4 ±1.8	0.35 ±0.04	N.D.	N.D.	N.D.	N.D.

[1] Values of total carbon (TC) concentration in the studied BCs were determined in a study by Mancinelli *et al.* (2016)
BCs at 450 and 700 °C from wood chips, lignin and digested sewage sludge are BCWC450, BCWC700, BCLG450, BCLG700, BCDSS450, and BCDSS700, respectively
DOC – dissolved organic carbon. DW – dry weight. N.D. – not determined. w:v –weight to volume ratio. L_i/S – liquid to solid (L/S) ratio equal to 1, 3, and 5 l/kg DW for i equal to 1, 2, and 3, respectively. $\Sigma L_i/S$ – cumulative L/S ratio

(Liaw, Wu 2015) or due to the swelling of BC in water (Ahmad *et al.* 2014). Therefore, organic matter, which was entrapped in the char matrix, may become more accessible to leaching. It is likely that the leaching condition applied in the up-flow percolation test enhanced the swelling of BCWC in water, thus resulting in more DOC as a percentage of TC released with increasing liquid to solid ratios compared with BCLG.

As regards the temperature-related trend of the cumulative released quantity of DOC expressed as a percentage of the respective TC content, with an increase in pyrolysis temperature BCWC and BCLG showed a decrease of about 22 and 67%, respectively (Table 2).

With an increase in pyrolysis temperature, results of the up-flow percolation test and batch test showed a similar trend for BCWC with a decrease by about 22 and 71%, respectively (Table 2). Comparison of batch and up-flow percolation test results showed an opposite trend for BCLG, with an increase by about 8% and a decrease by about 67%, respectively (Table 2).

2.3. Trace metal speciation biochar eluates predicted by WHAM

Trace metal speciation predictions were obtained assuming various active DOC fractions of FA (0.55–0.90) and HA (0–0.1) covering a range of assumptions reported in the literature about WHAM modelling for trace metal speciation in waters regarding active and inert fractions of DOM.

Speciation model results of Cd, Cu and Pb showed their affinity to form complexes in all the eluates collected at various L/S ratios from BCWC and BCLG, being the sum of the fractions of these trace metals bound to colloidal FA and HA equal to the respective total dissolved concentrations for all the assumptions taken in consideration for WHAM modelling (i.e., various fractions of DOC considered as active FA and HA). According to Unsworth *et al.* (2006), Fe, Al, and Mn can compete with Cd, Cu, and Pb for binding with HA and FA, thus affecting their speciation.

The high proportion of metal-FA and metal-HA complexes predicted for all the modelled eluate types may be due to the observed characteristics, such as alkaline pH levels, high DOC concentrations and the absence of elements that can compete for binding with HA and FA, such as major cations (i.e., Ca^{2+} and Mg^{2+}), Fe, Al, and Mn.

Comparisons between the free ion activity of Cd, Cu, and Pb considering 35% inert fractions of DOC and no inert fractions of DOC (Table 1) showed between 2.3 and 3 times higher free ion activity of the analysed trace metals (Figs 4–6).

Regarding the free ion activity of the analysed trace metals considering 35% DOC as inert fractions (Table 1), the modelling results of the eluates collected from BCWC and BCLG exhibited a different trend. Regardless of the L/S ratios, the modelling results of the eluates collected from BCWC exhibited higher free ion activity for Cd (between 11.4 and 23.3%), Cu (between 13.6 and 25.8%), and Pb (between 11.6 and 23.3%) considering 65% DOC acting as active FA compared to 55% DOC acting as active FA and 10% DOC acting as active HA.

Except for the eluate collected from BCLG700 at the final L/S ratio, in all the eluates collected from BCLG, the free ion activity was higher for Cd (between 0.8 and 16.8%), Cu (between 3 and 22.4%), and Pb (between 1 and 16.9%) considering 65% DOC acting as active FA compared to 55% DOC acting as active FA and 10% DOC acting as active HA. Conversely, the eluate collected from BCLG700 at the final L/S ratio showed a decrease by about 3.4–3.7 times in free ion activity.

In the work by Chakraborty and Chakrabarti (2006), a decrease in electrical conductivity and hence in ionic strength in diluted mine effluents was associated with stronger bonds in metal-DOC complexes resulting from a higher electrostatic attraction between positively charged metal ions, such as Cd^{2+} and Cu^{2+}, and the negatively charged sites of DOC.

In the present study, pH values of all the BC eluates were in the alkaline range. Furthermore, electrical conductivity greatly decreased with increasing L/S ratios. In addition, for all the modelled eluate types a strong affinity of Cd, Cu, and Pb to form DOC complexes was predicted. Therefore, it is likely that the trace metals released from BCWC and BCLG formed a strong bond with DOC, thus not representing a risk of impairment for the receiving water bodies.

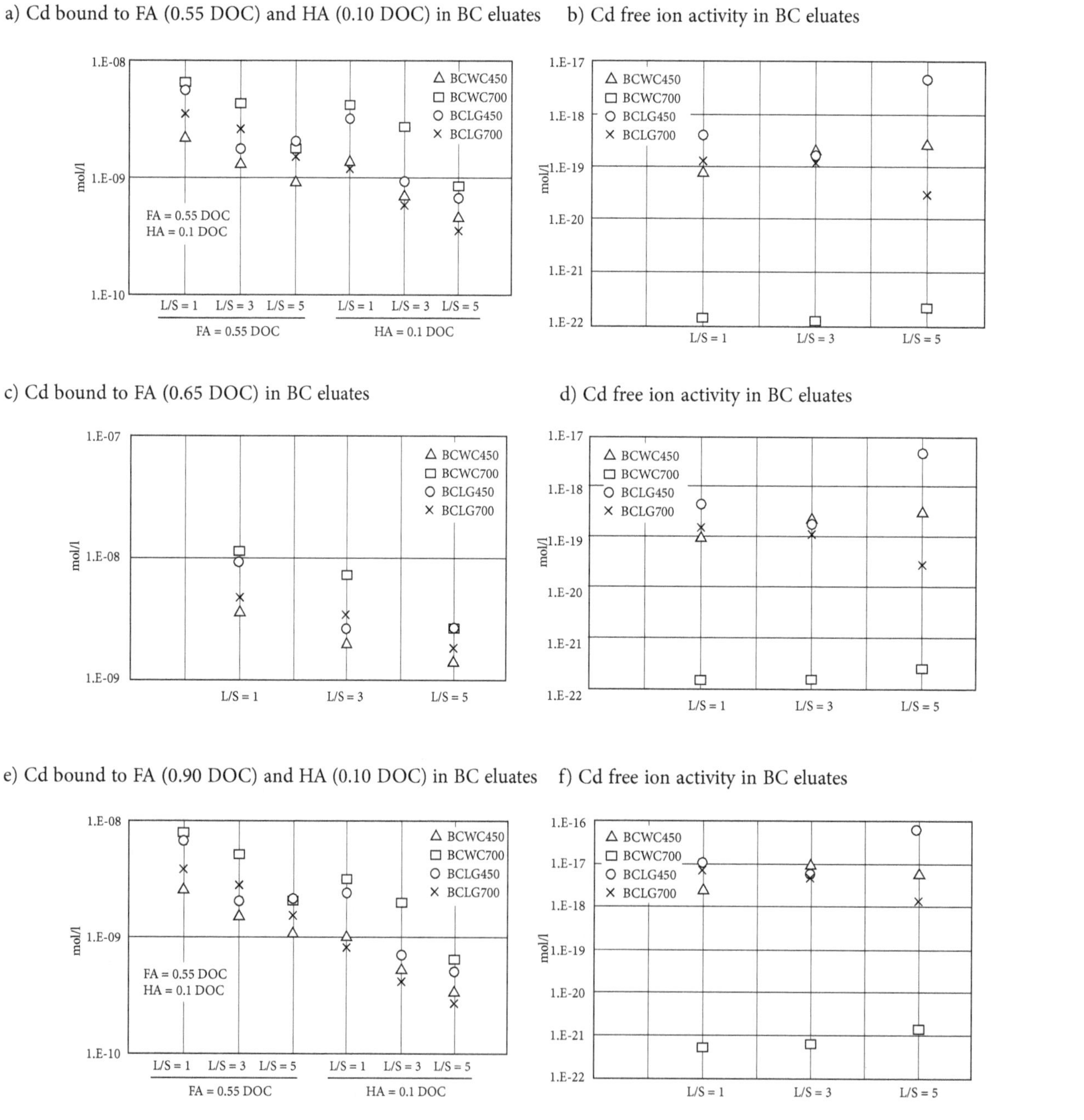

Fig. 4. Fraction bound to colloid fulvic acid (FA) and humic acid (HA) (a, c, e) and free ion activity (b, d, f) of Cd in biochar (BC) eluates collected at various liquid to solid (L/S) ratios, predicted by WHAM. DOC – dissolved organic carbon. BC at 450 and 700 °C from wood chips and lignin and are BCWC450, BCWC700, BCLG450, and BCLG700, respectively

Conclusions

1. For each type of feedstock (i.e., wood chips (WC) and lignin (LG)), analysis of the up-flow percolation test showed the same temperature related trend in dissolved organic carbon (DOC) leachability from biochars (BCs), with high temperature (700 °C) BCs releasing lower cumulative amounts of DOC compared with low temperature (450 °C) BCs, which were in the range 0.02–0.07% and 0.06–0.09% of total carbon (TC) content, respectively. Under the adopted pyrolysis conditions, the increase in pyrolysis temperature from 450 to 700 °C was more effective in lowering DOC leachability from BCs from LG (BCLG) compared with BCs from WC (BCWC), with a decrease by about 67 and 22% in the cumulative released quantity of DOC expressed as a percentage of TC content, respectively.

2. Regarding the influence of feedstock selection for BC production on DOC leachability, for both pyrolysis temperatures, BCWC exhibited a tendency to release more DOC compared with BCLG, as it was the cumulative

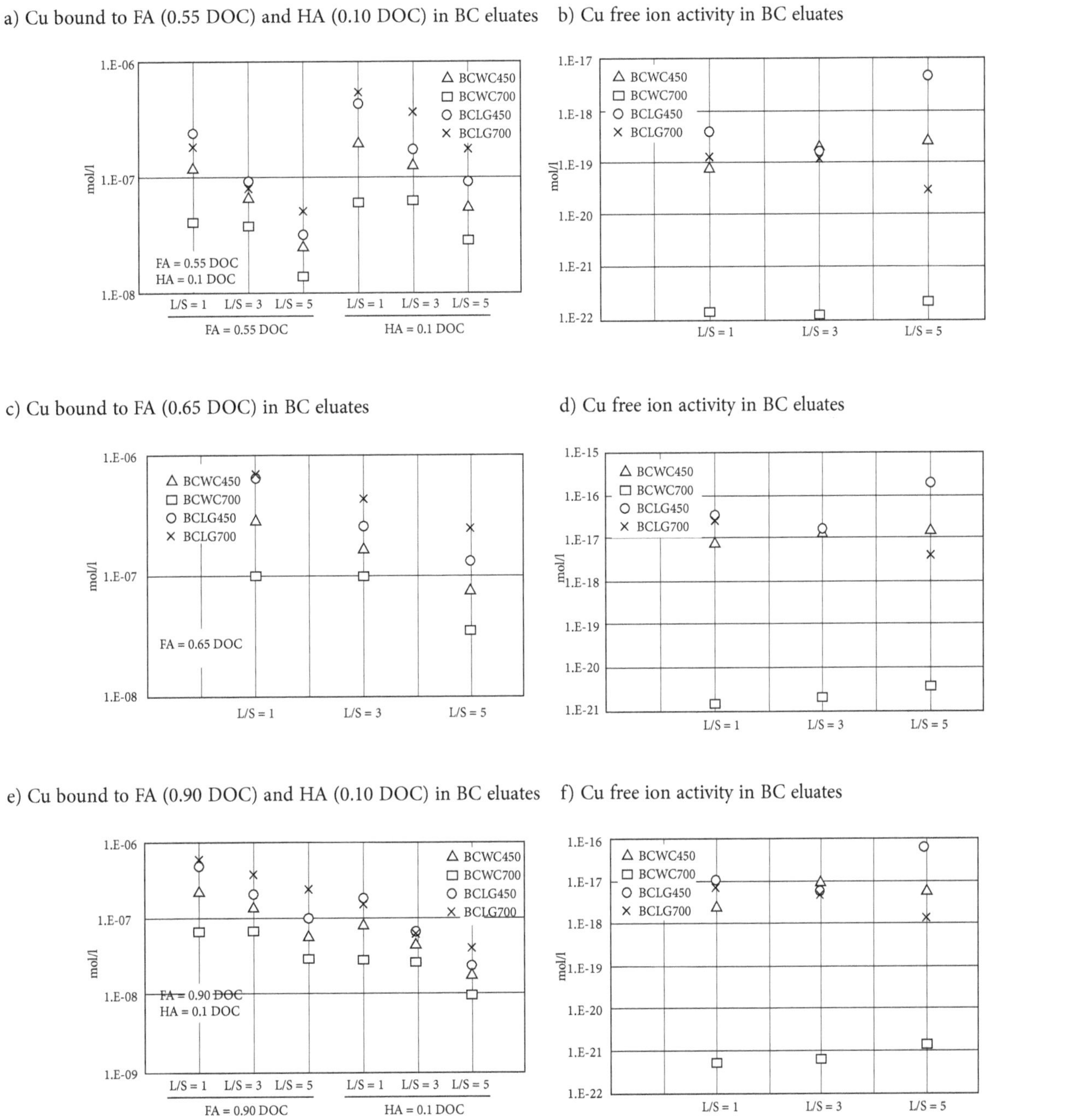

Fig. 5. Fraction bound to colloid fulvic acid (FA) and humic acid (HA) (a, c, e) and free ion activity (b, d, f) of Cu in biochar (BC) eluates collected at various liquid to solid (L/S) ratios, predicted by WHAM. DOC – dissolved organic carbon. BC at 450 and 700 °C from wood chips and lignin and are BCWC450, BCWC700, BCLG450, and BCLG700, respectively

released quantity of DOC of BCWC expressed as a percentage of TC content 1.5 times higher at 450 °C and 3.5 times higher at 700 °C than the respective quantity of DOC calculated for BCLG. It is likely that the leaching condition applied in the up-flow percolation test enhanced the swelling of BCWC in water by opening closed or blocked pores, thus increasing accessibility to leach the organic matter that was entrapped in the char matrix during the pyrolysis process.

3. Trace metal speciation modelling results showed the predominance of fulvic acid (FA) and humic acid (HA) complexes of Cd, Cu, and Pb in all the eluates collected at various liquid to solid (L/S) ratios from BCWC and BCLG irrespective of the inert fractions (0 and 35%) of DOC or the different fractions of active FA (55, 65, 90% DOC) or HA (0 and 10% DOC) considered. The high proportion of metal-FA and metal-HA complexes predicted for all the modelled eluate types may be due to the observed characteristics, such as alkaline pH values, high DOC concentrations and the absence of elements that can compete for binding with HA and FA.

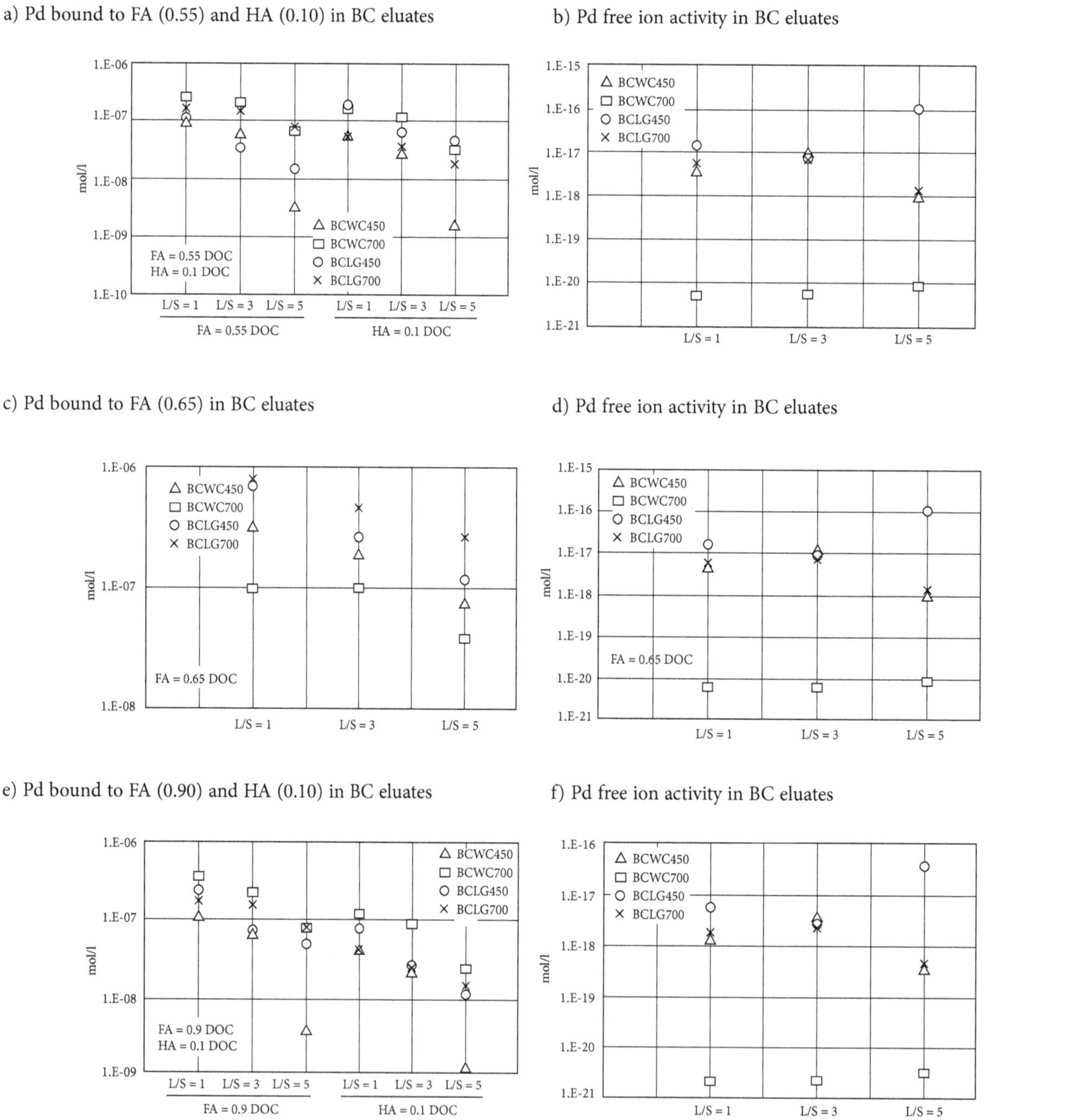

Fig. 6. Fraction bound to colloid fulvic acid (FA) and humic acid (HA) (a, c, e) and free ion activity (b, d, f) of Pd in biochar (BC) eluates collected at various liquid to solid (L/S) ratios, predicted by WHAM. DOC – dissolved organic carbon. BC at 450 and 700 °C from wood chips and lignin and are BCWC450, BCWC700, BCLG450, and BCLG700, respectively

4. If BCWC or BCLG are intended for water application, such as a filter media, the amount of DOC that may be emitted in surface water as a result of BC release in order to avoid high load of BC-borne DOC should be carefully estimated. It is likely that the trace metals released from BCWC and BCLG formed a strong bond with DOC, thus not representing a risk of impairment for the receiving water bodies.

References

Ahmad, M.; Rajapaksha, A.; Lim, J.; Zhang, M.; Bolan, N.; Mohan, D.; Vithanage, M.; Lee, S. S.; Ok, Y. 2014. Biochar as a sorbent for contaminant management in soil and water: a review, *Chemosphere* 99: 19–33. https://doi.org/10.1016/j.chemosphere.2013.10.071

Ahmed, I.; Hamilton–Taylor, J.; Lofts, S.; Meeussen, J.; Lin, C.; Zhang, H.; Davison, W. 2013. Testing copper-speciation predictions in freshwaters over a wide range of metal–organic matter ratios, *Environmental Science & Technology* 47(3): 1487–1495. https://doi.org/10.1021/es304150n

Alburquerque, J. A.; Sánchez, M. E.; Mora, M.; Barrón, V. 2014. Slow pyrolysis of relevant biomasses in the Mediterranean basin. Part 2. Char characterisation for carbon sequestration and agricultural uses, *Journal of Cleaner Production* 120: 191–197. https://doi.org/10.1016/j.jclepro.2014.10.080

Al-Wabel, M.; Al-Omran, A.; El-Naggar, A.; Nadeem, M.; Usman, A. 2013. Pyrolysis temperature induced changes in characteristics and chemical composition of biochar produced from conocarpus wastes, *Bioresource Technology* 131: 374–379. https://doi.org/10.1016/j.biortech.2012.12.165

Baltrėnaitė, E.; Baltrėnas, P.; Lietuvninkas, A. 2016. *The sustainable role of the tree in environmental protection technologies*: monograph. Springer. https://doi.org/10.1007/978-3-319-25477-7

Baltrėnas, P.; Baltrėnaitė, E.; Kleiza, J.; Švedienė, J. 2016. A biochar-based medium in the biofiltration system: Removal efficiency, microorganism propagation, and the medium penetration modeling, *Journal of the Air & Waste Management Association* 66(7): 673–686. https://doi.org/10.1080/10962247.2016.1162227

Beesley, L.; Marmiroli, M. 2011. The immobilisation and retention of soluble arsenic, cadmium and zinc by biochar. *Environmental Pollution* 159(2): 474–480. https://doi.org/10.1016/j.envpol.2010.10.016

Chakraborty, P.; Chakrabarti, C. 2006. Chemical speciation of Co, Ni, Cu, and Zn in mine effluents and effects of dilution of the effluent on release of the above metals from their metal–dissolved organic carbon (DOC) complexes, *Analytica Chimica Acta* 571(2): 260–269. https://doi.org/10.1016/j.aca.2006.04.069

Christensen, J.; Christensen, T. 2000. The effect of pH on the complexation of Cd, Ni and Zn by dissolved organic carbon from leachate-polluted groundwater, *Water Research* 34(15): 3743–3754. https://doi.org/10.1016/S0043-1354(00)00127-5

Dee, K. T. 2016. *Dissolved organic carbon (DOC) characteristics in metal-rich waters and implications for copper aquatic toxicity*: Doctoral dissertation. Colorado School of Mines, Colorado.

Devi, P.; Saroha, A. 2014. Risk analysis of pyrolyzed biochar made from paper mill effluent treatment plant sludge for bioavailability and eco-toxicity of heavy metals, *Bioresource Technology* 162: 308–315. https://doi.org/10.1016/j.biortech.2014.03.093

Dias, B.; Silva, C.; Higashikawa, F.; Roig, A.; Sánchez-Monedero, M. 2010. Use of biochar as bulking agent for the composting of poultry manure: effect on organic matter degradation and humification, *Bioresource Technology* 10: 1239–1246. https://doi.org/10.1016/j.biortech.2009.09.024

Gaskin, J. W.; Steiner, C.; Harris, K.; Das, K. C. 2008. Effect of low-temperature pyrolysis conditions on biochar for agricultural use, *Transactions of the American Society of Agricultural and Biological Engineers* 51(6): 2061–2069. https://doi.org/10.13031/2013.25409

Graber, E.; Tsechansky, L.; Lew, B.; Cohen, E. 2014. Reducing capacity of water extracts of biochars and their solubilization of soil Mn and Fe, *European Journal of Soil Science* 65: 162–172. https://doi.org/10.1111/ejss.12071

Groenenberg, J.; Lofts, S. 2014. The use of assemblage models to describe trace element partitioning, speciation, and fate: a review, *Environmental Toxicology and Chemistry* 33(10): 2181–2196. https://doi.org/10.1002/etc.2642

Iqbal, H.; Garcia-Perez, M.; Flury, M. 2015. Effect of biochar on leaching of organic carbon, nitrogen, and phosphorus from compost in bioretention systems, *Science of The Total Environment* 521: 37–45. https://doi.org/10.1016/j.scitotenv.2015.03.060

ISO 7888: 1985. Water quality – Determination of electrical conductivity. International Organization for Standardization, Geneva, Switzerland, 1985. 6 p.

ISO 10523:2008. Water quality – Determination of pH. International Organization for Standardization, Geneva, Switzerland, 2008. 13 p.

Jamieson, T.; Sager, E.; Guéguen, C. 2014. Characterization of biochar-derived dissolved organic matter using UV–visible absorption and excitation–emission fluorescence spectroscopies, *Chemosphere* 103: 197–204. https://doi.org/10.1016/j.chemosphere.2013.11.066

Jones, D. L.; Murphy, D. V.; Khalid, M.; Ahmad, W.; Edwards-Jones, G.; DeLuca, T. H. 2011. Short-term biochar-induced increase in soil CO_2 release is both biotically and abiotically mediated, *Soil Biology and Biochemistry* 43(8): 1723–1731. https://doi.org/10.1016/j.soilbio.2011.04.018

Komkiene, J.; Baltrenaite, E. 2016. Biochar as adsorbent for removal of heavy metal ions [Cadmium (II), Copper (II), Lead (II), Zinc (II)] from aqueous phase, *International Journal of Environmental Science and Technology* 13(2): 471–482. https://doi.org/10.1007/s13762-015-0873-3

Li, J.; Li, Y.; Wu, Y.; Zheng, M. 2014. A comparison of biochars from lignin, cellulose and wood as the sorbent to an aromatic pollutant, *Journal of Hazardous Materials* 280: 450–457. https://doi.org/10.1016/j.jhazmat.2014.08.033

Liaw, S.; Wu, H. 2015. Tuning biochar properties via partial gasification: facilitating inorganic nutrients recycling and altering organic matter leaching, *Energy Fuels* 29: 4407–4417. https://doi.org/10.1021/acs.energyfuels.5b01020

Lin, Y.; Munroe, P.; Joseph, S.; Henderson, R.; Ziolkowski, A. 2012. Water extractable organic carbon in untreated and chemical treated biochars, *Chemosphere* 87(2): 151–157. https://doi.org/10.1016/j.chemosphere.2011.12.007

Lu, H.; Zhang, W.; Yang, Y.; Huang, X.; Wang, S. 2011. Relative distribution of Pb2+ sorption mechanisms by sludge-derived biochar, *Water Res* 46(3): 854–862. https://doi.org/10.1016/j.watres.2011.11.058

Luo, L.; Xu, C.; Chen, Z.; Zhang, S. 2015. Properties of biomass-derived biochars: Combined effects of operating conditions and biomass types, *Bioresource Technology* 192: 83–89. https://doi.org/10.1016/j.biortech.2015.05.054

Mancinelli, E.; Baltrėnaitė, E.; Baltrėnas, P.; Paliulis, D.; Passerini, G. 2016. Trace metals in biochars from biodegradable by-products of industrial Processes, *Water, Air, & Soil Pollution* 227(6): 1–21. https://doi.org/10.1007/s11270-016-2892-1

Mancinelli, E.; Baltrėnaitė, E.; Baltrėnas, P.; Paliulis, D.; Passerini, G.; Almås, Å. 2015. Trace metal concentration and speciation in storm water runoff on impervious surfaces, *Journal of Environmental Engineering and Landscape Management* 23(1): 15–27. https://doi.org/10.3846/16486897.2014.936441

Mueller, K.; Lofts, S.; Fortin, C.; Campbell, P. 2012. Trace metal speciation predictions in natural aquatic systems: incorporation of dissolved organic matter (DOM) spectroscopic quality, *Environmental Chemistry* 9(4): 356–368. https://doi.org/10.1071/EN11156

Mukherjee, A.; Zimmerman, A. R. 2013. Organic carbon and nutrient release from a range of laboratory-produced biochars and biochar-soil mixtures, *Geoderma* 193–194: 122–130. https://doi.org/10.1016/j.geoderma.2012.10.002

Peckyte, J.; Baltrenaite, E. 2015. Assessment of heavy metals leaching from (bio)char obtained from industrial sewage sludge, *Mokslas – Lietuvos Ateitis* 7(4): 399–406. https://doi.org/10.3846/mla.2015.811

Smith, C. R.; Hatcher, P. G.; Kumar, S.; Lee, J. W. 2016. Investigation into the sources of biochar water-soluble organic compounds and their potential toxicity on aquatic microorganisms, *ACS Sustainable Chemistry & Engineering* 4(5): 2550–2558. https://doi.org/10.1021/acssuschemeng.5b01687

Smith, C.; Buzan, E.; Lee, J. 2012. Potential impact of biochar water-extractable substances on environmental sustainability, *ACS Sustainable Chemistry & Engineering* 1(1): 118–126. https://doi.org/10.1021/sc300063f

Tipping, E.; Lofts, S.; Sonke, J. 2011. Humic ion-binding model VII: a revised parameterisation of cation-binding by humic substances. *Environmental Chemistry* 8: 225–235. https://doi.org/10.1071/EN11016

Tipping, E. 1994. WHAM – a chemical equilibrium model and computer code for waters, sediments and soils incorporating a discrete-site/electrostatic model for ion-binding by humic substances, *Computers & Geosciences* 20(6): 973–1023. https://doi.org/10.1016/0098-3004(94)90038-8

Trompowsky, P.; de Melo Benites, V.; Madari, B.; Pimenta, A.; Hockaday, W.; Hatcher, P. 2005. Characterization of humic like substances obtained by chemical oxidation of eucalyptus charcoal, *Organic Geochemistry* 36(11): 1480–1489. https://doi.org/10.1016/j.orggeochem.2005.08.001

Uchimiya, M.; Ohno, T.; He, Z. 2013. Pyrolysis temperature-dependent release of dissolved organic carbon from plant, manure, and biorefinery wastes, *Journal of Analytical and Applied Pyrolysis* 104: 84–94. https://doi.org/10.1016/j.jaap.2013.09.003

Unsworth, E. R.; Warnken, K. W.; Zhang, H.; Davison, W.; Black, F.; Buffle, J.; Cao, J.; Cleven, R.; Galceran, J.; Gunkel, P.; Kalis, E. 2006. Model predictions of metal speciation in freshwaters compared to measurements by *in situ* techniques, *Environmental Science & Technology* 40(6): 1942–1949. https://doi.org/10.1021/es051246c

Van der Sloot, H.; Dijkstra, J. 2004. *Development of horizontally standardized leaching tests for construction materials: a material based or release based approach?* Identical leaching mechanisms for different materials. Energy Research Center of The Netherlands, Report No-. ECN-C-04-060.

Wood, C.; Al-Reasi, H.; Smith, D. 2011. The two faces of DOC, *Aquatic Toxicology* 105(3): 3–8. https://doi.org/10.1016/j.aquatox.2011.03.007

Wu, H.; Yip, K.; Kong, Z.; Li, C.-Z.; Liu, D.; Yu, Y.; Gao, X. 2011. Removal and recycling of inherent inorganic nutrient species in mallee biomass and derived biochars by water leaching, *Industrial & Engineering Chemistry Research* 50: 12143–12151. https://doi.org/10.1021/ie200679n

Zhang, J.; Lü, F.; Zhang, H.; Shao, L.; Chen, D. 2015. Multiscale visualization of the structural and characteristic changes of sewage sludge biochar oriented towards potential agronomic and environmental implication, *Scientific Reports, 5*. https://doi.org/10.1038/srep09406

Enrico MANCINELLI. PhD student in Industrial Engineering at Marche Polytechnic University (UNIVPM, Italy). Master of Science in Civil Engineering (UNIVPM, 2013). Research interests: environmental protection, waste management, water pollution.

Edita BALTRĖNAITĖ. Dr, Assoc. Prof. at the Department of Environmental Protection, Vilnius Gediminas Technical University (VGTU). She gained her Environmental Engineering and Landscape Management PhD in 2007, the diploma of the Associate Professor in 2010 from Vilnius Gediminas Technical University (Lithuania). Publications: author/co-author of approx. 38 scientific papers, participant of about 20 international conferences. She is a deputy member of the Management Committee of COST TD1107 Biochar as Option for Sustainable Resource Management and FA0905 Mineral-Improved Crop Production for Healthy Food and Feed. Main research visits: Latvian State Institute of Wood Chemistry under FP7 project The Implementation of Research Potential of the Latvian State Institute of Wood Chemistry in the European Research Area (WOOD-NET) (FP7-203459)); University of Ljubljana (Slovenia) under FP7 project SPIRIT Support of Public and Industrial Research Using Ion Beam Technology; short-term scientific mission at Norwegian University of Life Sciences, in the framework of COST 859 action Phytotechnologies to Promote Sustainable Land Use and Improve Food Safety (COST-859-STSM-04824). Research interests: biogeochemistry, phytogeochemistry, bioindication.

Pranas BALTRĖNAS. Dr Habil, Prof. and Head of the Department of Environmental Protection, Vilnius Gediminas Technical University (VGTU). Doctor Habil of Science (air pollution), Leningrad Civil Engineering Institute (Russia), 1989. Doctor of Science (Air Pollution) Ivanov Textile Institute (Russia), 1975. Employment: Professor (1990), Associate Professor (1985), Senior Lecturer (1975), Vilnius Civil Engineering Institute (VISI, now VGTU). Publications: author of 13 monographs, 24 study-guides, over 320 research papers and 67 inventions. Honorary awards and membership: prize-winner of the Republic of Lithuania (1994), a corresponding Member of the Ukrainian Academy of Technological Cybernetics, a full Member of International Academy of Ecology and Life Protection. Placements in Germany and Finland. Research interests: air pollution, pollutant properties, pollution control equipment and methods.

Eglė MARČIULAITIENĖ. Dr, Assoc. Prof., Department of Environmental Protection, Vilnius Gediminas Technical University (VGTU). Doctor of Technological Sciences (environmental engineering and landscape management), VGTU, 2012. Master of Science (environmental protection engineering), VGTU, 2007. Bachelor of Science (environmental engineering), Siauliai University, 2005. Publications: 25 scientific publications. Research interests: waste management, odour management and research, sewage sludge management, monitoring.

Giorgio PASSERINI. Assoc. Prof. at Engineering Faculty of Marche Polytechnic University. M.Eng. in Electronic Engineering, PhD in Applied Physics. Employment: Associate Professor (2003), Researcher (1998) Marche Polytechnic University. Publications: 2 authored books, 4 edited books, over 90 papers. Research interests: air pollution, global warming assessment, energy conversion and management.

15

USE OF HYDROGEN PEROXIDE AND PERCARBONATE TO TREAT CHLORINATED AROMATIC HYDROCARBON-CONTAMINATED SOIL

Marika Viisimaa, Anna Goi

Department of Chemical Engineering, Tallinn University of Technology, Ehitajate tee 5, 19086 Tallinn, Estonia

Abstract. This study compared treatment methods that utilised a liquid carrier of hydrogen peroxide and a solid carrier, percarbonate, for p-dichlorobenzene, p-chloro-m-cresol and p-chlorothymol degradation in the soil. The targeted chlorinated aromatic contaminants in the soil degraded to a certain level when treated with the liquid hydrogen peroxide, but the removal efficacy was not dependent on the dosage. In contrast, an increase in the percarbonate dosage enhanced the contaminant removal. Supplementary ferrous iron was more effective for the treatment that employed the liquid carrier of hydrogen peroxide than the treatment employing the solid carrier. Although acidic pH conditions (initial pH of 2.5) favoured contaminant degradation using liquid hydrogen peroxide, the treatment involving percarbonate resulted in more effective contaminant removal without any soil pH pre-adjustment. Therefore, the solid carrier of hydrogen peroxide, percarbonate, was concluded to be an effective alternative to the liquid carrier, resulting in greater contaminant removal at natural soil pH values.

Keywords: soil cleaning technologies, chemical oxidation, hydrogen peroxide carrier, activation aid.

Reference to this paper should be made as follows: Viisimaa, M.; Goi, A. 2014. Use of hydrogen peroxide and percarbonate to treat chlorinated aromatic hydrocarbon-contaminated soil, *Journal of Environmental Engineering and Landscape Management* 22(1): 30–39. http://dx.doi.org/10.3846/16486897.2013.804827

Introduction

Hydrogen peroxide is one of the most popular chemicals used for contaminated soil treatment by chemical oxidation (Watts, Teel 2006). Oxidation with hydrogen peroxide occurs mostly through the generation of free hydroxyl radicals, and the most widespread activator used for the radical pathway is ferrous iron. Hydrogen peroxide activated by ferrous iron is known (Barb *et al.* 1951) as a classical Fenton's reagent, while any deviation from the classical mixture is known as a modified Fenton system. The modified Fenton system has been used to overcome apparent disadvantages in the use of the classical system, including the costly acidic range needed to maintain the radical reactions and the hydroxyl radicals' non-productive consumption, and it is based on the fundamental principles of Fenton chemistry. However, this system may utilise the application of different concentrations of H_2O_2 (2–20%) (Watts *et al.* 1999; Žukauskaitė *et al.* 2012) and carriers of hydrogen peroxide such as CaO_2 (Arienzo 2000; Northup, Cassidy 2008; Goi *et al.* 2011), and it may use chelating agents such as EDTA in addition to transition metals (Sun, Pignatello 1992; Vicente *et al.* 2011). While Fenton's reagent system refers to the iron-mediated hydroxyl radical production from H_2O_2 decomposition, most soil transition metals and ion metals containing minerals (Watts *et al.* 1992) can also activate the hydrogen peroxide oxidation.

The application of sodium percarbonate, $Na_2CO_3 \bullet 1.5H_2O_2$ (hereafter referred to as percarbonate), is another way to modify the Fenton system. Cravotto *et al.* (2007) and Rivas *et al.* (2010) have reported that percarbonate is a solid carrier of hydrogen peroxide with the same features as the liquid carrier, but having safer handling and environmentally friendly procedures. Percarbonate is a crystalline chemical compound that readily dissolves in water, yielding hydrogen peroxide and sodium carbonate (Eq. 1) and creating an alkaline, oxidative environment as a result (McKillop, Sanderson 1995). The actual name of this chemical compound, which reflects the structure of this chemical, is carbonate perhydrate (McKillop, Sanderson 1995).

$$2(Na_2CO_3 \bullet 1.5H_2O_2) \rightarrow 2Na_2CO_3 + 3H_2O_2. \quad (1)$$

It is believed (Rivas *et al.* 2010; McKillop, Sanderson 1995) that the dominant chemistry of the percarbonate oxidation system is that of hydrogen peroxide. Released hydrogen peroxide can oxidise contaminants through the

Corresponding author: Anna Goi
E-mail: anna.goi@ttu.ee

formation of hydroxyl radicals or any other oxidative species and is also decomposed to water and oxygen (Eq. 2). Due to the oxygen formation, percarbonate is also used (Vesper *et al.* 1994) to support the aerobic biodegradation of contaminants.

$$2H_2O_2 \rightarrow 2H_2O + O_2. \quad (2)$$

The sodium carbonate formed in Eq. 1 gives a basic reaction of the carbonate ion hydrolysis in water:

$$CO_3^{2-} + H_2O \rightarrow HCO_3^- + OH^-; \quad (3)$$

$$HCO_3^- \leftrightarrow CO_3^{2-} + H^+. \quad (4)$$

Rivas *et al.* (2010) have stated that the only potential drawback of $Na_2CO_3 \cdot 1.5H_2O_2$ could be the scavenging of hydroxyl radicals (Eqs 5–7) exerted by the release of (bi)carbonate anions (Eqs 3 and 4). Due to this scavenging, it is believed (Ku *et al.* 1996; Glaze *et al.* 1995) that (bi)carbonate anions have an overall negative impact on the treatment efficacy.

$$HCO_3^- + OH^\bullet \rightarrow HCO_3^\bullet + OH^-; \quad (5)$$

$$HCO_3^\bullet \leftrightarrow H^+ + CO_3^{-\bullet}; \quad (6)$$

$$CO_3^{2-} + OH^\bullet \rightarrow CO_3^{-\bullet} + OH^-. \quad (7)$$

However, the carbonate radical anions ($CO_3^{-\bullet}$) generated by reactions 6 and 7 (Umschlag, Hermann 1999) have been reported to yield a reduction potential ($CO_3^{-\bullet}$, H^+/HCO_3^-) of 1.63 V at pH 8.4 (Zuo *et al.* 1999) and have been found (Umschlag, Hermann 1999) to be effective for the degradation of various aromatic compounds, such as benzene, p-xylene, toluene, chlorobenzene, nitrobenzene and benzonitrile. In addition, it has been found (Flangan *et al.* 1986) that bicarbonate anions can also react with hydrogen peroxide to generate peroxymonocarbonate ions (HCO_4^-) (Eq. 8) with an estimated redox potential of 1.8 V (Yao, Richardson 2000) and several other active oxygen species, such as superoxide ions (O_2^-) and singlet oxygen ($^1O_2$).

$$HCO_3^- + H_2O_2 \rightarrow HCO_4^- + H_2O. \quad (8)$$

A bicarbonate-activated hydrogen peroxide system has been found (Flangan *et al.* 1986; Yao, Richardson 2000; Regino, Richardson 2007) to be capable of oxidising of a variety of organic compounds in the liquid phase. Moreover, the reactivity of HCO_4^- towards organic sulphides has been found (Regino, Richardson 2007) to be higher than that of H_2O_2 by a factor of 100–500 depending on the specific substrates. Thus, the chemistry of the percarbonate oxidation system is more complicated than previously suggested and may allow for the destruction of a wide range of resistant contaminants in soil and water.

There has been only one published study (Cravotto *et al.* 2007) on the application of percarbonate to contaminated soil treatment. This study combined sodium percarbonate and microwaves, resulting in the complete degradation of the aromatic compounds (4-chloronaphtol and 2,4-dichlorophenoxyacetic acid) in the soil. However, unaccompanied applications of microwaves and ultrasound (Duong *et al.* 2010) also had the potential to degrade chlororganic contaminants in soil. As a result, there are no significant studies comparing the efficacy of contaminated soil treatment methods that utilise an unaccompanied solid hydrogen peroxide carrier (percarbonate) and a liquid carrier (hydrogen peroxide).

Chlorobenzenes (CBs) are used in engine-block cleaners, solvents, pesticides, pharmaceutical intermediates and for the synthesis of chlorophenols and disinfectants. Chlorothymols (CTh) are applied as effective fungicide components and chlorocresols (CCr) can be found in metalworking fluids, pharmaceutical products and agricultural pesticides. The extensive use of these compounds has led to their substantial release into the environment (Gomez-Belinchon *et al.* 1991; Jan *et al.* 1994; Kot-Wasik *et al.* 2003; Feidieker *et al.* 1994), necessitating effective treatment methods. The application of liquid hydrogen peroxide has been found (Watts *et al.* 1997, 2008) to be effective for CB-contaminated soil treatment, indicating the need to test the solid hydrogen peroxide carrier, percarbonate, as a potential alternative to the liquid carrier. Although it is known that CBs can degrade in the presence of hydrogen peroxide in soil (Watts *et al.* 2008), the degradation efficacy of CBs, CCr and CTh when utilising hydrogen peroxide in the presence of carbonates/bicarbonates has not been studied.

Therefore, this study performed a comparative assessment of the treatment methods utilising percarbonate and liquid hydrogen peroxide for p-dichlorobenzene (p-DCB), p-chloro-m-cresol (p-C-m-Cr) and p-chlorothymol (p-CTh) removal in soil. The testing of certain activation aids was executed (pH pre-adjustment and supplemental ferrous iron addition) because the activator efficiency depends not only on the target contaminant but also on the remedial chemical used. The influence of chemical dosage on the degradation of selected chlorinated aromatic contaminants was also studied.

1. Methods

1.1. Soil characterisation

Natural topsoil (0–20 cm) was dried overnight at 30 °C in a circulating air drying oven before being spiked and sieved through a 2.0-mm sieve using a Retsch (AS 200) digital sieve shaker. Several characteristics of the soil are presented in Table 1. The soil pH was measured according to EPA method 9045C (1995) using a digital pH meter (CG-840, Schott) equipped with a Mettler Toledo InLaB 412 electrode. The total extractable iron

Table 1. Soil properties

Parameter (unit)	Value (mean ± standard deviation)
pH	5.8
Ferrous iron (g kg^{-1} of soil)	1.9±0.5
Total extractable iron (g kg^{-1} of soil)	12.1±0.9
Ion-exchangeable Fe(II) fraction (mg kg^{-1} of soil)	2.0±0.3
Organic carbon (mg kg^{-1} of soil)	460±30
Sand (%)	45.5
Silt (%)	52
Clay (%)	2.5

in the soil was extracted using the method in Heron *et al.* (1994). Ferrous iron and the ion-exchangeable Fe(II) fraction were extracted according to the procedure presented by Tessier *et al.* (1979) and the iron in the extracts was measured photometrically at 492 nm using the phenanthroline method (Merck 1974). The soil's organic carbon was determined using sulphochromic oxidation (ISO 14235 1998) and the texture of the soil was identified as sandy silt. The soil identification was based on the principles established by ISO 14688-1, 2 (2002, 2004) using a laser scattering particle size distribution analyser (LA-950, Horiba).

The initial concentrations of each individual contaminant in the soil were 0.77±0.03 g p-DCB kg^{-1} of soil, 0.59±0.06 g p-C-m-Cr kg^{-1} of soil and 2.50 ± 0.14 g p-CTh kg^{-1} of soil. Thus, the total concentration of p-DCB, p-C-m-Cr and p-CTh in the untreated soil was 3.86±0.08 g kg^{-1} of soil. The initial concentrations of each target compound were verified by the GC-FID analysis of six replicates.

1.2. Soil treatment

Sodium percarbonate ($Na_2CO_3 \cdot 1.5H_2O_2$, powder) and hydrogen peroxide (35%) were obtained from Aldrich for this experiment. The comparative assessment of the treatment methods' efficacy when utilising percarbonate and liquid H_2O_2 for the removal of chlorinated hydrocarbons, p-DCB, p-C-m-Cr and p-CTh, was performed by applying similar dosages of H_2O_2. The concentration of H_2O_2 was photometrically verified at 254 nm and the molar extinction coefficient for the hydrogen peroxide at 254 nm taken for the calculation of the H_2O_2 concentration was 19.6 M^{-1} cm^{-1} (Baxendale, Wilson 1957). In the experiments using a supplemental activator addition, $FeSO_4 \cdot 7H_2O$ salt (solution) was used as a ferrous iron source. The ferrous iron concentration in the stock solution was photometrically verified at 492 nm using the complex reaction with 1,10-phenanthrolinium chloride (Merck 1974).

The treatment of p-DCB-, p-C-m-Cr- and p-CTh-contaminated soil with different chemical/contaminant weight (w/w) ratios was carried out in batch mode. The standard procedure entailed a slurry of 10 g soil with 10 mL of liquid (bi-distilled water and remedial chemical solution) being treated in a 0.2-L cylindrical glass reactor using vigorous (300 rpm) magnetic stirring. The treatment time of 24 h was selected to ensure the complete decomposition of hydrogen peroxide. There was no residual hydrogen peroxide after 24 h of treatment. The flasks were sealed with a laboratory film (Parafilm) to reduce volatilisation losses. In the experiments using metal ion-activated hydrogen peroxide oxidation, ferrous iron solution was added and the reaction was initiated by a single addition of the chemical (hydrogen peroxide or percarbonate) solution. The soil was treated without pH pre-adjustment. In the experiments testing the pH effect on the treatment performance, the pH of the soil slurry was pre-adjusted to a value of 2.5 using H_2SO_4.

All of the experiments were carried out at 20±1 °C and in triplicate.

1.3. Extraction and determination of contaminants

The solid and liquid phases were not separated after treatment. Ten mL of hexane were added to the slurry and left overnight on an orbital shaker. Next, a vortex (Genius3, IKA) extraction procedure was employed two times (with 10 mL of n-hexane) for 2 min each. Concentrated hexane extracts were mixed (1/1) with the internal standard, which was $C_{16}H_{34}$ dissolved in n-hexane with a concentration of 0.235 g L^{-1}. The measurement of the contaminant concentration in the extracts was performed using a gas chromatograph (FocusGC, Finnigan, Thermo Electron Corporation) with a flame-ionisation detector (GC-FID). 2 µL of the extract were injected in a crossbond 100% dimethylpolysiloxane capillary column RTX-1 (30.0 m × 320 µm id × 0.25 µm film thickness) at an injector temperature of 260 °C. The GC-FID temperature program included the following steps. The initial temperature was held at 50 °C for 1 min and then increased at a rate of 8 °C min^{-1} to 200 °C with a 3-min hold. The detector temperature was 330 °C and a splitless injection mode was used. The velocity of the carrier gas (nitrogen) was 1 mL min^{-1} and the coefficient of variation for the GC-FID method was 1.5%. The external standard was prepared by dissolving p-DCB, p-C-m-Cr and p-CTh in n-hexane, and the method's detection limit was 11.3 mg p-DCB, 9.0 mg p-C-m-Cr and 7.8 mg p-CTh per L of solvent, which corresponded to 34 mg p-DCB, 27 mg p-C-m-Cr and 26 mg p-CTh per kg of soil. The GC-FID analyses of the un-spiked soil showed no indication of contaminants.

1.4. Soil filtrate analyses

The residual hydrogen peroxide in the soil filtrate treated by titanium sulphate with pertitanic acid formation was determined photometrically at 410 nm (Eisenberg 1943). The pH after the treatment was measured according to EPA method 9045C (1995). The ion content in the soil filtrate was measured using an ion chromatograph (761 Compact IC, Metrohm) equipped with a suppressed conductivity detector and a Metrosep A Supp 5 (150 mm × 4.0 mm id) analytical column.

2. Results and discussion

2.1. Removal of contaminants with the non-supplemented hydrogen peroxide treatment

p-DCB, p-C-m-Cr and p-CTh in soil degraded with the addition of hydrogen peroxide only (Fig. 1), indicating the ability of the natural activators present in soil to activate the oxidation. The potential for naturally occurring minerals in the soil and some transition metals to activate the hydrogen peroxide oxidation process has been indicated in several previous studies (Matta *et al.* 2008; Watts *et al.* 1997; Kong *et al.* 1998; Teel *et al.* 2001). Nevertheless, the degradation of the contaminants in the soil with liquid hydrogen peroxide was uncompleted, resulting in the same level of contaminant removal regardless of the chemical dosage used (Fig. 1). The removal of p-DCB, p-C-m-Cr and p-CTh with the liquid carrier averaged 10.1 ± 0.7% of the 0.77 ± 0.03 g of p-DCB, 5.9 ± 0.9% of the 0.59 ± 0.06 g of p-C-m-Cr and 22.7 ± 0.8% of the 2.50 ± 0.14 g of p-CTh targeted in each kg of soil.

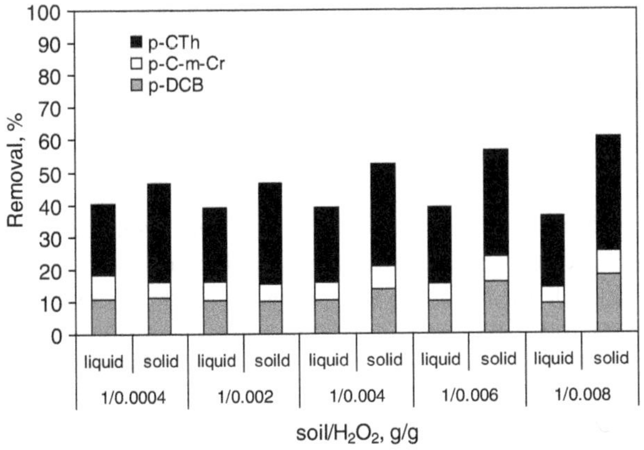

Fig. 1. p-DCB, p-C-m-Cr, and p-CTh removal (% of initial content) in the soil after a 24-h treatment with liquid hydrogen peroxide and percarbonate, without the addition of supplemental ferrous iron and utilising different weight ratios of soil to H_2O_2

A lack of available activators in soil may limit the effective production of hydroxyl radicals and could be a reason for the oxidation reaction termination. The loss of the hydrogen peroxide without the degradation of the contaminants could be explained by competing reactions, such as non-productive chemical decomposition without the generation of oxidants, reaction with the soil's organic matter, etc. (Watts, Teel 2006).

Although the treatment with percarbonate in a study by Rivas *et al.* (2010) has indicated that metoprolol in aqueous solution has the same oxidising power as the liquid hydrogen peroxide, a 24-h soil treatment with percarbonate in the present study led to the improved degradation of p-DCB, p-C-m-Cr and p-CTh compared to that obtained by the treatment with liquid hydrogen peroxide (Fig. 1). The soil treatment with percarbonate at soil/hydrogen peroxide ratios ranging from 1/0.0004 to 1/0.008 allowed for the removal of 10–18% of p-DCB, 5–7% of p-C-m-Cr and 30–36% of p-CTh. Thus, some increase in the removal of each targeted compound was obtained with the application of percarbonate.

The contaminants removal by the percarbonate treatment increased somewhat as the percarbonate dosage increased (Fig. 1). This scenario assumed that several other reactions besides the hydrogen peroxide oxidation were able to proceed. For example, in a study by Regino and Richardson (2007), kinetic and spectroscopic studies of the bicarbonate-catalysed oxidation of organic sulphides have strongly supported the role of peroxymonocarbonate (HCO_4^-) as the oxidant in bicarbonate-activated hydrogen peroxide solution. In a study by Xu *et al.* (2011), bicarbonate anions have also been identified as efficient activators of the hydrogen peroxide oxidation of some organic pollutants, including 4-chlorophenol, providing an efficient transition-metal-free treatment method.

Moreover, certain chemical-physical properties (hydrophobicity, pKa) of the studied compounds, caused primarily by their aromatic nature and functional groups attached to the ring (Table 2), may also have influenced the efficacy. CCr and CTh are weak acids, and their solubility to polar solvents (water) is higher in basic conditions when they are dissociated. The pH increase (Table 3) with the rise in the percarbonate dosage could have resulted in the increased solubility of the target contaminants, which may have influenced the efficacy of the treatment. Increased solubility results in the increased availability of the contaminant to the oxidant, as it is known (Miller, Valentine 1999) that the oxidation reactions proceed mainly in the liquid phase.

The reactivity of the compounds to both applied chemicals followed the order p-C-m-Cr < p-DCB < p-CTh, which could be explained by their chemical structure that influence the rate of oxidizing species attack on the ring, chemical-physical properties of the

Table 2. Chemical-physical properties (molecular weight, water solubility, log K_{ow} and pK_a) of the target compounds

Chemical name and abbreviation	p-dichlorobenzene, 1,4-dichlorobenzene, p-DCB	p-chloro-m-cresol, 4-chloro-3-methylphenol, p-C-m-Cr	p-chlorothymol, 4-chloro-2-isopropyl-5-methylphenol, p-CTh
Structural formula		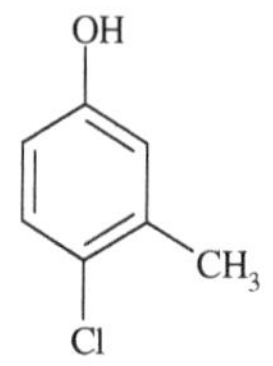	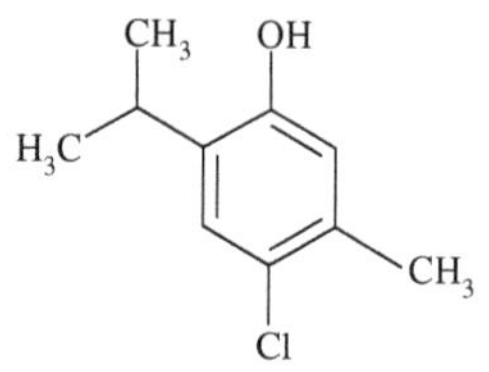
Molecular weight (g mol^{-1})	147.0[a]	142.6[a]	184.7[a]
Water solubility at 25 °C (mg L^{-1})	76[a]	3830[a]	89.2[a]
pK_a	–	9.55[a]	9.98[a]
log K_{ow}	3.38[b]	3.10[c]	3.76[d]

[a] Howard *et al.* (1997)
[b] Mackay and Boethling (2000)
[c] Wilfried (2005)
[d] Finizio *et al.* (1997)

contaminants (Table 2) and the difference in the initial concentrations of the target compounds that were also known to have a strong influence on the efficacy of contaminant removal during the Fenton treatment.

2.2. Removal of contaminants with the ferrous iron-supplemented hydrogen peroxide treatment

Higher contaminants degradations, compared to those obtained by unaccompanied liquid hydrogen peroxide, were achieved with the addition of supplemental ferrous iron (Fig. 2a), suggesting a lack of available activators. The increasing Fe^{2+}/H_2O_2 weight ratio, from 0.05/1 to 0.3/1 at lower dosages of liquid hydrogen peroxide (contaminant/H_2O_2 = 1/0.1, g/g), did not improve the degradation and even diminished the contaminant removal in the soil (Fig. 2a). This effect was explained by the excessive iron that provided conditions for the increased quenching of oxidising agents (Eq. 9) (Barb *et al.* 1951), reducing the contaminant degradation efficacy as a result.

$$Fe^{2+} + OH^{\bullet} \rightarrow Fe^{3+} + OH^{-}. \quad (9)$$

The increase in the supplemental activator/hydrogen peroxide ratio from 0.05/1 to 0.2/1 at a higher (contaminant/H_2O_2 = 1/2, g/g) dosage of hydrogen peroxide improved the degradation of the contaminants from 51.3 ± 3.3 to 73.2 ± 1.8%, respectively. The supplemental ferrous iron addition had a similar effect on the removal of the contaminants during the soil treatment with percarbonate (Fig. 2b). While the increase in the Fe^{2+}/H_2O_2 ratio at a lower ratio of contaminant/H_2O_2 = 1/0.1 (w/w) did not improve the degradation, the increase in that ratio at a higher ratio of contaminant/H_2O_2 = 1/1 (w/w) considerably increased the removal efficacy. The changes in the contaminant removal efficacy with the addition of the ferrous iron in the percarbonate treatment indicated that the transition metals could also be involved as activators of the oxidation process. The increase in the Fe^{2+}/H_2O_2 weight ratio, from 0.1/1 to 0.3/1, increased the removal of the contaminants from 58.0 ± 0.03 to 80.8 ± 4.3%, respectively. Thus, it was important to optimise not only the hydrogen peroxide/contaminant ratio but also the hydrogen peroxide/transition metal activator ratio for effective contaminant degradation with either carrier of hydrogen peroxide.

In spite of the fact that a supplemental ferrous iron addition could improve the degradation of the contaminants with both carriers of H_2O_2 applying certain ratio of contaminant/H_2O_2/Fe^{2+}, the enhancement of the treatment efficacy was more obvious with the liquid hydrogen peroxide (Figs 2a and 2b). The addition of supplementary ferrous iron to the liquid hydrogen peroxide resulted in a higher removal of all

Table 3. pH values of the soil slurry after 24 h of treatment with liquid hydrogen peroxide and percarbonate, with and without a supplemental ferrous iron addition

Contaminant/H_2O_2/Fe^{2+}, w/w/w	pH after 24 h of treatment
Liquid hydrogen peroxide	
1/0.1/0	5.8
1/0.1/0.005	5.7
1/2/0	4.9
1/2/0.1	4.4
Percarbonate	
1/0.1/0	6.2
1/0.1/0.01	5.6
1/1/0	8.5
1/1/0.1	7.8

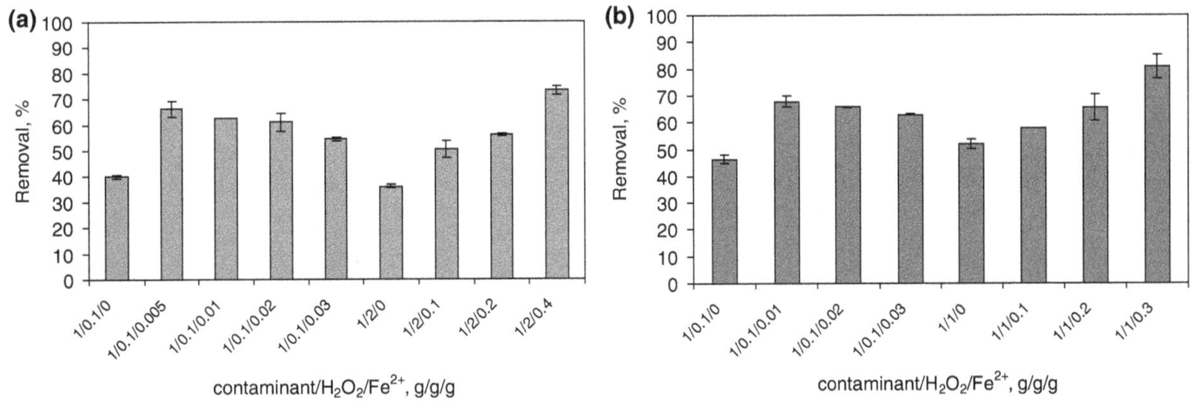

Fig. 2. Total contaminant removal (% of initial content) in the soil after a 24-h treatment with a) hydrogen peroxide and b) percarbonate, with and without the addition of supplemental ferrous iron, at different weight ratios of contaminant-to-H_2O_2. The error bars represent the standard deviation of the replicates (n = 3)

the target compounds compared to that obtained with a single chemical application (Figs 1 and 3). For example, while the removal of p-DCB, p-C-m-Cr and p-CTh with the treatment involving liquid hydrogen peroxide supplemented by a ferrous iron addition was 15–18%, 9–10% and 31–35%, respectively (Fig. 3), the removal of p-DCB, p-C-m-Cr and p-CTh with the treatment involving unaccompanied liquid hydrogen peroxide averaged 10.1±0.7%, 5.9±0.9% and 22.7±0.8%, respectively (Fig. 1).

Increased removal of p-CTh obtained by the treatment with percarbonate supplemented by ferrous iron compared to that achieved by the treatment with a single chemical was noted only. The soil treatment with percarbonate supplemented by ferrous iron allowed for the removal of 47–55% of p-CTh, depending on the chemical dosages used (Fig. 3). The removal of p-CTh with the treatment involving unaccompanied liquid hydrogen peroxide (soil/H_2O_2 = 1/0.0004, g/g) averaged 30% (Fig. 1). On the other hand, the percarbonate application supplemented by ferrous iron showed decreased removal of p-DCB and no influence on the

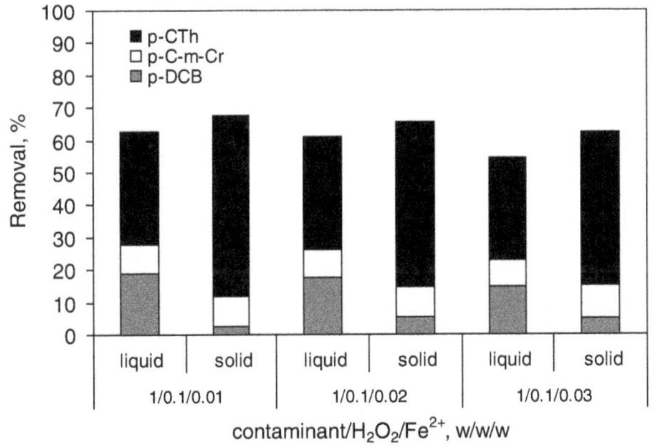

Fig. 3. p-DCB, p-C-m-Cr, p-CTh removal (% of initial content) in the soil after 24 h of treatment with liquid hydrogen peroxide and percarbonate with the addition of supplemental ferrous iron

p-C-m-Cr removal (the removal was 4–5% in both cases). While the removal of p-DCB with a single percarbonate treatment was 11% (Fig. 1), the treatment involving the percarbonate-supplementary ferrous iron system removed only 2–4% (Fig. 3). The rates of hydroxyl radical attack on the ring of target compound might be very different depending on the type and degree of substitution. Carbonates may also act as scavengers of the hydroxyl radicals formed in excess during the rapid decomposition of hydrogen peroxide achieved by the addition of a supplementary activator. Thus, unproductive consumption of hydroxyl radicals in side reactions (Watts, Teel 2006) could occur in place of their reaction with the specific target contaminant during the treatment of soil with percarbonate supplemented by ferrous iron. This can result in p-DCB degradation decrease with the addition of supplementary ferrous iron system. Ferrous iron also has had no activation effects in a past study by Cravotto *et al.* (2007) during the percarbonate treatment of 4-chloronaphthol, 2,4-dibromophenol, 2,4-dichlorophenoxyacetic acid and p-nonylphenol contaminated soil.

Thus, the improvement in contaminants total degradation with supplemental ferrous iron addition was more substantial during the treatment with liquid hydrogen peroxide (Figs 2a, b) than with the percarbonate. Lower enhancement in the treatment efficacy with percarbonate could also have occurred because the ferrous iron activation efficacy could be substantially reduced due to the pH increase (Table 3) achieved during the soil treatment with percarbonate. It is known (Sylva 1972) that the Fenton reaction rate tends to decrease as pH increases due to the hydrolysis of iron depending on counter ion, ionic strength, and total iron concentration, culminating in the precipitation of amorphous ferric oxyhydroxides. Thus, the application of a soluble ferrous iron as an activation aid was more effective for the oxidation system utilising liquid hydrogen peroxide.

The results for the soil slurry pH values measured after a 24-h treatment with either liquid hydrogen peroxide or percarbonate are presented in Table 3. The application of a lower dosage of hydrogen peroxide (contaminant/H_2O_2 = 1/0.1, g/g) did not influence the soil's pH. The soil's buffering capacity was usually sufficient to offer resistance to pH changes during the decomposition of hydrogen peroxide applied in moderate dosages. However, the treatments with elevated hydrogen peroxide and/or activator dosages have been found capable (Liang *et al.* 2008) of exceeding the buffering capacity of soil. As Table 3 shows, the decrease in soil pH was achieved during the treatment with a twenty-fold increase in the hydrogen peroxide dosage (weight ratio of contaminant/H_2O_2 changed from 1/0.1 to 1/2). A slightly higher decrease in pH, due to the presence of sulphate ions from $FeSO_4$ dissociation, was observed in the experiments involving the addition of a ferrous iron activator. The increased SO_4^{2-} concentration (130–250 mg $SO_4^{2-}L^{-1}$ of the soil filtrate depending on the added ferrous iron sulphate dosage) in comparison with initial value (40 mg $SO_4^{2-}L^{-1}$ of the soil filtrate) was detected in the soil filtrate after the treatment with a supplementary ferrous iron system. On the other hand, the pH of the soil slurry after the treatment with a solid hydrogen peroxide carrier, percarbonate, increased substantially and was greater during the application of higher dosages of percarbonate. A supplementary ferrous iron addition slightly reduced the pH from that achieved by the treatment with a single chemical.

2.3. Influence of soil pH pre-acidification on the removal of contaminants with the hydrogen peroxide treatment

The initial pH of the soil (Table 1) is higher than the optimum pH range of 2–4 identified for the chemical oxidation of chlorinated aromatics using the hydrogen peroxide oxidation process in water (Sedlak, Andren 1991; Ghaly *et al.* 2001). Thus, the treatment of the soil pre-acidified to a pH of 2.5 was tested. Such pre-acidification allowed for the improvement of the contaminant degradation process when using a liquid carrier of hydrogen peroxide (Fig. 4).

The reason for this improvement in contaminant degradation could be that acidic conditions sustain the solubility of natural soil metals mobilised during strong chemical oxidation conditions, involving them as the activators of the oxidation process (Villa *et al.* 2008; Dahnami *et al.* 2006). Moreover, only in acidic conditions the hydroxyl radical is predominant oxidant (Neyens, Baeyens 2003). On the contrary, the percarbonate treatment of the soil at an initial pH of 2.5 resulted in no improvement in the contaminant degradation. Percarbonate dissolves in water and may yield not only hydroxyl radicals but also several other oxidation species (HCO_4^-, $CO_3^{\cdot-}$) that may be more stable at near-neutral

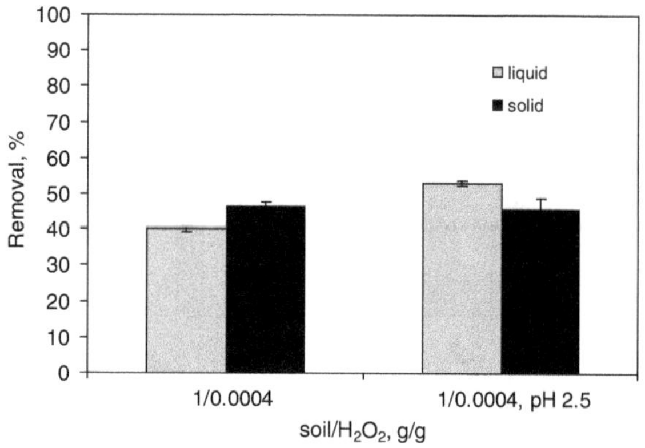

Fig. 4. Total contaminant removal (% of initial content) in soil after 24 h of treatment with liquid hydrogen peroxide and percarbonate at both a natural soil pH and a pH pre-adjusted to 2.5. The error bars represent the standard deviation of the replicates (n = 3)

pH values (Zuo *et al.* 1999; Yao, Richardson 2000; Regino, Richardson 2007). For example, in a study by Yao and Richardson (2000) on bicarbonate-activated hydrogen peroxide, a maximum catalytic efficacy for the oxidation of organic sulphides was observed at a pH range of 7–9.

In summary, the solid carrier of hydrogen peroxide, percarbonate, was employed as an effective alternative to the liquid carrier for the treatment of chlorinated aromatic hydrocarbon-contaminated soil, leading to a higher contaminant degradation level at natural soil pH values. However, further research on the chemistry of the percarbonate process and optimisation of the treatment performance will be required before full-scale treatment of contaminated soil.

Conclusions

This study performed a comparative assessment on the efficacy of using liquid and solid (percarbonate) carries of hydrogen peroxide to treat p-DCB-, CTh- and p-C-m-Cr-contaminated soil. The following conclusions were drawn from the experimental data.

1. The application of liquid hydrogen peroxide to the treatment of chlorinated aromatic hydrocarbon-contaminated soil resulted in 38 ± 2% total removal of the contaminants, independent of the chemical dosage used. However, increasing the dosage of percarbonate increased the contaminant degradation when allowing several other reactions besides hydrogen peroxide oxidation to proceed.
2. The addition of supplementary ferrous iron increased the contaminant removal efficacy during treatment with both carriers of hydrogen peroxide. However, the improvement in contaminant degradation was more substantial during the treatment with liquid hydrogen peroxide.

3. Pre-acidifying the soil to a pH of 2.5 increased the contaminant degradation during the treatment involving liquid hydrogen peroxide, but did not influence the treatment efficacy of the percarbonate.
4. The results of this study showed that percarbonate could be utilised as an effective alternative to hydrogen peroxide for the treatment of chlorinated aromatic hydrocarbon-contaminated soil at natural soil pH values.

Acknowledgments

The financial support of the Estonian Research Council (project JUT 1–7) and the Estonian Science Foundation (grant ETF7812) is gratefully acknowledged. We would like to thank MSc degree student Jevgenia Malinina for the experimental assistance.

References

Arienzo, M. 2000. Degradation of 2,4,6-trinitrotoluene in water and soil slurry utilizing a calcium peroxide compound, *Chemosphere* 40: 331–337. http://dx.doi.org/10.1016/S0045-6535(99)00212-X

Barb, W. G.; Baxendale, J. H.; George, P.; Hargrave, K. R. 1951. Reactions of ferrous and ferric ions with hydrogen peroxide. Part I. The ferrous ion reaction, *Transactions of the Faraday Society* 47: 462–500. http://dx.doi.org/10.1039/tf9514700462

Baxendale, J. H.; Wilson, J. A. 1957. The photolysis of hydrogen peroxide at high light intensities, *Transactions of the Faraday Society* 53: 344–356. http://dx.doi.org/10.1039/tf9575300344

Cravotto, G.; Carlo, S. D.; Ondruschka, B.; Tumiatti, V.; Roggero, C. M. 2007. Decontamination of soil containing POPs by the combined action of solid Fenton-like reagents and microwaves, *Chemosphere* 69: 1326–1329. http://dx.doi.org/10.1016/j.chemosphere.2007.05.078

Dahnami, M. A.; Huang, K.; Hoag, G. 2006. Sodium persulfate oxidation for the remediation of chlorinated solvents (USEPA Superfond Innovative Technology Evaluation Program), *Water Air and Soil Pollution* 6: 127–141. http://dx.doi.org/10.1007/s11267-005-9002-5

Duong, P. T.; Shrestha, R. A.; Sillanpää, M.; Virkutyte, J. 2010. Ultrasound-assisted treatment of kaolin artificially contaminated with phenanthrene, fluoranthene and hexachlorobenzene, *Journal of Environmental Engineering and Landscape Management* 18: 251–258. http://dx.doi.org/10.3846/jeelm.2010.29

Eisenberg, G. M. 1943. Colorimetric determination of hydrogen peroxide, *Industrial & Engineering Chemical Research* 15: 327–328.

EPA Method 9045C 1995. *Soil and Waste pH* [online], [cited 15 April 2012]. Available from Internet: http://www.epa.gov

Flangan, J.; Jones, D. P.; Griffith, W. P.; Skapski, A. C.; West, A. P. 1986. On the existence of peroxocarbonates in aqueous solution, *Journal of the Chemical Society, Chemical Communications* 1: 20–21. http://dx.doi.org/10.1039/c39860000020

Feidieker, D.; Kampfer, P.; Dott, W. 1994. Microbiological and chemical evaluation of a site contaminated with chlorinated aromatic compounds and hexachlorocyclohexanes, *FEMS Microbiology Ecology* 15: 265–278. http://dx.doi.org/10.1111/j.1574-6941.1994.tb00250.x

Finizio, A., Vighi, M., Sandroni, D. 1997. Determination of n-octanol/water partition coefficient (Kow) of pesticide, *Chemosphere* 34: 131–161. http://dx.doi.org/10.1016/S0045-6535(96)00355-4

Ghaly, M. Y.; Härtel, G.; Mayer, R.; Haseneder, R. 2001. Photochemical oxidation of p-chlorophenol by UV/H_2O_2 and photo-Fenton process. A comparative study, *Water Management* 21: 41–47. http://dx.doi.org/10.1016/S0956-053X(00)00070-2

Glaze, W. H.; Lay, Y.; Kang, J. W. 1995. Advanced oxidation processes: a kinetic model for the oxidation of 1,2-dibromo-3-chloropropane in water by the combination of hydrogen and UV radiation, *Industrial & Engineering Chemical Research* 34: 2314–2323. http://dx.doi.org/10.1021/ie00046a013

Goi, A; Viisimaa, M.; Trapido, M.; Munter, R., 2011. Polychlorinated biphenyls-containing electrical insulating oil contaminated soil treatment with calcium and magnesium peroxides, *Chemosphere* 82: 1196–1201. http://dx.doi.org/10.1016

Gomez-Belinchon, J. I.; Grimalt, J. O.; Albaiges, J. 1991. Volatile organic compounds in two polluted rivers in Barcelona (Catalonia, Spain), *Water Research* 25: 577–589. http://dx.doi.org/10.1016/0043-1354(91)90130-I

Heron, G.; Christensen, T. H.; Tjell, J. C. 1994. Oxidation capacity of aquifer sediments, *Environmental Science and Technology* 28: 153–158. http://dx.doi.org/10.1021/es00050a021

Howard, P. H.; Meylan, W. M.; Fake, F. F. 1997. *Handbook of physical properties of organic chemicals.* 1st ed. Boca Raton: CRC Press/Lewis Publishers.

Jan, J.; Zupancic, K. L.; Kralj, B.; Marsel, J. 1994. The influence of exposure time and transportation routes on the pattern of organochlorines in plants from a polluted region, *Chemosphere* 29: 1603–1610. http://dx.doi.org/10.1016/0045-6535(94)90308-5

ISO 14235 1998. *Soil quality — determination of organic carbon by sulfochromic oxidation.* International Organization for Standardization, Geneva, Switzerland.

ISO 14688-1 2002. *Geotechnical investigation and testing - identification and classification of soil. Part 1: identification and description.* International Organization for Standardization, Geneva, Switzerland.

ISO 14688-2 2004. *Geotechnical investigation and testing - identification and classification of soil. Part 2: principles for a classification.* International Organization for Standardization, Geneva, Switzerland.

Kong, S.-H.; Watts, R. J.; Choi, J.-H. 1998. Treatment of petroleum-contaminated soils using iron mineral catalyzed hydrogen peroxide, *Chemosphere* 37: 1473–1482. http://dx.doi.org/10.1016/S0045-6535(98)00137-4

Kot-Wasik, A.; Dezbska, J.; Namiesnik, K.; Pacyna, J. 2003. Physical, chemical, and biological changes in the gulf of Gdansk ecosystem (Southern Baltic Sea), *Reviews of Environmental Contamination and Toxicology* 179: 1–36. http://dx.doi.org/10.1007/0-387-21731-2_1

Ku, Y.; Su, W. J.; Shen, Y. S. 1996. Decomposition kinetics of ozone in aqueous solution, *Industrial & Engineering Chemistry Research* 35: 3369–3374. http://dx.doi.org/10.1021/ie9503959

Liang, C.; Lee, I.-L.; Hsu, I.-Y.; Liang, C.-P.; Lin, Y.-L. 2008. Persulfate oxidation of trichloroethylene with and without iron activation in porous media, *Chemosphere* 70: 426–435. http://dx.doi.org/10.1016/j.chemosphere.2007-06.077

Mackay, D.; Boethling, R. S. 2000. *Handbook of property estimations methods for chemicals: environmental and health sciences.* New York, USA: CRC Press.

Matta, R.; Hanna, K.; Kone, T.; Chiron, S. 2008. Oxidation of 2,4,6-trinitrotoluene in the presence of different iron-bearing minerals at neutral pH, *Chemical Engineering Journal* 144: 453–458. http://dx.doi.org/10.1016/j.cej.2008.07.013

McKillop, A.; Sanderson, W. R. 1995. Sodium perforate and sodium per carbonate: cheap, safe and versatile oxidizing agents for organic synthesis, *Tetrahedron* 51: 6145–6166. http://dx.doi.org/10.1016/0040-4020(95)00304-Q

Merck, E. 1974. *The testing of water.* 19th ed. Merck, Darmstadt.

Miller, C. M.; Valentine, R. L. 1999. Mechanistic studies of surface catalyzed H_2O_2 decomposition and contaminant degradation in the presence of sand, *Water Research* 33: 2805–2816. http://dx.doi.org/10.1016/S0043-1354(98)00500-4

Neyens, E.; Baeyens, J. 2003. A review of classic Fenton's peroxidation as an advanced oxidation technique, *Journal of Hazardous Materials* 98: 33–50. http://dx.doi.org/10.1016/S0304-3894(02)00282-0

Northup, A.; Cassidy, D. 2008. Calcium peroxide (CaO_2) for use in modified Fenton chemistry, *Journal of Hazardous Materials* 152: 1164–1170. http://dx.doi.org/10.1016/j.jhazmat.2007.07.096

Regino, C. A.; Richardson, D. E. 2007. Bicarbonate-catalyzed hydrogen peroxide oxidation of cysteine and related thiols, *Inorganica Chimica Acta* 360: 3971–3977. http://dx.doi.org/10.1016/j.ica.2007.05.020

Rivas, F. J.; Gimeno, O.; Borralho, T.; Carbajao, M. 2010. UV-C radiation based methods for aqueous metoprolol elimination, *Journal of Hazardous Materials* 179: 357–362. http://dx.doi.org/10.1016/j.jhazmat.2010.03.013

Sedlak, D. L.; Andren, A. W. 1991. Oxidation of chlorobenzene with Fenton's reagent, *Environmental Science and Technology* 25: 777–782. http://dx.doi.org/10.1021/es00016a024

Sun, J.; Pignatello, J. J. 1992. Chemical treatment of pesticide wastes. Evaluation of Fe(III) chelates for catalytic hydrogen peroxide oxidation of 2,4-D at circumneutral pH, *Journal of Agricultural and Food Chemistry* 40: 322–327. http://dx.doi.org/10.1021/jf00014a031

Sylva, R. N. 1972. The hydrolysis of iron (III), *Reviews of Pure and Applied Chemistry* 22: 115–130.

Teel, A. L.; Warberg, C. R.; Atkinson, D. A.; Watts, R. J. 2001. Comparison of mineral and soluble iron Fenton's catalysts for the treatment of trichloroethylene, *Water Research* 35: 977–984. http://dx.doi.org/10.1016/S0043-1354(00)00332-8

Tessier, A.; Campbell, P. G. C.; Bisson, M. 1979. Sequential extraction procedure for the speciation of particulate trace metals, *Analytical Chemistry* 51: 844–851. http://dx.doi.org/10.1021/ac50043a017

Umschlag, T.; Hermann, H. 1999. The carbonate radical ($HCO_3^{\bullet}/CO_3^{\bullet-}$) as a reactive intermediate in water chemistry: kinetics and modelling, *Acta Hydrochimica et Hydrobiologica* 27: 214–222. http://dx.doi.org/10.1002/(SICI)1521-401X(199907)27:4214::AID-AHEH2143.0.C O;2-6

Vesper, S. J.; Lawrence, C. M.; Hayes, S.; Davis-Hoover, W. J. 1994. Solid oxygen source for bioremediation in subsurface soils, *Journal of Hazardous Materials* 36: 265–274. http://dx.doi.org/10.1016/0304-3894(94)85019-4

Vicente, F.; Rosas, J. M.; Santos, A.; Romero, A. 2011. Improvement soil remediation by using stabilizers and chelating agents in a Fenton-like process, *Chemical Engineering Journal* 172: 689–697. http://dx.doi.org/10.1016/j.cej.2011.06.036

Villa, R. D.; Trovó, A. G.; Nogueira, R. F. P. 2008. Environmental implications of soil remediation using the Fenton process, *Chemosphere* 71: 43–50. http://dx.doi.org/10.1016/j.chemosphere.2007.10.043

Wilfried, P. 2005. *Directory of microbicides for the protection of materials. A handbook.* Netherlands: Springer–Verlag.

Watts, R. J.; Udell, M. D.; Leung, S. W. 1992. *Treatment of contaminated soils using catalysed hydrogen peroxide.* Chemical Oxidation Technologies for the Nineties, Lancaster: WA Technomic Publishing Company.

Watts, R. J.; Jones, A. P.; Chen, P.-H.; Kenny, A. 1997. Mineral-catalyzed Fenton-like oxidation of sorbed chlorobenzenes, *Water Environment Research* 69: 269–275. http://dx.doi.org/10.2175/106143097X125443

Watts, R. J.; Foget, M. K.; Kong, S. H.; Teel, A. L. 1999. Hydrogen peroxide decomposition in model subsurface systems, *Journal of Hazardous Materials* 69: 229–243. http://dx.doi.org/10.1016/S0304-3894(99)00114-4

Watts, R. J.; Teel, A. L. 2006. Treatment of contaminated soils and groundwater using ISCO, *Practice Periodical of Hazardous, Toxic, and Radioactive Waste Management* 10: 2–9. http://dx.doi.org/10.1061/(ASCE)1090-025X(2006)10:1(2)

Watts, R. J.; Haeri-McCarroll, T. M.; Teel, A. L. 2008. Effect of contaminant hydrophobicity in the treatment of contaminated soils by catalyzed H_2O_2 propagations (modified Fenton's reagent), *Journal of Advanced Oxidation Technologies* 11: 354–361.

Xu, A.; Li, X.; Xiong, H.; Yin, G. 2011. Efficient degradation of organic pollutants in aqueous solution with bicarbonate-activated hydrogen peroxide, *Chemosphere* 82: 1190–1195. http://dx.doi.org/10.1016/j.chemosphere.2010.11.066

Yao, H.; Richardson, D. E. 2000. Epoxidation of alkenes with bicarbonate-activated hydrogen peroxide, *Journal of*

American Chemical Society 122: 3220–3221. http://dx.doi.org/10.1021/ja993935s

Zuo, Z.; Cai, Z.; Katsumura, Y.; Chitose, N.; Muroya, Y. 1999. Reinvestigation of the acid-base equilibrium of the (bi)carbonate radical and pH dependence of its reactivity with inorganic reactants, *Radiation Physics and Chemistry* 55: 15–23. http://dx.doi.org/10.1016/S0969-806X(98)00308-9

Žukauskaitė, A.; Jakubauskaitė, V.; Ambrazaitienė, D.; Zabukas, V.; Paulauskienė, T. 2012. The impact of chemical additives on the process of biodegradation of oil products, *Journal of Environmental Engineering and Landscape Management* 20: 17–26. http://dx.doi.org/10.3846/16486897.2011.633336

Marika VIISIMAA. Doctoral Student, MSc (chemical and environmental technology, 2010), is the author of 17 scientific publications, 4 of which are ISI Web of Science Publications. Viisimaa underwent research training in the field of soil remediation in Finland and is currently involved in the implementation of several projects on soil remediation and wastewater purification in Estonia. Research interests: chemical treatment of soil, modified Fenton treatment of wastewater.

Anna GOI. Senior Research Scientist, PhD (chemistry and chemical engineering, 2005). Dr Anna Goi is the author of over 70 publications, 21 of which are ISI WEB of Science Publications. Research interests: advanced oxidation processes, combined chemical-biological treatment of soil, wastewater treatment.

16

ENHANCING COPPER AND LEAD BIOACCUMULATION IN RAPESEED BY ADDING HEMP SHIVES AS SOIL NATURAL AMENDMENTS

Corneliu TANASE[a], Irina VOLF[b], Valentin I. POPA[c]

[a]*Faculty of Biology, Al. I. Cuza University of Iasi, 20 A Bd. Carol I, 700505 Iasi, Romania*
[b]*Environmental Engineering and Management Department, Gheorghe Asachi Technical University of Iasi, 71 Bd. Prof. D. Mangeron, 700050 Iasi, Romania*
[c]*Institute of Research, Development and Innovation in Technical and Natural Sciences, Aurel Vlaicu University of Arad, 77 Bd Revoluției, 310130 Arad, Romania.*

Abstract. The current study reveals the results of a phytoremediation process applied to a multi metal contaminated soil, located in close proximity to an energy power plant. Phytoremediation process was studied using the rapeseed cultivation, both in the presence and absence of hemp shives, considered as potential natural soil amendments. The physiological responses of the rapeseed plants, such as variations in length and accumulation of biomass, as well as the content of assimilating pigments were investigated. The concentrations of Cu (II) and Pb (II) ions have also been determined in soil and plants aiming at locating the metal ions in different organs of the plants. By using hemp shives as amendments, bioaccumulation of copper and lead ions in plant increases. The plants cultivated in the presence of high amounts of heavy metal ions have been characterized by an inhibition of growth and development that consisted in decrease of root and stem lengths, as well in the amount of biomass accumulated in different vegetative organs.

Keywords: heavy metals, soil contamination, phytoremediation, rapeseed, amendments, hemp shives.

Introduction

The concentrations of heavy metals such as copper and lead in the environment are currently increasing, mainly as a result of industrial activities associated with processes as metal plating and manufacture of batteries, pigments, plastics and fossil fuels burning. Any of these metals, in sufficiently high concentrations, can cause severe damage to physiological and biochemical activities of plants and humans. This damage can be intensified when two or more metals present in soil act synergistically (Jankaite, Vasarevičius 2005).

High concentrations of copper in the growth medium affect the different parameters of plant metabolism including dry biomass production, water content, macro and micro-nutrients balance (Jankaite, Vasarevičius 2005). They also inhibit root growth and block photosynthetic electron transport, thus reducing the content of chlorophyll pigments (Bernal *et al.* 2007; Kovacik *et al.* 2006). The inhibition of plant growth in the presence of this metal ion may be the result of degradation of chlorophyll pigments during the photosynthesis process (Upadhyay, Panda 2009).

Humans are exposed to copper via inhalation of particulate copper by occupational exposure or consumption of copper-contaminated water and food. In human body, copper is maintained in homeostasis (Jesse, Mary 2004). If the intake of copper exceeds the limit of human body tolerance, it may lead to toxic effects including hemolysis, jaundice and even death. The overload of common copper in vivo may result in hepatic cirrhosis (Bjorn *et al.* 2003), changes in lipid profile, oxidative stress, renal dysfunction (Galhardi *et al.* 2004) and stimulation of mucous membrane of alimentary canal.

Studies on the effects of lead in children have demonstrated a relationship between exposure to lead and a variety of adverse health effects. These effects include impaired mental and physical development, decreased heme biosynthesis, elevated hearing threshold, and decreased serum levels of vitamin D. The neurotoxicity of lead is of particular concern, because evidence from prospective longitudinal studies has shown that neurobehavioral effects, such as impaired academic performance and deficits in motor skills, may persist even after Pb (II) levels have returned to normal (Needleman 2004).

Corresponding author: Irina Volf
E-mail: iwolf@tuiasi.ro

In order to ensure the protection of soil, as a mean of increasing soil resources and environmental protection, different methods and technologies of bioremediation were designed to neutralize or block the flow of pollutants and to obtain an efficient and proper protection for a desired quality of soil (Stingu *et al.* 2010).

Phytoremediation is a cost effective and eco-friendly "green" remediation technology for environmental clean-up. Among the phytoremediation techniques, the phyto-extraction consists in removal of trace elements from soil through their uptake and accumulation by plants (Nascimento, Xing 2006). By harvesting the resulted plant biomass that contains heavy metals, these are removed from the site. This technique is effective only if the plants accumulate high concentrations of metals/metalloids in shoots (Stingu *et al.* 2011) and a reasonable amount of biomass is produced (McGrath, Zhao 2003). Phytoextraction is best suited for the remediation of diffusively polluted areas, where pollutants occur superficially in a relatively low concentration. Phytostabilisation is considered to be a particular phytoremediation technique, which aims at establishing a vegetation cover and causing in situ inactivation of trace elements by combining the use of metal-tolerant plants with soil amendments. These result in decrease of mobility and toxicity of pollutants concomitant with increase of soil fertility and improvement of plant establishment. The plants recommended for phytostabilisation should retain the metals at the root level restricting their transport to aerial parts thus avoiding the further transfer into the food chain (Wenzel *et al.* 1999).

One approach to increase the uptake of heavy metals is the use of chelators to mobilize the metals to the root zone (Madrid *et al.* 2003). Synthetic chelates have been successfully used to induce accumulation of metals by high biomass plants but they pose serious environmental drawbacks regarding excessive amount of metals solubilized and being poorly biodegraded in the soils (Luo *et al.* 2006). Increasing interest in the replacement of synthetic chelators has determined to research into natural sources of chelating agents, especially using plants as raw materials (Volf *et al.* 2012). This will allow a possibility to valorize waste industrial products which represent an important challenge of sustainable development based on environmentally friendly processes correlated with complex processing of biomass. Another approach is the use of soil amendments: material added to a soil to improve its physical properties as water retention, permeability, water infiltration, drainage, aeration and structure, with a view to provide a better environment for roots development (Davis, Wilson 2005). Hemp shives can be used as soil amendments. When the hemp stocks are processed, about 70–80% of the initial material is rejected as shives, which are not used conveniently evermore; sometimes used as low quality construction materials (tiles that contain fiber, gypsum, cement), as bedding for composting in agriculture, or are converted into fuel briquettes. However, the accessibility of these products led currently to their usage as soil amendment in phytoremediation.

Bioactive compounds existing in aqueous extracts modulated the bioaccumulation process of cadmium ions in Avena sativa depending on heavy metal concentrations and polyphenolic extracts compositions (Stingu *et al.* 2011). *Picea abies* bark and *Asclepia syriaca* plant extracts, both of them containing polyphenolic compounds, stimulated cadmium bioaccumulation, promoting the translocation of heavy metals to the aerial part of the plant. The conclusion of Stingu *et al.* (2011) study was that polyphenolic extracts could be properly used in bioremediation.

In this context, the current study examined the possibility to enhance copper and lead tolerance/bioaccumulation by adding natural products as hemp shives in soil. The relation between natural products and heavy metal tolerance/bioaccumulation was investigated for *Brassica napus* L. plant, cultivated on multi-metal polluted soils which were sampled from the neighborhood of an energy power plant.

The possibility of using natural amendments in heavy metals phytoremediation was investigated through all specific methods for plants (biometric measurements and quantitative determination of plant biomass and content of photoassimilating pigments: chlorophyll *a*, chlorophyll *b* and carotenoids) and soils analysis.

1. Methodology

1.1. Soil samples

The soil samples (S1, S2, S3) taken at different distances from the energy power plant (using coal as fuel) (Table 1) were characterized in terms of water content, pH value and content of heavy metal ions as copper, lead, iron, cadmium and zinc. For the content of heavy metal ions analysis, the soil samples were first digested with a mixture of hydrochloric and nitric acids (10.5 mL HCl 12 mol/L and 3.5 mL HNO_3 15.8 mol/L) for 12 hours and, subsequently, they were kept on a hot plate for another 2 hours. The concentrations of heavy metal ions were determined using a GBC 2008 Avanta Atomic Absorption Spectrophotometer, Germany (Stingu *et al.* 2009).

Table 1. Location of soil sampling sites

Soil sample	Location of soil sampling sites
S1	500 m E from the slag and ash storage yard and 1.5 km from the flue gas stack
S2	700 m SE from the slag and ash storage yard and 1.3 km from the flue gas stack
S3	1.8 km NE from the slag and ash storage yard and 300 m from the flue gas stack

1.2. Characterization of hemp shives

The hemp shives used in this study are wastes resulted after bast fibers separation. The vegetal material was randomly collected, while the species was identified and authenticated. The vegetal material was shade-dried at room temperature for two weeks, a reference sample (control) being kept for comparison. For chemical characterization, 20 grams of ground dried material with particle size of 0.5–1 mm were extracted consecutively for three times with 125 mL distilled water on a water bath, for 45 minutes, at 80 °C, and finally, the extracts were cumulated to a final volume of 500 mL using distilled water. Dry matter content in the extracts was determined by evaporation of 25 mL extract on water bath and drying at 105 °C, until a constant mass is reached. After that the crucible containing the dry matter was placed into a furnace at 600 °C, in order to establish the mineral and organic matter content. Folin Ciocalteu (FC) method was used to determine the total polyphenolic content (TPC) in aqueous extract and the concentrations were expressed as the number of equivalents of gallic acid (GAE) (Bao *et al.* 2005; Ignat *et al.* 2011). For the HPLC analysis, the cumulated aqueous extracts were concentrated under vacuum to 10 mL. Before HPLC characterization the concentrated sample was subjected to a fractionation step by successive liquid-liquid extractions with ethyl acetate. The organic phases were evaporated to dryness, diluted in methanol and further subjected to HPLC analysis.

A reversed-phase high-performance liquid chromatographic technique was used to identify and quantify the phenolic compounds. The HPLC analysis was carried out using a Dionex UltiMate 3000 chromatograph LC Systems, Germany, coupled to a PDA detector. Separations were performed with a Zorbax RX C18 (4.6×250 mm, particle size 5 μm) column, operating at 30 °C with a flow rate of 1.2 mL/min, and the injection volume was 5 μL. The mobile phase used was 1% acetic acid in water (A) versus methanol (B) for a total run time of 40 min, and the gradient changed as follows: solvent B started at 10% and increased immediately to 40% in 40 min. For quantification, standards for external calibration were used.

1.3. Experimental assay

Three different samples of contaminated soils were tested in different conditions, both in the absence (S1, S2, S3) and presence (S11, S22, S33) of hemp shives (1g/30cm^2). The plants have been cultivated in greenhouse conditions using those three samples of contaminated soils (100 g/pot). *Brassica napus* L. seeds were directly sown into pots. Each sample was replicated in ten pots, and three uniform plants have been spaced evenly in each pot and were allowed to grow.

The cultivated soils were wetted daily with 15 mL tap water for one week, until germination of the plants was observed. From this point, rapeseed plants were wetted every two days with the same amount of water. After 45 days from the beginning of the experiments, the rapeseed plants were separated into roots, coleoptiles and leaves which were characterized in order to evidence the different effects of the cultivation conditions on plants growth and development (Fig. 1).

1.4. Plantlet analysis

Plant growth and development. To evaluate the influence of natural amendment on plant growth and development in multi-metal stress conditions, after ten days, the rapeseed plants were separated into roots, stems and primary leaves, followed by biometric measurements of plant elongation and quantitative determinations of biomass. Samples of separated plant have been oven dried at 70 °C until constant values of mass were reached and the dry biomass was quantified.

Determination of heavy metal concentration in plants. For heavy metals ions analysis, plant tissues (root, stem and leaves) were digested with a mixture of nitric and

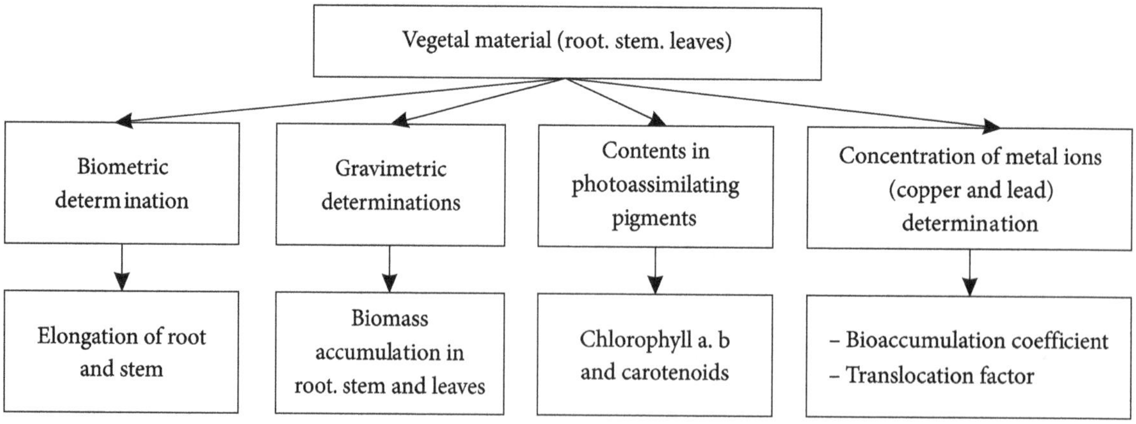

Fig. 1. Research methodology on plant material (*Brassica napus* L.)

hydrochloric acids similar to the case of soil samples. The concentrations of copper were determined by calibration curves obtained using standard solutions of pure copper and lead ions (Fisher Scientific). Calibration curves were generated using three replicates per metal concentration (0.02–5.00 μg/mL) by reading the absorbance at fixed wavelength.

The following parameters have been calculated:

Translocation factor (TF) = metal concentration in shoots (mg/kg) / metal concentration in roots (mg/kg) (Sun *et al.* 2009).

Bioaccumulation coefficient = (metal concentration mg/kg dry plant tissue) / (metal concentration mg/kg soil samples) (Stingu *et al.* 2009).

Photoassimilating pigments assay. The chlorophyll was extracted in 80% acetone and determined spectrophotometrically by measuring the absorbance at fixed wavelengths of 470 (A_{470}), 646 (A_{646}) and 663 nm (A_{663}).

The concentrations of chlorophyll pigments (chlorophyll *a* and *b*: [chl a] and [chl b]) were calculated by using the specific coefficients (Lichtenthaler, Wellburn 1983).

Chlorophyll *a* (μg/mL) = $12.21 \times A_{663} - 2.81 \times A_{646}$;

Chlorophyll *b* (μg/mL) = $20.13 \times A_{646} - 5.03 \times A_{663}$;

Carotenoids (μg/mL) = ($100A_{470}$ – 3.27 [chl a] – 104 [chl b])/ 227.

Statistical analysis. All the results are expressed as mean±standard error where $n = 3$. Comparison of the means was performed by the Fisher least significant

Table 2. Characterization of the soil samples

Sample	Water content (%)	pH value	Lead ions concentration (mg/kg)	Copper ions concentration (mg/kg)	Iron ions concentration (mg/kg)	Cadmium ions concentration (mg/kg)	Zinc ions concentration (mg/kg)
S1	19±1.03 [b]	7.53±0.14 [b]	549±13.67 [a]	330±9.93 [b]	36216±24.32 [c]	0.11±0.00 [a]	74±1.43 [a]
S2	15±0.89 [b]	7.43±0.12 [b]	301±8.86 [c]	464±11.42 [a]	51268±43.21 [a]	0.09±0.00 [b]	65±0.92 [b]
S3	25±1.23 [a]	7.72±0.21 [a]	242±5.49 [d]	344±10.78 [b]	24175±21.33 [d]	0.08±0.00 [b]	55±0.59 [c]

Different letters within columns indicate significant differences ($p < 0.05$).

Table 3. Characteristics of hemp shives aqueous extract

Dry mater content (g/L extract)	Organic matter content (g/L extract)	pH (at 25 °C)	TPC (g/100g)	Catechins (g/100g)	Caffeic acid (g/100g)	p - coumaric acid (g/100g)	Ferulic acid (g/100g)
0.51±0.003	0.42±0.003	4.7±0.2	16.41±0.62	5.21±0.23	9.68±0.35	0.80632±0.02	0.32598±0.005

Results reported represent the average values of triplicate determinations (n =3)±standard error. TPC – total polyphenolic content.

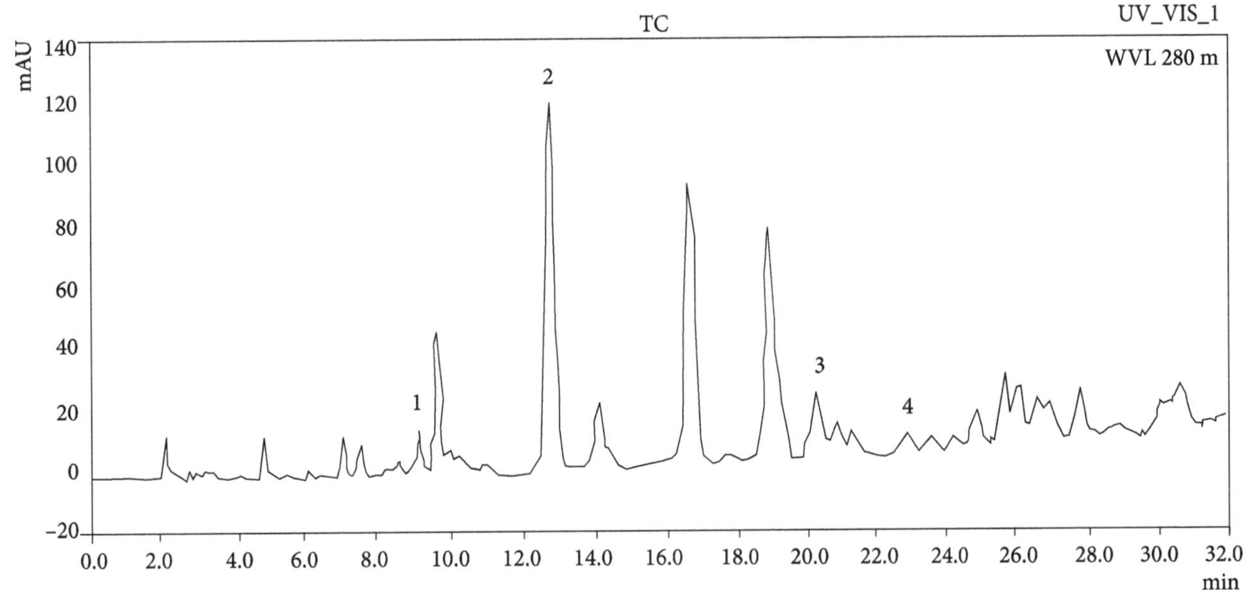

Fig. 2. Chromatographic profile of hemp shives aqueous extract: 1: catechin; 2 – caffeic acid; 3 – p-coumaric acid; 4 – ferulic acid

difference (LSD) test ($p \le 0.05$) after ANOVA analysis using program PAST 2.14. Sampling and chemical analyses were examined in triplicate, in order to decrease the experimental errors and to increase the experimental reproducibility.

2. Results and discussion

2.1. Characterization of the soil samples

The characterisation of soil samples showed a multi-metal polluted soil. As concerns the water content (Table 2), the highest value is recorded in sample S3. The specific humidity of S3 soil sample, taken at 1.8 km NE from the slag and ash storage yard and 300 m from the flue gas stack exhibits a slight increase compared to other areas as a result of the fact that this soil belongs to swampy areas.

The pH of the three sampling sites is slightly alkaline. The pH value has a direct influence on the mobility and accessibility of nutrients into soil. Optimal pH domain for plant nutrition is ranged between 6.0 and 6.5, since between these values the solubility and optimum accessibility of most nutrients at soil level are ensured. In general, for a normal development, the plants need a soil with pH values ranged between 6.5 and 7.2. Alkaline soil reaction is unfavorable for most crops and causes a low fertility of soils (Wang *et al.* 2004).

The concentrations of the studied heavy metals ions (Pb (II), Cu (II)) in the three soils samples tested (S1, S2, S3), have exceeded the alert threshold laid down in regulations (MWFEP 1997), excepting the value of lead ions in S3, which is very close to the limit of alert allowed. The highest concentration of lead ions was found in the soil sample S1, located at the nearest distance from slag and ash storage yard and 1.5 km from the flue gas stack. The content of lead ions has decreased as the distance to storage yard has increased. Pb(II) and Cu(II) were chosen for the trials due to high risk and transferability, though other heavy metal ions (iron, cadmium, zinc) were also determined in soil samples.

High concentration of potentially mobile Pb(II) and Cu(II) in soil run the risk of their downward migration and possible contamination of groundwater.

2.2. Characterization of the hemp shives extract

The total polyphenols content of hemp shives aqueous extract was 16.41±0.62 g/100g. The chromatographic analysis has allowed identification and quantification of components as catechin and phenolic acids, (Fig. 2 and Table 3), along with other compounds. Thus, the amount of catechins was 5.21 g/100g plant material; caffeic acid was presented in a higher concentration (9.68 g/100g) compared with those of other identified phenolic acids.

2.3. Rapeseed plantlet growth and development

The rape was chosen because it is considered as a plant with a high potential to be used for polluted soil remediation. The sorption of metals from the soil by plants is influenced by a variety of factors, including pH, temperature, soil ions, the cation exchange capacity of the soil, organic matter content of the soil, the type and concentration of metal, and the species of plant. The metals enter the root in the form of dissolved ions and move with the inflow of water apoplastically through the root hairs and into the cortex and are then translocated to other parts of plants. The metals interact with the physiological and biochemical activities of plants to reduce their vigour and in extreme cases can completely inhibit growth.

Lead has been shown to have toxic effects on a variety of metabolic processes essential to plant growth and development, including photosynthesis, transpiration, DNA synthesis, and mitotic activity. Of these processes, photosynthesis was found to be the most sensitive to lead contamination. Lead occurs in soils in the form of insoluble minerals and soluble compounds. It tightly binds to organic soil particles which may decrease the mobility of lead in most soils and may reduce uptake by plants.

Copper, unlike lead, is necessary for the growth of plants. It functions as a structural and catalytic component of several enzymes and proteins. Despite this requirement, excess copper when absorbed by plants can be toxic, leading to reduced growth, chlorosis, and malformation of roots.

The stem and root elongation of rapeseed plants (Fig. 3) which have been cultivated in the presence of amendments (S11, S22, S33) is usually significantly lower compared to the variants without amendments (S1, S2, S3). A special situation is recorded for root of rapeseed plants cultivated in S33 variant, where the elongation is significantly higher than the control (S3). Biomass accumulated in vegetative organs is presumed to be directly proportional to their length increase.

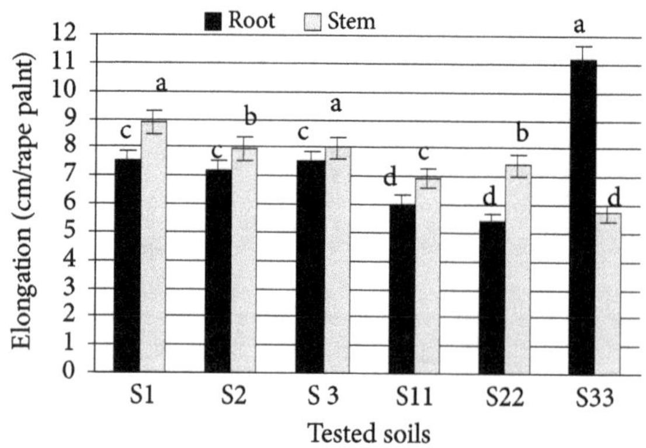

Fig. 3. Elongation of vegetative organs of rapeseed plants. Bars showing the same letter are not significantly different at $p \le 0.05$. Error bars represent the standard error of means ($n = 3$)

An inhibition in accumulation of the dry biomass in stems and leaves was obvious in experimental variants containing hemp shives as amendment, but in roots the opposite tendency prevailed, especially in S33 (Fig. 4 and Fig. 5). The results could be correlated with the characteristics features of soils samples. The sample S3 from swampy area possibly contains more and different microorganisms which can be involved in the interactions with the other components of the system (e.g. pollutants, polyphenols, products resulting in biomass transformation).

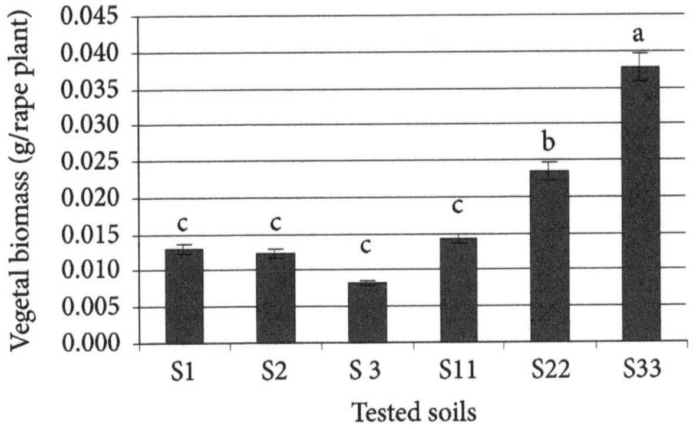

Fig. 4. Accumulation of biomass in roots of rapeseed plants under multi-metal stress conditions. Bars showing the same letter are not significantly different at $p \leq 0.05$. Error bars represent the standard error of means (n = 3)

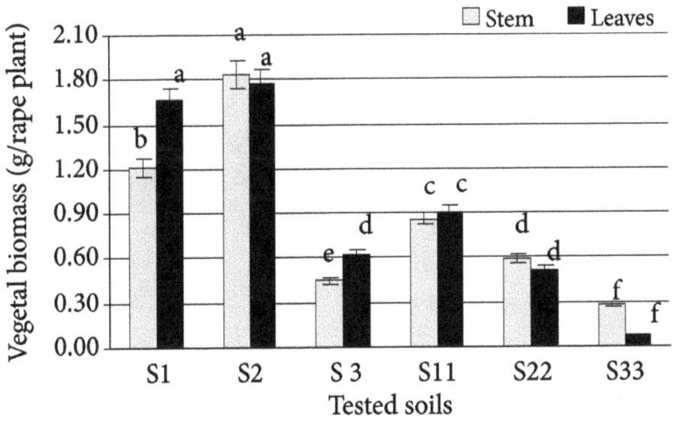

Fig. 5. Accumulation of biomass in stem and leaves of rapeseed plants under multi-metal stress conditions. Bars showing the same letter are not significantly different at $p \leq 0.05$. Error bars represent the standard error of means (n = 3)

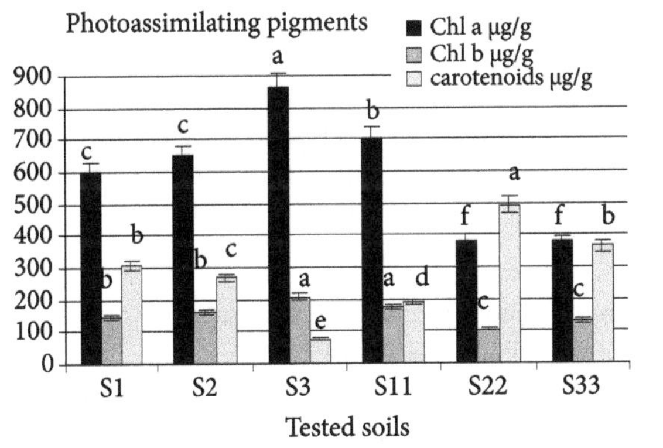

Fig. 6. Content of photo assimilating pigments in rapeseed leaves. Bars showing the same letter are not significantly different at $p \leq 0.05$. Error bars represent the standard error of means (n = 3)

2.4. Photoassimilating pigments

The photosynthetic apparatus may be especially sensitive to damage in the presence of heavy metals. An excess of heavy metals could negatively affect photoassimilating pigments biosynthesis.

Analyzing the variation of content of assimilating pigments (Fig. 6), a difference between the experimental variants was observed, that seems to be influenced by peculiarities of rapeseed plants and characteristic features of the soil samples. Thus the quantity of chlorophyll *a* (Chl a) in S11 variant has significantly higher value than those recorded in the case of the plants that have been developed in the absence of hemp shives. On the contrary, for S22 and S33 variants, the values are significantly lower than those recorded in case of the plants that have been developed in the absence of hemp shives.

The same trend is recorded for chlorophyll *b* (Chl b). So in most cases the leaves of rapeseed plants were significantly affected by copper and lead ions, which resulted in a decline of chlorophyll content. Copper and lead ions inhibit chlorophyll biosynthesis and retard the incorporation of these pigments in photosynthesis. Concerning the accumulation of carotenoids, the situation is opposite (Fig. 6). Significantly higher amounts of carotenoids were observed in S22 and S33 variants compared to values recorded for the plants cultivated in the variants S2 and S3. The leaves of rapeseed plants were significantly affected by copper and lead ions, which resulted in a decline of chlorophyll content. Copper and lead ions inhibit chlorophyll and carotene biosynthesis, and retard the incorporation of these pigments in photosynthesis.

From these results it seems that at the level of the soil, in the presence of humidity an extraction of polyphenols from hemp shives is possible. At the same time, an interaction of polyphenols with soil microorganisms and heavy metals could influence the plants development and their metabolism. It is known that polyphenols are also involved in metal complexing, plant growth regulation (oxidative phosphorylation) and reactive oxygen species inhibition. When a plant is exposed to more than one pollutant, interactions between those pollutants may occur. Although heavy metals may have antagonistic, additive, or synergistic effects in plants, synergistic interactions are found to be most common. The presence of hemp shives and its decomposition products could also influence this complex process.

2.5. Heavy metal ions content in plants

The concentrations of copper ions accumulated in all vegetative organs of rapeseed plants have shown that the presence of hemp shives in soil often determines an improvement in the process of bioaccumulation of metal ions (Fig. 7). The presence of natural amendments (hemp shives) in a multi-metal polluted soil has often induced the stimulation of the Cu(II) bioaccumulation process in plant. The Pb (II) content accumulated in different parts of rapeseed plants has revealed a similar conclusion (Fig. 8). The presence of hemp shives as amendment in soil usually increases the capacity of bioaccumulation both for copper and lead ions. The accumulation of metals in various parts of plants is often accompanied by an induction of variety of cellular changes, some of which directly contribute to metal tolerance capacity of plants. The high concentration of Cu(II) in soil (S2) probably caused a significant decrease of bioaccumulation of this element in in the plant leaves. Rapeseed has obviously higher bioaccumulation capacity of Pb(II) compared with Cu(II).

In cases of higher metal ion concentration, a decrease in plant growth and development processes were also observed, expressed by significant decrease of biomass accumulated in stems and leaves and usually significant decrease in root and stem lengths. The translocation factor reveals that the presence of hemp shives in heavy metals contaminated soil causes a decrease (usually significant) of metal ions transport to the upper parts of plants for S11 and S22 variants, but a significant increase of their transport for variant S33 (Fig. 9).

These phenomena could be explained by complex interaction between heavy metals, compounds liberated from hemp shives by aqueous elution or action of microorganisms and plants. These mechanisms will be elucidated in our further studies.

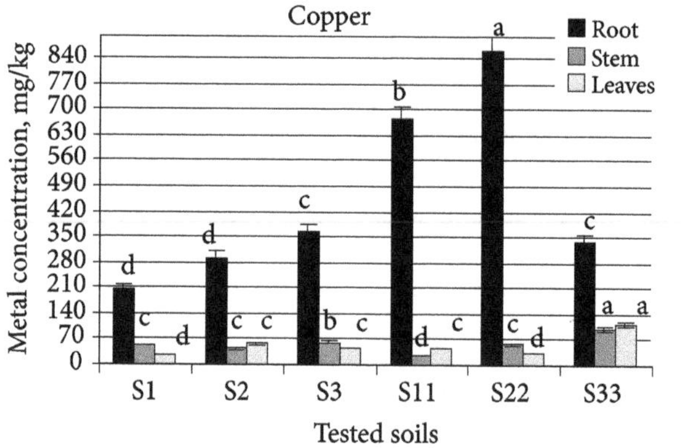

Fig. 7. Bioaccumulation coefficient of Cu (II) in vegetative organs of rapeseed plants. Bars showing the same letter are not significantly different at $p \leq 0.05$. Error bars represent the standard error of means ($n = 3$)

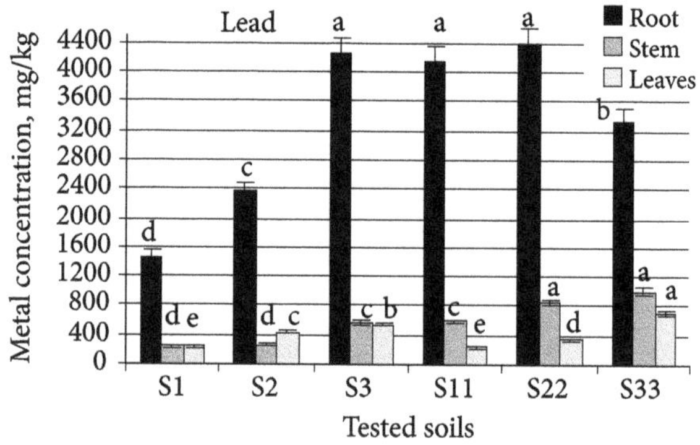

Fig. 8. Bioaccumulation coefficient of Pb (II) in vegetative organs of rapeseed plants. Bars showing the same letter are not significantly different at $p \leq 0.05$. Error bars represent the standard error of means ($n = 3$)

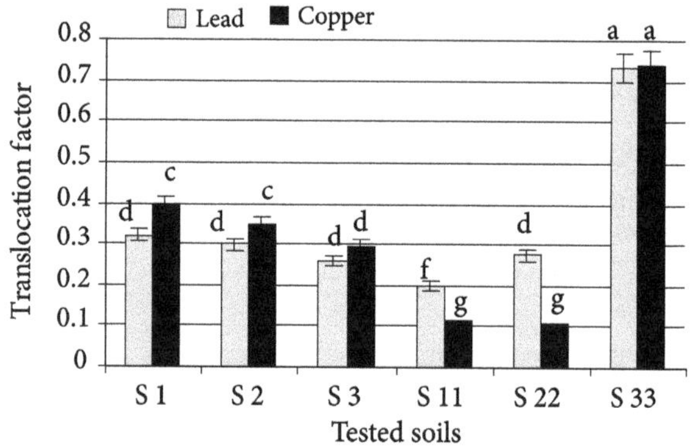

Fig. 9. Translocation factor of rapeseed plants. Bars showing the same letter are not significantly different at $p \leq 0.05$. Error bars represent the standard error of means ($n = 3$)

Conclusions

A multi-metal contaminated soil from a site located close to an energy power plant was used to study a bioremediation process using rape seed plants, in both the absence and presence of hemp shives as natural amendments. The development of rapeseed plants cultivated is influenced by the presence of heavy metals ions and hemp shives and the following effects were observed:

1. Prevalence of an inhibitory effect on biomass accumulation in plants demonstrated by significant decrease in lengths of stems and in the amount of biomass accumulated in the stems and leaves.
2. Prevalence of significant decrease in chlorophyll *a* and *b* contents and significant increase of carotenoids compared to the values recorded for plants that have grown in the absence of hemp shives.
3. Modulation effects of hemp shives on Cu (II) and Pb (II) bioaccumulation in rapeseed plant; the translocation factor calculated indicated a decreased transport of metal ions to the upper parts of plants for S11 and S22 variants and a promotion of heavy metals to the aerial parts of plant for S33 variant.
4. Bioactive compounds existing in hemp shives aqueous extracts can modulate the bioaccumulation process of lead and copper ions in *Brassica napus*, depending on heavy metal concentrations in soil. At low concentrations in soil the presence of hemp shives stimula-

tes lead and copper bioaccumulation, promoting the translocation of heavy metals to the aerial part of the plant. In this case the polyphenolic compounds eluted from hemp shives as natural amendments, increase the solubilisation of heavy metals and improve the phytoextraction process. At higher concentration in soil another effect may take place: reducing Pb(II) and Cu(II) mobility, decreasing thus translocation and determining in situ inactivation of heavy metal ions being suitable in phytostabilisation. Presumably, other components of lignocellulose are involved in this process.

5. The prevalence of significant increase of Pb(II) and Cu (II) bioaccumulation in different vegetative organs of rapeseed plants after adding hemp shives indicates that they can be recommended to be used as amendments in bioremediation of polluted soils. The results obtained in this study will be further developed with the aim to elucidate the mechanism of the complex interactions of components of the system.

Acknowledgements

This paper was realized with the support of Project POSCCE – "Bast fibers plants – strategic renewable resources for European economy" – BASTEURES and CUANTUMDOC "Doctoral studies for European performances in research and innovation" projects funded by the European Social Found and Romanian Government.

References

Bao, J. S.; Cai, Y.; Sun, M.; Wang, G. Y.; Corke, H. 2005. Anthocyanins, flavonols, and free radical scavenging activity of Chinese bayberry (*Myrica rubra*) extracts and their color properties and stability, *Journal of Agricultural and Food Chemistry* 53: 2327–2332. http://dx.doi.org/10.1021/jf048312z

Bernal, A. J.; Jensen, J. K.; Harholt, J.; Sørensen, S.; Moller, I.; Blaukopf, C.; Johansen, B.; de Lotto R.; Pauly, M.; Schelle, H. V. 2007. Disruption of ATCSLD5 results in reduced growth reduced xylan and homogalacturonan synthase activity and altered xylan occurrence in Arabidopsis, *Plant Journal* 52: 791–802. http://dx.doi.org/10.1111/j.1365-313X.2007.03281.x

Bjorn, P. Z.; Hermann, H. D.; Max, L.; Heide, S.; Barabara, K. G.; Hartmut, D. 2003. Epidemiological investigation on chronic copper toxicity to children exposed via the public drinking water supply, *Science of Total Environment* 302: 127–144. http://dx.doi.org/10.1016/S0048-9697(02)00399-6

Davis, J. G.; Wilson, C. R. 2005. Choosing a soil amendment, fact sheet, *Gardening Series Basics* [online], [cited August 2013], Colorado State University No. 7235. Available from Internet: http://www.ext.colostate.edu/pubs/garden/07235.html

Galhardi, C. M.; Diniz, Y. S.; Faine, L. A.; Rodrigues, H. G.; Burneiko, R. C.; Ribas, B. O.; Novelli, E. L. 2004. Toxicity of copper intake: lipid profile, oxidative stress and susceptibility to renal dysfunction, *Food Chemistry and Toxicology* 42: 2053–2060. http://dx.doi.org/10.1016/j.fct.2004.07.020

Ignat, I.; Volf, I.; Popa, V. I. 2011. A critical review of methods for characterization of polyphenolic compounds in fruits and vegetables, *Food Chemistry* 126: 1821–1835. http://dx.doi.org/10.1016/j.foodchem.2010.12.026

Jankaite, A.; Vasarevičius, S. 2005. Remediation technologies for soils contaminated with heavy metals, *Journal of Environmental Engineering and Landscape Management* 13(2): 109–113. http://dx.doi.org/10.1080/16486897.2005.9636854

Jesse, B.; Mary, R. L. 2004. Maintaining copper homeostasis regulation of copper-trafficking proteins in response to copper deficiency or overload, *Journal of Nutritional Biochemistry* 15: 316–322. http://dx.doi.org/10.1016/j.jnutbio.2004.02.004

Kovacik, J.; Tomko, J.; Backor, M.; Repcak, M. 2006. *Matricaria chamomilla* is not a hyperaccumulator, but tolerant to cadmium stress, *Journal of Plant Growth Regulation* 50: 239–247. http://dx.doi.org/10.1007/s10725-006-9141-3

Lichtenthaler, H. K.; Wellburn, A. R. 1983. Determinations of total carotenoids and chlorophylls a and b of leaf extracts in different solvents, *Biochemical Society Transactions* 11: 591–593.

Luo, C.; Shen, Z.; Lou, L.; Li, X. 2006. EDDS and EDTA enhanced phytoextraction of metals from artificially contaminated soils and residual effects of chelant compounds, *Environmental Pollution* 144: 862–871. http://dx.doi.org/10.1016/j.envpol.2006.02.012

Madrid, F.; Liphadzi, M. S.; Kirkham, M. B. 2003. Heavy metal displacement in chelate-irrigated soil during phytoremediation, *Journal of Hydrology* 271: 107–119. http://dx.doi.org/10.1016/S0022-1694(02)00258-5

McGrath, S. P.; Zhao, F. J. 2003. Phytoremediation of metals and metalloids from contaminated soils, *Current Opinion in Biotechnology* 14: 277–282. http://dx.doi.org/10.1016/S0958-1669(03)00060-0

MWFEP. 1997. Ministry of Waters, Forests and Environment Protection, Order No. 184 – September 21st, 1997 on the approval of the procedure for the carrying out of environmental balance sheets (Annex 4).

Nascimento, C. W. A.; Xing, B. 2006. Phytoextraction: a review on enhanced metal availability and plant accumulation, *Sciencia Agricola* 63: 299–311. http://dx.doi.org/10.1590/S0103-90162006000300014

Needleman, H. 2004. Lead poisoning, *Annual Review Medicine* 55: 209–22. http://dx.doi.org/10.1146/annurev.med.55.091902.103653

Stingu, A.; Volf, I.; Popa, V. I. 2009. Study of copper and cadmium accumulation by bean, *Environmental Engineering and Management Journal* 5: 1247–1252.

Stingu, A.; Volf, I.; Popa, I. V. 2010. Spruce bark extract as modulator in rape plant copper bioaccumulation, *Cellulose Chemistry and Technology* 45: 281–286.

Stingu, A.; Volf, I.; Popa, V. I.; Gostin, I. 2011. New approaches concerning the utilization of natural amendments in cadmium phytoremediation, *Industrial Crops and Products* 35: 53–60. http://dx.doi.org/10.1016/j.indcrop.2011.06.005

Sun, Y.; Zhou, Q.; Liu, W.; An, J.; Xu, Z.; Wang, L. 2009. Joint effects of arsenic and cadmium on plant growth and metal bioaccumulation: a potential Cd hyperaccumulator and As-excluder *Bidens pilosa* L, *Journal of Hazardous Materials* 161: 808–814. http://dx.doi.org/10.1016/j.jhazmat.2008.04.030

Upadhyay, R. K.; Panda, S. K. 2009. Copper induced growth inhibition, oxidative stress and ultrastructural alterations in freshly grown water lettuce (*Pistia stratiotes* L.), *Comptes Rendus Biologies* 332: 623–632. http://dx.doi. org/10.1016/j.crvi.2009.03.001

Volf, I.; Stingu, A.; Popa, V. I. 2012. New natural chelating agents with modulator effects on copper phytoextraction, *Environmental Engineering and Management Journal* 11: 487–491 [online], [cited August 2013]. Available from Internet: http://omicron.ch.tuiasi.ro/EEMJ/isues/vol11/vol11no2.htm

Wang, C.; Isoda, A.; Wang, P. 2004. Growth and yield performance of some cotton cultivars in Xinjiang China, an arid area with short growing period, *Journal of Agronomy and Crop Science* 190: 177–183. http://dx.doi.org/10.1111/j.1439-037X.2004.00090.x

Wenzel, W. W.; Lombi, E.; Adriano, D. 1999. Biogeochemical processes in the rhizosphere: role in phytoremediation of metal-polluted soils, in N. Prasad; J. Hagemeyer (Eds.). *Heavy metal stress in plants – from molecules to ecosystems.* Heidelberg: Springer Verlag, 273–303. http://dx.doi.org/10.1007/978-3-662-07745-0_13

Corneliu TANASE. Doctor of Science (Biology and Chemical Engineering), "Al. I. Cuza" University of Iasi (Romania) 2012. Author of 1 book, author of 7 peer-reviewed journal articles indexed in International Data Bases, 12 international conference proceeding publications. Phd, Gheorghe Asachi Technical University of Iasi. Research interests: plant physiology, bioremediation of degraded land and poor crop production based on vegetal material, regulators of plant growth based on natural products.

Irina VOLF. Associate Professor, Gheorghe Asachi Technical University of Iasi, Environmental Engineering and Management Department, Senior Researcher. Doctor of Science (Chemical Engineering), Gheorghe Asachi Technical University of Iasi (Romania), 2002. Publications: author of 6 books, author/co-author of 5 book chapters, 40 ISI peer-reviewed journal articles, 35 peer-reviewed journal articles indexed in international databases, 35 international conference proceeding publications, 1 patent. Internal and international scientific projects: 24. Research interests: biomass conversion, natural treatment systems: bioremediation of degraded land and poor crop production based on vegetal material (peat, sawdust, lignocellulose and polyphenolic products) as amendments, phytoremediation, regulators of plant growth based on natural products, industrial biotechnology.

Valentin I. POPA. Professor of Wood chemistry and Biotechnology. Publications: books and chapters of book 43, 600 papers, 6 patents. Editor of books in the fields of polysaccharides and polymeric biomaterials (Smithers Rapra and CRC Pess). Supervisor of PhD program-thesis defended: 30. Internal and international scientific projects: 60. Research interest: chemical and structural composition of biomass (cellulose, hemicelluloses, lignin secondary compounds), biorefining, biological properties of natural compounds with aromatic structure, biosynthesis and biodegradation of chemical compounds of biomass, bioremediation. Editor in chief of Cellulose Chemistry and Technology journal.

17

THE ROLE OF PH IN HEAVY METAL CONTAMINATION OF URBAN SOIL

Asta KAZLAUSKAITĖ-JADZEVIČĖ[a], Jonas VOLUNGEVIČIUS[b], Virginija GREGORAUSKIENĖ[c], Saulius MARCINKONIS[d]

[a]*Traku Voke Branch of the Lithuanian Research Centre for Agriculture and Forestry, Žalioji a. 2, Trakų Vokė, LT-02232 Vilnius, Lithuania*

[b]*Faculty of Environmental Sciences, Vilnius University, M. K. Čiurlionio g. 21/27, LT-03101 Vilnius, Lithuania*

[c]*Lithuania Geological Survey, S. Konarskio g. 35, LT-03123 Vilnius, Lithuania*

[d]*Vilniaus kolegija/University of Applied Sciences, Saltoniškių 58, LT-08105 Vilnius, Lithuania*

Abstract. The aim of this study to assess interdependence between urban soil pH and its accumulation of heavy metals. The article meant to be a contribution to a better knowledge of peculiarities and diagnostics of urban soil and its anthropogenic transformation. The hypothesis assumes that-relationship between urban soil pH and its accumulation of heavy metals may be determined by the origin and age of parent material as well as the nature and degree of the anthropogenic impact. The spatial variability of topsoil pH level was performed in 100 points in eldership of Šnipiškės of the city Vilnius. Laboratory analysis was based on ISO 10390:2005. Samples were collected from 20 cm topsoil layer in the same sampling points where have been analysed concentrations of topsoil chemical elements using optical atomic emission spectrophotometry. The contamination of urban soils exhibits somewhat different compared to agricultural soils. In contradiction to earlier studies in Lithuanian agricultural soils where strong correlation between soil pH and Cr, Cd, Pb, Ni, Cu and Z_d found, the conducted analysis shows a statistically reliable, but very weak (<0.3) correlation between the soil pH and concentration of contaminants. The proof to this correlation is provided by an existing relationship between pH and the concentration of copper ($r = 0.20$), mercury ($r = 0.15$), strontium ($r = -0.12$) and the overall contamination index ($r = 0.12$). The applied statistical analysis, however, failed to reveal the nature of interdependence between the soil pH and its contamination with studied heavy metals there concentration of contaminant chemical elements depends on the pH range of the soil and, conversely, the chemical reactivity of the soil changes affect on the concentration of studied chemical elements.

Keywords: soil pH, soil contamination, correlation analysis.

Introduction

One of the many environmental problems is a growth in extent to which the soil is being anthropogenised, subsequently causing the soil properties and its chemical and physical composition to change. The ever-growing rate of soil anthropogenesis (human impact on the environment, in this case resulting from the use of soil for human needs) expands the gap between the chemical, physical and biological properties of the soil found in urban territories and more natural soils. Nowadays Europe is facing an on-going loss of healthy soils reporting approximately 3.5 million sites potentially contaminated, resulting in failure to provide ecosystem services (Thematic Strategy ... 2006). In search of the ways how the contamination problem of the urban territories can be solved, the constant monitoring of the soil state as well as documentation of any changes observed in it are highly efficient. An assessment of the environmental risk due to soil pollution is of particular importance for urban areas, because of heavy metals direct and indirect effects on human health (Grzebisz *et al.* 2002). The inclusion of ecological risk assessment in soil quality standards shows an increasing interest in many EU Member States (Swartjes *et al.* 2008).

Individual concentration rates of chemical elements K_k (HN 60:2004; Taraškevičius 1996) as well as the overall index of contamination with these chemical elements Z_d (HN 60:2004; Taraškevičius 1996) and criteria of the

Corresponding author: Asta Kazlauskaitė-Jadzevičė
E-mail: asta.kaz@gmail.com

maximum permissible contents usually are used for soil pollution assessment. Studies of trace element dynamics in long-term field experiments have been shown to be useful complementary tool to assess soil accumulation as well (Jones *et al.* 1987).

Within last decades in Lithuania, significant research interest has been shown in the problem of soil contamination with heavy metals in agricultural soils (Marcinkonis 2008; Sunkieji metalai … 2001). The extensive overview on long-term agricultural impact on accumulation of heavy metals in topsoil and subsoil using selected reference elements and reference horizons has been reported by S. Marcinkonis (2008). The sorption of the heavy metals from polluted after and their migration in the system soil-tree has been provided by D. Butkus *et al.* (2004). Overview for the analysis of the correlation between Lithuanian soils pH and concentration of soluble heavy metals found in the agricultural soils has been provided in monograph by Sunkieji metalai… (2001).

The extensive studies of urban soil contamination with heavy metals been reported by different Lithuanian scientists (Taraškevičius, Zinkutė 2011; Zinkutė *et al.* 2011; Kumpienė *et al.* 2011; Jankauskaitė *et al.* 2008; Gregorauskienė 2006; Gregorauskienė, Kadūnas 2006; Taraškevičius 1996). But due to high variation of heavy metals in urban soils mainly long-term statistically significant changes are detectable not only in Lithuania (Vasarevičius, Greičiūtė 2004; Sunkieji metalai … 2001) but in other countries as well (Reimann, Caritat 2012; Irha *et al.* 2009; Lado *et al.* 2008; Imperato *et al.* 2003).

Up to the present, there are limited studies of interactions and interdependence of soil pH and heavy metals not only in Lithuania (Eidukevičienė *et al.* 2010) but in other countries as well (Paartel *et al.* 2004). Soil resistance/sensitivity to contamination with chemical elements depending its physicochemical properties including pH range been overviewed at European scale by R. G. Burau and R. J. Zasoski (2002), by Kadūnas *et al.* (2005). In general the scientific papers providing the relevant information on the role of pH in heavy metal contamination are often regionally fragmented. From various countries both, the heavy metals found in soil and its pH range reported by V. Cappuyns and R. Swennen (2008), Tong-Bin Chen *et al.* (2005), T. L. Woodard *et al.* (2007).

Up to the present, there are limited Pedological studies of soil contamination with heavy metals in urban Lithuanian soils. A possible relationship between the soil pH and urban soil contamination (in Vilnius city) with heavy metals is represented in this article. This study contributes to the body of applied sciences by investigating issues of plants in urban territories, it is important for specific planting work and selection of recreation sites. It represents more comprehensive study with the aim to assess interdependence between urban soil pH and its accumulation of heavy metals and meant to be a contribution to a better knowledge of peculiarities and diagnostics of urban soil and its anthropogenic transformation; as well as the basis provided for the geochemical index of assessing the soil state.

1. Materials and methods

1.1. Site

The eldership of Šnipiškės, chosen for the analysis, is one of the oldest elderships found in the city of Vilnius. Eldership occupies part of the city centre and the north-western part of Vilnius (Fig. 1). From the geomorphological point of view, the eldership of Šnipiškės lies in the sandy terraces of the river Neris valley. In its central part, low-rise private houses are found predominant. In the northern, eastern and north-eastern districts of the eldership, however, the most dominant are multi-storey blocks of flats, and the eldership is surrounded by flows of heavy traffic on its western, northern and eastern boundaries.

It well represents the diversity of anthropogenic impact on the soil that is observed in Vilnius as well as the most typical state of anthropogenised soil that is characteristic of urban territories. The most common source of pollution results from traffic flows and industrial activity. Therefore this eldership can best represent the common state of the anthropogenised soil found in the city of Vilnius. The distinctive feature of the eldership is a large number of former and present pollution sources: organizations of construction sector, transport maintaining companies, boiler-houses, repair workshops, the Institute of Thermal Insulation, toy factory, companies producing metal equipment, furniture factories, garages used for car repairs, marketplace, residential houses (multi-storey blocks of flats and quarters of poor and neglected low-rise houses) and streets of heavy traffic.

1.2. Sampling

The soil samples were collected in an "envelope's" principle, taking 5 samples in equal amounts from the soil surface horizon A on specific locations as shown in Figure 1. The topsoil horizon A was chosen for the analysis because it undergoes the biggest impact of versatile origin – mechanical, physical, chemical and biological. Due to the extent of the chemical impact, soil obtains its unique composition and its properties that are only typical to it (e.g. a shift in the pH range found in the surface of the soil).

Soil pH samples were collected in the same locations as the soil samples for chemical element concentration detection. Sampling points were decided by the Vilnius city municipality monitoring study "Urban and recreational territories soil pollution level evaluation" (Taraškevičius 2007). Sampling locations were chosen with consideration for the territories identified by Vilnius city municipality for being areas of intense pollution and differing pollution sources: transport hubs, marketplaces, teaching institutions, areas of mass assembly or residential areas. Therefore the choosing of sampling locations was based on three principles: technogenic pollution diversity, accessibility to data about soil pollution by chemical elements and possibility of comparison, comparing results of the soil pH measurements with results of soil pollution by chemical elements.

The analysed attribute meanings meet the normal eligibility requirements.

The sampling was formed to reflect the diversity of technogenisation patterns of the chosen territories, as well as to incorporate all geomorphological structures in the territory. Therefore while formulating the sampling it was aimed at being representative of the technogenisation diversity and the geomorphological situation.

Samples were collected in standard Urban conditions, during summer time (June 2009). For collection of samples a day without rain was chosen, in meteorological conditions which are common for Vilnius.

Sampling area fully represents regular approbation network. Soil samples for the pH analysis were collected in 2009 in the same soil sampling area, in Šnipiškės neighbourhood, Vilnius. The same soil sampling area was used in 1996 for determining the chemical element concentration coefficients.

1.3. Laboratory analyses

The collection and preparation of the soil samples for the laboratory analysis were based on methods as they are found in recommendations provided in international standards (ISO 10390:2005) and widely employed by Lithuanian geologists in their geochemical research. The analysis was conducted in the laboratory of General and Inorganic Chemistry Department, at Vilnius University, for which the pH meter of "Mettler Toledo MP220 pH Meter" was used.

Taking the measurements of the pH range in soil, the samples were prepared in ratio of 1 part of soil surface material (2 g) to 5 parts (10 ml) of boiled distilled water. In an attempt to obtain a more precise estimation of pH range, the solution was filtrated through a usual neutral filter paper. For calibration, the standard buffer solutions (pH 7.00 and pH 4.00) were used. The samples were analysed three times.

1.4. Data mining and statistical procedures

All of the data on sample collection and analysis were stored into the geographic database (GIS). The statistical analysis of contamination rate of individual chemical elements was carried out taking the data of Vilnius Municipality, provided in R. Taraškevičius's report of 1996. Along with the pH data, individual concentration rates of chemical elements K_k (HN 60:2004; Taraškevičius 1996) as well as the overall index of contamination with these

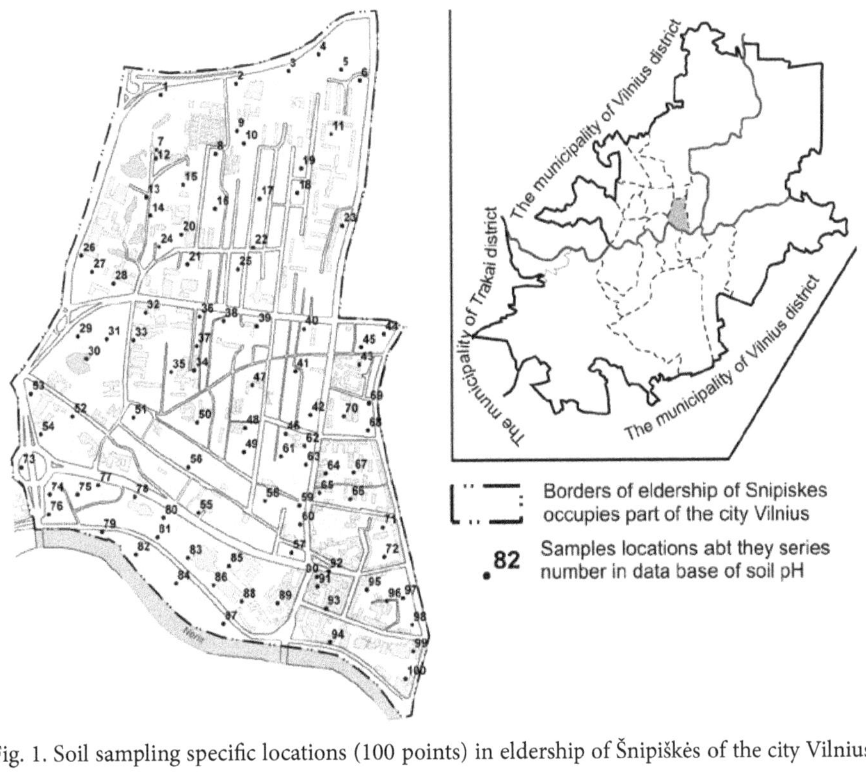

Fig. 1. Soil sampling specific locations (100 points) in eldership of Šnipiškės of the city Vilnius

chemical elements Z_d (HN 60:2004; Taraškevičius 1996) and background individual concentration rates of chemical elements (HN 60:2004; Taraškevičius 1996) were used. According to the data submitted in 2007 by Vilnius city council (Taraškevičius 2007) the major part of the soil chemical element concentration during the timeframe from 1996 to 2006 did not change and remained the same.

The statistical analyses (Fig. 3) were executed using the *"Statistica 6"* program. In reference to the collected data of the investigated samples, contour maps of topsoil pH range (Fig. 2) were prepared with the use of *"ArcGis 9.1"* program.

2. Results and discussion

What sets apart the investigated territory of the eldership of Šnipiškės from other neighbouring more natural territories is that it is only distinguished by the intensity of the conducted anthropogenic activity, which in the long term has caused the pH of the soil to increase (Fig. 2), i.e. the surface horizon of the soil has turned alkaline. As much as 98.04% of the surface horizon A of the investigated soil turned out to have the pH which is typically found in the alkaline surface of soils (≥7.0). Only 1.96% of the territory could be described as having the close to neutral pH (6.6–6.9).

Undoubtedly, one of the most significant factors influencing the reactivity of the soil surface is the geomorphological features of the location. Background values are important when analysing anthropogenic soil pH. The pH of the soil depends on physical, chemical and microbiological soil qualities a more visible reaction change during a relatively short soil formation period is caused only by liming of soil or soil dumping. The fluvio-glacial (Flg) and fluvial (Fl) deposits that shape the landform of the eldership of Šnipiškės had strong influence on the formation of the background values of the soil pH found in the territory.

The tendency that was observed, while investigating the territory, suggests that greater pH values (pH > 8) are found as typical of younger deposited sediment (Fl s-I) rather than the older deposits (Flg II, Flg III), where the value of pH ranges from 7 to 7.5 (Fig. 2). Within the eldership of Šnipiškės, the topsoil have a tendency of being more acid with their samples taken from the younger fluvial deposits (Fl s-I) in the south and less acid

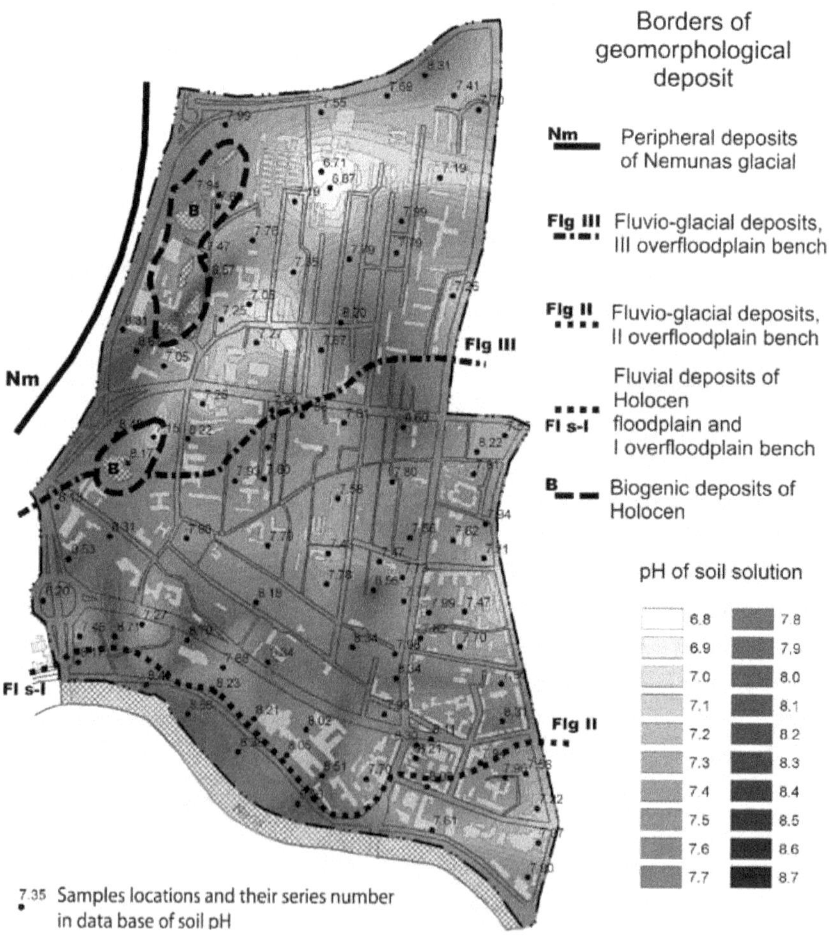

Fig. 2. Specific soil samples locations and soil pH values in eldership of Šnipiškės of the city Vilnius

with samples taken from the older fluvio-glacial deposits (Flg II, Flg III) in the north.

I–III terrace as well as B biogenic structure pH measurement massifs: for Flg I the average pH is 7.98, Flg II the average pH is 7.98 as well. The average for Flg III pH is 7.67 and for B the average pH is 7.72.

In the north-western part of the eldership of Šnipiškės, there are a few small lakes of thermal karst origin that have turned into a wetland. Because of the fenny edges of the lakes, biogenic deposited sediment (B) is found prevailing in the lowland. These deposits commonly have a greater value of topsoil pH due to the strong sorption of the alkaline contaminant elements caused by organic matter.

The north-western part of the investigated territory clings to the slope of the river Nėris valley formed by the peripheral deposit (Nm) brought from the river Nemunas in the glacial period. This deposit is found beyond the borders of the investigated territory and therefore has no significant impact on reactivity of topsoil.

According to the pH map of background soils published by J. Grybauskas (1978), the predominant soils in the territory of Vilnius were soils with the pH value ranging between 4.5 and 6.1. However, under the influence of morphological processes that took place in Vilnius and the slowdown of these processes, – the environment of the time was gradually shaped by human action and anthropogenic contamination that came along.

Studies in different environments (agricultural, urban and transition land-use zones) demonstrate that in the acid soils, heavy metals are more mobile than in the alkaline one, and the alkaline environment better sustains metals (Sunkieji metalai … 2001). The solubility of heavy metals decreases with increasing pH and vice versa, therefore accumulation of heavy metals often observed in the alkaline environment. On other hand, contamination of soil with heavy metals can stimulate rising of soil pH level.

The results achieved in the cluster analysis of soil contamination with heavy metals (Fig. 3) showed that the strongest association within the investigated territory is made up of Ba-Co-Ni-Zn, and the weaker one by Ag-Cr-Pb and Cd-Hg-Sn, which in most cases appear in the soil because of air pollution and, taken together, form the index of the overall contamination Z_d.

Certain regularities emerge when analysing the soil contamination with individual chemical elements in topsoil. Statistically, the strongest accumulation association is formed by Ba-Co-Ni-Zn, among which a strong linear interdependence is observed and which can be related with both clay minerals of organic origin and pollution levels in the city. In this association, two elements: barium (Ba) and cobalt (Co) are closely linked. The weakest genetic links of these metals connect them with zinc (Zn) that is found in their association.

The accumulation association of boron (B) and

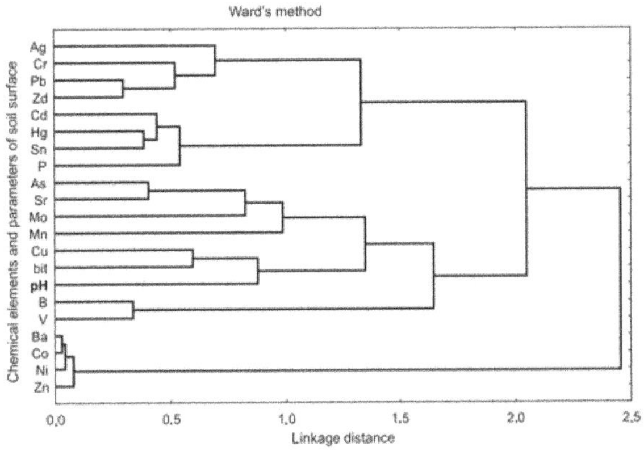

Fig. 3. Relationships and linkages of heavy metals and rare elements in soil (Šnipiškės eldership, Vilnius) by cluster analysis Ward's minimum variance method

vanadium (V) is related to associations of Cu-Bi and As-Sr-Mo-Mn via correlative links. Molybdenum (Mo) and manganese (Mn) are closely related to clayey soils, and an interdependent association between them usually shows up in areas of high pollution levels caused by traffic. Except for B-V, which are related to clay minerals found in the soil and commonly regarded as typical elements of power plat pollution, and Sr-As, which are linked to carbonaceous soil material of large scale broken rock, these elements and parameters form pseudo-associations based on weak links.

With reference to the book "Soil and water chemistry" (Burau, Zasoski 2002), it becomes possible to explain a stronger correlation between lithogenous elements of arsenic (As), strontium (Sr), molybdenum (Mo), manganese (Mn) and the pH value of the topsoil. Copper (Cu) is yet another lithogenous trace element found in soil, but the increased levels of it, as they are observed in the topsoil, can be put down on anthropogenic activity due to the intensive use of copper in industry. In terms of the city's environment, nickel (Ni) and zinc (Zn) are more likely to be taken as metals of technogenic origin and their concentration in soil is related to carbonaceous factors. They poorly correlate with the pH value, although they are known to have strong correlative links between themselves (Ni-Zn).

Results of correlation analysis suggest that the topsoil pH has weakest impact on technogenic origin metals.

The interdependence between soil reactivity and its contamination with heavy metals is hypothetical. A possible interdependence between the two issues suggests that the concentration of pollutant elements found in the soil depends on the pH value of the soil and, likewise, the soil reactivity changes affection the concentration of the pollutant elements. In the course of investigation, the conducted correlation analysis between the topsoil pH and the concentration of contaminant elements (Ag, Pb, Sn, Zn, Hg, Cd, Sr, Co, Ba, As, Ni, Cr, Cu, Mn, Mo, V, B, P) found in the soil showed very weak linear interdependence ($r < 0.3$). High variation in the content often reported to make it difficult to analyse their degree of contamination without taking into account the spatial variability of metals and their content in the parent materials of soils, which is the natural background (Grzebisz *et al.* 2002). Studies in Moscow region showed that distribution of heavy metals by the fractions primarily depended on the chemical properties of the heavy metals themselves; the pollution level; and, to a lesser degree, on the soil properties (Plyaskina, Ladonin 2009).

Copper (Cu) concentration found having very weak linear correlation with pH ($r = 0.2$), but that mainly based on the three ultimate points (out of 100 possible ones) (Fig. 4).

The mercury (Hg) linear correlation with pH $r = 0.15$ was found even weaker than that of copper (Cu) (Fig. 5). The weak negative, but statistically reliable association was indicated between pH and strontium (Sr), $r = -0.12$ (Fig. 6).

In the course of investigation, it was also established that there is a very weak direct relationship between topsoil pH value and the index of the overall contamination Z_d ($r - 0.12$) (Fig. 7).

The conducted analysis determined the interdependence between the index (Z_d) of overall contamination with some contaminants (Cu, Hg, Sr) and the topsoil pH value.

The observed results confirmed that the contamination of urban soils exhibits somewhat different compared to agricultural soils. In contradiction to earlier studies in Lithuanian agricultural soils where strong correlation between soil pH and Cr, Cd, Pb, Ni, Cu and Z_d found (Sunkieji metalai … 2001), the conducted urban soils analysis shows a statistically reliable, but very weak (<0.3) correlation between the soil pH and concentration of contaminants (Figs 4, 5, 6, 7).

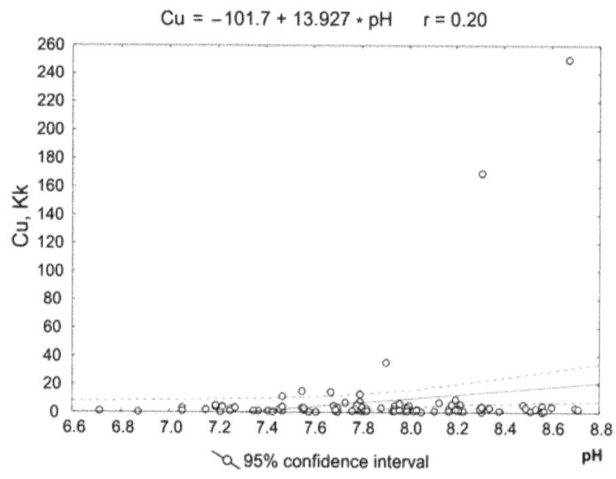

Fig. 4. Correlation (r) between concentration coefficient of copper (Cu) and topsoil pH in eldership of Šnipiškės of the city Vilnius

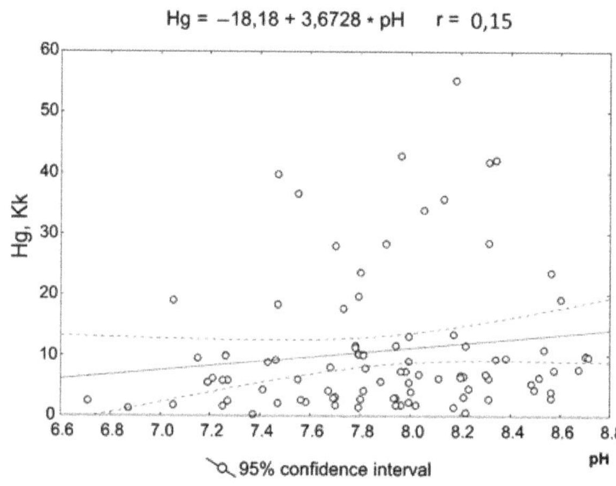

Fig. 5. Correlation (r) between concentration coefficient of mercury (Hg) and topsoil pH in eldership of Šnipiškės of the city Vilnius

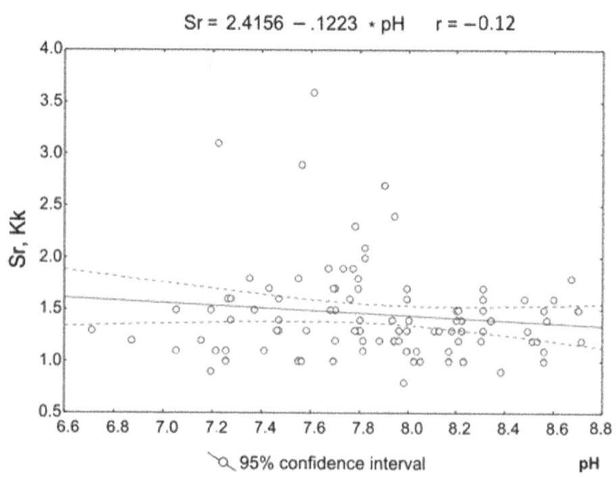

Fig. 6. Correlation (r) between concentration coeficient of stroncium (Sr) and topsoil pH value in eldership of Šnipiškės of the city Vilnius

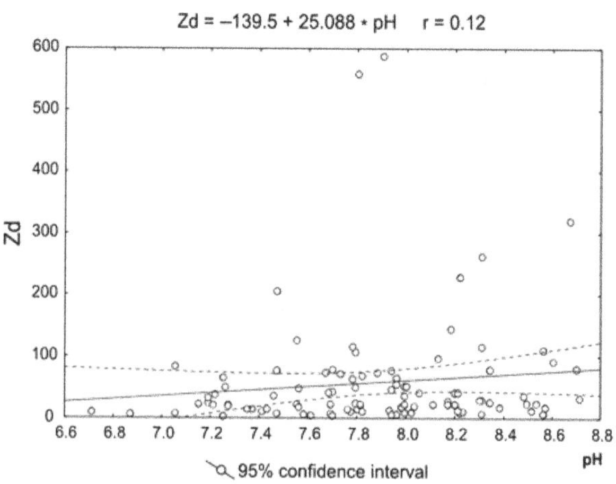

Fig. 7. Correlation (r) between the index of overall contamination (Z_d) and topsoil pH value in eldership of Šnipiškės of the city Vilnius

Conclusions

Anthropogenic impact determines the alkalinity of the urban soil (pH → 9) and, simultaneously, the intensity of sorption of specific contaminants (silver (Ag), lead (Pb), tin (Sn), zinc (Zn), mercury (Hg), bitumoide, cadmium (Cd), strontium (Sr), cobalt (Co), barium (Ba), arsenic (As), nickel (Ni), chromium (Cr), copper (Cu), manganese (Mn), molybdenum (Mo), vanadium (V), boron (B), phosphorus (P) the overall contamination (Z_d).

The morphological features of the territory influence the pH value of the soil. Greater pH values ($pH > 8$) are more characteristic of younger deposited sediment than of the older deposits.

The analysed spatial premises, by which the pH shift in the soil and concentration of heavy metals in contaminated areas as well as the interdependence of the two issues are determined, confirm the claim for the existence of this interdependence.

According to analysis results specific heavy metals form specific accumulation associations that indicate different contamination sources.

The conducted analysis showed a statistically reliable, yet very weak correlation between the soil pH and concentration of heavy metals (Ag, Pb, Sn, Zn, Hg, Cd, Sr, Co, Ba, As, Ni, Cr, Cu, Mn, Mo, V, B, P) found in the soil ($r < 0.3$). Between specific heavy metals and the topsoil pH, a strongest statistically reliable relationships were observed between the soil pH and copper (Cu) as well as the soil pH and the index of overall contamination (Z_d).

The conducted analysis does not reveal the nature of the interdependence between the soil pH and its contamination with heavy metals, i.e. it remains unproven if the soil reactivity influences its contamination or, conversely, if the soil contamination causes changes to its reactivity.

References

Burau, R. G.; Zasoski, R. J. 2002. *Soil and water chemistry.* California. 153 p.

Butkus, D.; Paliulis, D.; Baltrenaite, E. 2004. Sorption of heavy metals from polluted water and their migration in the system soil-tree, *Journal of Environmental Engineering and Landscape Management* 12(4): 120–125.

Cappuyns, V.; Swennen, R. 2008. The application of pH_{stat} leaching tests to assess the pH-dependent release of trace metals from soil, sediments and waste materials, *Journal of Hazardous Materials* 158(1): 185–195. http://dx.doi.org/10.1016/j.jhazmat.2008.01.058

Eidukevičienė, M.; Volungevičius, J.; Marcinkonis, S.; Tripolskaja, L.; Karčiauskienė, D.; Fullen, M. A.; Booth, C. A. 2010. Interdisciplinary analysis of soil acidification hazard and its legacy effects in Lithuania, *Natural Hazards and Earth System Sciences* 10: 1477–1485. http://dx.doi.org/10.5194/nhess-10-1477-2010

Gregorauskienė, V.; Kadūnas, V. 2006. Vertical distribution patterns of trace and major elements within soil profile in Lithuania, *Geological Quarterly* 50(2): 229–237.

Gregorauskienė, V. 2006a. Mapping of geochemical contamination in urban areas of Lithuania, *Journal of Environmental Engineering and Landscape Management* 14(1): 52–57.

Grybauskas, J. 1978. *Lietuvos dirvožemio pH žemėlapis* [Map of Lithuania soil pH]. M 1:3000 000 [in Lithuanian].

Grzebisz, W.; Cieoela, L.; Komisarek, J.; Potarzycki, J. 2002. Geochemical assessment of heavy metals pollution of urban soils, *Polish Journal of Environmental Studies* 11(5): 493–499.

HN 60:2004. Lithuanian hygienic norm HN 60:2004 Maximum permitted concentrations of hazardous chemical substances in soil, *Valstybės Žinios* [State news], 41-1357 (in Lithuanian).

Imperato, M.; Adamo, P.; Naimo, D.; Arienzo, M.; Stanzione, D.; Violante, P. 2003. Spatial distribution of heavy metals in urban soils of Naples city (Italy), *Environmental Pollution* 124(2): 247–256. http://dx.doi.org/10.1016/S0269-7491(02)00478-5

Irha, N.; Steinnes, E.; Kirso, U.; Petersell, V. 2009. Mobility of Cd, Pb, Cu, and Cr in some Estonian soil types, *Estonian Journal of Earth Sciences* 58(3): 209–214. http://dx.doi.org/10.3176/earth.2009.3.05

ISO 10390:2005. 2010. *Soil quality. Determination of pH.* 7 p.

Jankauskaitė, M.; Taraškevičius, R.; Zinkutė, R.; Veteikis, D. 2008. Relationship between landscape self-regulation potential and topsoil additive contamination by trace elements in Vilnius city, *Journal of Environmental Engineering and Landscape Management* 16(1): 5–14. http://dx.doi.org/10.3846/1648-6897.2008.16.5-14

Jones, K. C.; Symon, C. J.; Johnston, A. E. 1987. Retrospective analysis of an archived soil collection I, *Science of the Total Environment* 61: 131–144. http://dx.doi.org/10.1016/0048-9697(87)90363-9

Kadūnas, V.; Katinas, V.; Radzevičius, A.; Taraškevičius, R.; Zinkutė, R. 2005. Geochemical atlas of Lithuania – a basis for detailed geochemical investigations of long-term ecological research sites, *Acta Zoologica Lituanica* 15(2): 131–135. http://dx.doi.org/10.1080/13921657.2005.10512389

Kumpienė, J.; Brännvall, E.; Taraškevičius, R.; Aksamitauskas, Č.; Zinkutė, R. 2011. Spatial variability of topsoil contamination with trace elements in preschools in Vilnius, Lithuania, *Journal of Geochemical Exploration* 108(1): 15–20. http://dx.doi.org/10.1016/j.gexplo.2010.08.003

Lado, L. R.; Hengl, T.; Reuter, H. I. 2008. Heavy metals in European soils: a geostatistical analysis of the FOREGS Geochemical database, *Geoderma* 148(2): 189–199. http://dx.doi.org/10.1016/j.geoderma.2008.09.020

Marcinkonis, S. 2008. Assessing trace element accumulation and depletion in agricultural soils in Lithuania, *Acta Agriculturae Scandinavica, Section B – Plant Soil Science* 58(2): 114–123.

Paartel, M.; Helm, A.; Ingerpuu, N.; Reier, U.; Tuvi, E. L. 2004. Conservation of Northern European plant diversity: the correspondence with soil pH, *Biological Conservation* 120(4): 525–531. http://dx.doi.org/10.1016/j.biocon.2004.03.025

Plyaskina, O.; Ladonin, D. 2009. Heavy metal pollution of urban soils, *Eurasian Soil Science* 42(7): 816–823. http://dx.doi.org/10.1134/S1064229309070138

Reimann, C.; Caritat, P. 2012. New soil composition data for Europe and Australia: demonstrating comparability, identifying continental-scale processes and learning lessons for global geochemical mapping, *Science of the Total Environment* 416: 239–252. http://dx.doi.org/10.1016/j.scitotenv.2011.11.019

Mažvila, J. (Ed.). 2001. *Sunkieji metalai Lietuvos dirvožemiuose ir augaluose:* monografija [Heavy metals in soils and plants of Lithuania: monograph]. 343 p. (in Lithuanian).

Swartjes, F. A.; Carlon, C.; de Wit, N. H. S. M. 2008. The possibilities for the EU-wide use of similar ecological risk-based soil contamination assessment tools, *Science of the Total Environment* 406(3): 523–529. http://dx.doi.org/10.1016/j.scitotenv.2008.07.034

Taraškevičius R.; Zinkutė R. 2011. Urban geochemical anomalies of Lithuania and their spread, *Baltica* 24: 163–168.

Taraškevičius R. 2007. *Vilniaus miesto dirvožemio (grunto) būklė ir kaita.* [Soil (subsoil) assessment and variation of Vilnius city] [online], [cited 28 May 2013]. Available from Internet: http://aplinka.vilnius.lt/pdf/vilniaus_miesto_dirvozemio_bukle_ir_kaita.pdf (in Lithuanian).

Taraškevičius, R. 1996. Šnipiškių (Vilnius) ekogeocheminis įvertinimas [Ekogeochemical assessment of Šnipiškės (Vilnius)], *Geologijos institutas* [Institute of Geology]. 62 p. (in Lithuanian).

Thematic Strategy for Soil Protection. 2006. Brussels: Commission of the European Communities. (COM (2006) 231 final). 12 p.

Chen, T.-B.; Zheng, Y.-M.; Lei, M.; Huang, Z.-C.; Wu, H.-T.; Chen, H.; Fan, K.-K.; Yu, K.; Wu, X.; Tian, Q.-Z. 2005. Assessment of heavy metal pollution in surface soil of urban parks in Beijing, *Chemosphere* 60(4): 542–551. http://dx.doi.org/10.1016/j.chemosphere.2004.12.072

Vasarevičius, S.; Greičiūtė, K. 2004. Investigation of soil pollution with heavy metals in Lithuanian military grounds, *Journal of Environmental Engineering and Landscape Management* 12(4): 132–137. http://dx.doi.org/10.1080/16486897.2004.9636834

Woodard, T. L.; Amarasiriwardena, D.; Shrout, K.; Xing, B. 2007. Roadside accumulation of heavy metals in soils in Franklin County, Massachusetts, and surrounding towns, *Communications in Soil Science and Plant Analysis* 38(7–8): 1087–1104. http://dx.doi.org/10.1080/00103620701280308

Zinkutė, R.; Taraškevičius, R.; Želvys, T. 2011. Major elements as possible factors of trace element urban pedochemical anomalies, *Central European Journal of Chemistry* 9(2): 337–347. http://dx.doi.org/10.2478/s11532-011-0012-z

Asta KAZLAUSKAITĖ-JADZEVIČĖ. She is a PhD student, working at the Lithuanian Research Centre for Agriculture and Forestry. Her research interests include renaturation, GIS, and soil carbon modelling.

Jonas VOLUNGEVIČIUS. He is a Doctor of Physical Sciences (Geography) and works as a Docent at the Department of Geography and Land Management, Faculty of Natural Sciences in Vilnius University, and also is a President of Lithuanian Soil Science Society at the Lithuanian Academy of Science Agricultural and Forest Sciences Department. He is the author and co-author of ~14 scientific papers and other scientific publications and has attended 2 international and 11 local conferences. His research interests include geography of landscape management, soil geography, environmental policy, and application of GIS in geography.

Virgilija GREGORAUSKIENĖ. She is a Doctor of Physical Sciences (Geography), works as a Senior technician at the Subdivision of Engineering Geology and Processes, Geological Survey of Lithuania, and is also a Lector at Mykolas Riomeris University. Her research interests include environmental geochemistry, soil science, soil contamination, soil quality standards, air borne contamination, methods of geochemical mapping at different scales, and medical geology.

Saulius MARCINKONIS. He is a Doctor of Biomedical Sciences (Agronomy), working as a Head of Strategic Development Department at the University of Applied Sciences (Vilniaus kolegija). He is the author and co-author of ~33 scientific papers and has attended 40 conferences. His research interests include soil chemical degradation, liming, fertilization, soil pollution, renaturation, and GIS.

NOVEL MATHEMATICAL MODELS FOR PREDICTION OF MICROBIAL GROWTH KINETICS AND CONTAMINANT DEGRADATION IN BIOREMEDIATION PROCESS

Azadeh KHOSHDEL, Behrooz MAHMOODZADEH VAZIRI

Department of Chemical Engineering, Quchan Branch, Islamic Azad University, Quchan, Iran

Abstract. Bioremediation is defined as a process, which involves decomposition of organic pollutant compounds available in soil and water resources into safe and eco-friendly materials, like water and CO2, by the microorganisms. In the present article, mathematical modeling of the bioremediation process was conducted comprehensively, and new models proposed for the microbial growth kinetics and substrate consumption (contaminant degradation). Accordingly, six kinetic models were suggested for the biomass growth and six models for the substrate consumption. Moreover, two models were considered for specific growth rate constant of the microorganisms. Then, model predictions were compared to and validated by the available experimental data in the literature. According to the obtained results, the microbial growth kinetic model, entitled as "*MVKH2*", the substrate (contaminant) consumption model, entitled as "*MVKH2s*", and the Aiba specific growth rate constant model had the best performance and the least error value in predicting the bioremediation process. Results achieved from this study are a promising beginning for practical and experimental works.

Keywords: bioremediation, environmental processes modeling, simulation, microbial growth kinetic, contaminant degradation.

Introduction

Bioremediation is defined as a process, which involves decomposition of organic pollutant compounds available in soil and water resources into safe and eco-friendly materials by the microorganisms. The process usually takes place extensively for cleaning coastal ecosystems, after leakage of oil compounds, cleaning of contaminated areas with heavy metals (such as uranium, arsenic, aluminum, tin and zinc), and also extraction of metals from their stone mine (Cristina Souza *et al.* 2014; Sheoran *et al.* 2010). Various physical and chemical methods could be employed to eliminate contaminant materials, but they are mostly expensive and incompatible with the environment. Therefore, bioremediation could be an appropriate strategy for removing pollutants, especially in places where cleanup is not physically or chemically feasible (e.g. in areas with low level of pollution) (Jeyasingh *et al.* 2011; Moscoso *et al.* 2012).

Practically, bioremediation process depends on the number and type of microorganisms. Selecting microbes also depends on the nature, chemical structure of the contaminant, and the environmental condition. Different microorganisms including bacteria, fungi, yeasts, and algae are used in bioremediation process for pollutant decomposition. Among them, bacteria, due to more adaptability with the environment, are highly considered. Contaminants are usually comprised of petroleum compounds and its derivatives, solvents, and pesticides (Khataee *et al.* 2011; Wang *et al.* 2008). Bioremediation is often a slow process that could be expedited through addition of nutrients, inoculation with bacterial consortium, and selection of suitable bioreactor. The successful completion of bioremediation depends on preparation of optimum physical, chemical and biological conditions in the polluted area (Nikakhtari *et al.* 2008).

Recently, the bioremediation process has attracted the attention of many researchers, however, few works have been conducted on modeling of bioremediation process. The following are some works done in the field of bioremediation process and modeling:

Corresponding author: Behrooz Mahmoodzadeh Vaziri
E-mails: vaziri@iauq.ac and behrouz.vaziri@gmail.com

Kindred, Celia (1989) considered simple first order kinetic for growth of biomass. Nakamura, Sawada (2000) suggested a mathematical model for prediction of chrome (VI) reduction, and its conversion to chrome (III) with less toxicity. Their model has desirable prediction only when the amount of substrate is low for bacterium growth. Tarighian *et al.* (2003) showed that glucose was a better co-substrate relative to phenol for the co-metabolic removal of 4-chlorophenol from wastewater by *Pseudomonas putida*. They mentioned that both co-substrates were metabolized by the biomass according to Monod model. Moussa *et al.* (2005) presented a model to study the effect of increasing sludge age, and the role of predators on the biomass composition in the sequenced batch reactors. Furthermore, the interaction between nitrifiers, heterotrophs and predators in wastewater treatment was considered in their model. Jankaite and Vasarevicius (2005) reviewed various remediation techniques for removing heavy metals from the contaminated soil. They pointed out that the selection of suitable soil remediation technology depends on concentration, type of pollutant, site characteristic and the final use of a contaminated medium. Van Ginneken *et al.* (2007) applied phytoremediation to clean the soils contaminated with heavy metals. They stated that uptake capacity of heavy metals can be promoted by plants via the addition of biodegradable physico-chemical agents, and stimulating the microbial community in and around the plant. Furthermore, they discussed in more detail to convert the harvested biomass crops into the biodiesel. Abbasi, Shquirat (2008) showed that the value of microbial specific growth rate constant (μ) is related to the microbial concentration and increasing that will lead to enhancement of this value. The overall bioremediation rate was modeled with the two limiting cases of zero and first order kinetics to describe biodegradation of a volatile organic compound (VOC) in a fluidized bioreactor by Clarke *et al.* (2009). Their model represented that maximum elimination capacity is limited by the microorganism growth rate and the microbial population. Also, the overall degradation of a substrate was a zero order process in the bioreactor at maximum elimination capacity. Guo *et al.* (2010) considered a first order kinetic model for the biomass growth in bioremediation of oil hydrocarbons.

Fallgren *et al.* (2010) used respirometric measurements of CO_2 due to microbial activity in a soil contaminated with diesel to derive empirical models. The presented empirical equations described the effects of environmental factors such as temperature, water content, nitrogen and phosphorus content on microbial hydrocarbon degradation via CO_2 data analysis.

Vaezihir *et al.* (2012) presented transport model and a field-scale three-dimensional numerical flow to simulate the destiny and transport of benzene, toluene, ethylbenzene and xylenes (BTEX) from six source regions of light nonaqueous-phase liquids (LNAPLs). They investigated the effect of natural debilitation and enhanced remediation procedures on the destiny of BTEX. Their results indicated that the remediation time for BTEX was more than 60 years, under natural debilitation only, but using enhanced bioremediation methods and LNAPL removal could decrease this time to about 30 years for BTEX.

Das Saha (2013) studied the effects of operational variables such as adsorbent dosage, contact time, dye concentration, initial pH, and agitation on the dye removal from wastewater using rice husk in a stirred tank reactor. She applied response surface methodology (RSM) to investigate the interaction between various process parameters, and the optimization of the variables on dye removal. In addition, the artificial neural network was used to model process parameters in her work.

Mohajeri *et al.* (2013) concluded that the bioremediation kinetic modeling of contaminated soil is complicated due to the existence of numerous factors. Despite this fact, they have used the first order kinetic to predict the biological decomposition of oil hydrocarbons.

Vasiliadou *et al.* (2013) applied a kinetic model to predict the effect of biomass growth and nutrients consumption on the sorption and bioremediation of the pollutants (pharmaceutical compounds) in individual and simultaneous treatment. In their work, the bioremediation process was modeled by a pseudo-first-order kinetic type with a double-Monod kinetic expression. Koreiviene *et al.* (2014) applied microalgae consortia for bioremediation of wastewater and their biomass potential to produce the biofuel. Their results revealed that: a) *Chlorella/Scenedesmus* consortium removed up to 99.7–99.9% of inorganic phosphorus and up to 88.6–96.4% of inorganic nitrogen from the wastewater within three weeks, b) The ammonium elimination was more efficient than that of nitrate, c) *Chlorella* algae grew better in diluted, while *Scenedesmus* in the concentrated wastewater, d) The consortium treated wastewater more efficiently than a single species.

Since the experimental study of bioremediation for evaluating different aspects of the process (such as microbial growth and contaminant degradation trends), is extremely time-consuming and expensive, process modeling and simulation are greatly contributed to achieve the desired objectives. In the present article, comprehensive modeling of microbial growth trend and substrate consumption, as two key parameters in bioremediation were accomplished. In this regard, novel kinetic models have been proposed and compared to predict the biomass growth behavior and the substrate consumption (pollutant degradation). Moreover, different models were considered to determine the value of μ.

1. Modeling of bioremediation process

Modeling is a useful tool for design, scale-up, and process control in bioremediation. Also, it could cause the realization of effective strategies and removing of misguided policies. Some factors such as biomass concentration and the amount of substrate (contaminant) have an important role in optimum performance of the bioremediation. Thus, various models have been proposed and compared for predicting the biomass concentration, the amount of substrate consumption, and the value of μ in the present study. In proposed biomass concentration models, both growth and death kinetics of microorganisms were considered, because the amount of biomass initially increase by passing of the time and substrate consumption and then will decrease due to reduction of substrate and cells' death.

1.1. Microbial growth kinetic models

As can be seen in Table 1, various kinetic models are suggested to predict the behavior of microbial growth through the bioremediation process (growth kinetic models identified by *MVKH*). In all developed growth models, the related terms of growth and death of microorganisms were considered (death of microorganisms will occur after the stationary phase).

In existing mathematical models in Table 1, X is the biomass concentration, μ the microbial specific growth rate constant, γ stationary constant, k_d death or biomass degradation constant, and X_m the maximum biomass concentration. In these models the term $k_d \left|\int_0^t X(t)dt\right|$ describes death kinetic of the cells.

1.2. Substrate consumption models

As can be seen in Table 2, different mathematical models have been designed to predict the rate of substrate consumption (contaminant degradation) in the bioremediation process (substrate consumption models defined by *MVKHs*).

In presented models of Table 2, S is the substrate (contaminant) concentration and, Y is the yield factor. It is worth mentioning that in substrate consumption models the corresponding terms of growth and stationary phases from growth curve were only considered, and the related terms of death kinetics have been ignored, because the amount of substrate consumption (contaminant biodegradation) is negligible in the death phase.

1.3. Models of microbial specific growth rate constant (μ)

The value of specific growth rate constant of microorganisms has a significant role in correct prediction of growth kinetic model. In most previous studies on bioremediation, the amount of this parameter (μ) was assumed as constant. So, the effect of contaminant concentration on μ will be ignored with this assumption (Mohajeri *et al.* 2013; Geng *et al.* 2013; Li *et al.* 2013; Louati *et al.* 2013). In order to overcome this shortcoming, two models were considered for predicting the value of μ in the present study. These models are shown in Table 3. In Table 3 μ_{max} is the maximum specific growth rate constant, K_s biomass affiliation to the substrate, and K_i substrate dissociation constant.

Table 1. The microbial growth kinetic models proposed in the present study

Model	Name
MVKH1	$\frac{dX}{dt}=\mu X(1-\gamma X)+k_d\left|\int_0^t X(t)dt\right|$
MVKH2	$\frac{dX}{dt}=\mu-\gamma+k_d\left|\int_0^t X(t)dt\right|$
MVKH3	$\frac{dX}{dt}=\mu X\left(1-\frac{X}{X_m}\right)-\gamma+k_d\left|\int_0^t X(t)dt\right|$
MVKH4	$\frac{dX}{dt}=\mu X\left(1-\frac{X}{X_m}\right)-\gamma X+k_d\left|\int_0^t X(t)dt\right|$
MVKH5	$\frac{dX}{dt}=\mu X-\gamma+k_d\left|\int_0^t X(t)dt\right|$
MVKH6	$\frac{dX}{dt}=\mu X-\gamma X+k_d\left|\int_0^t X(t)dt\right|$

Table 2. The substrate consumption models proposed in the present study

Model	Name
MVKH1s	$\frac{dS}{dt}=\mu X(1-\gamma X)\ /Y$
MVKH2s	$\frac{dS}{dt}=\mu-\gamma\ /Y$
MVKH3s	$\frac{dS}{dt}=\mu X\left(1-\frac{X}{X_m}\right)-\gamma\ /Y$
MVKH4s	$\frac{dS}{dt}=\mu X\left(1-\frac{X}{X_m}\right)-\gamma X\ /Y$
MVKH5s	$\frac{dS}{dt}=\mu X-\gamma\ /Y$
MVKH6s	$\frac{dS}{dt}=\mu X-\gamma X)\ /Y$

Table 3. Microbial specific growth rate constant models

Model	Name
Haldan (*P1*)	$\mu=\dfrac{\mu_{max}S}{S+K_s+\dfrac{S^2}{K_i}}$
Aiba (*P2*)	$\mu=\dfrac{\mu_{max}S\exp\left(\dfrac{-S}{K_i}\right)}{K_s+S}$

2. Process simulation

In order to evaluate the performance of the proposed models of the previous section, growth behavior simulation of prokaryotic communities were carried out on aliphatic and polycyclic aromatic hydrocarbons (as contaminant) in two cases. In the first case, each microbial growth kinetic model was simultaneously solved with the related substrate consumption (contaminant degradation) model, and Haldan's specific growth rate constant model. Similarly, each microbial growth kinetic models with corresponding substrate consumption model were solved through the second case by specific growth rate constant model of Aiba, simultaneously. Then, the obtained results from simulations were compared with the experimental data of Beolchini *et al.* (2010) for growth of prokaryotic cells on aliphatic and polycyclic aromatic hydrocarbon contaminants. As it will be shown in the next section, the predictions of some models are extremely in line with the experimental results, which is indicative of high reliability of the proposed models.

Table 4. The parameters value calculated by nonlinear regression in the first simulation case

Parameter	Value	Unit
K_s	38.5	μM
K_i	9840.2	μM
γ	0.04	–
k_d	–0.006	day^{-2}

Table 5. The values of μ_{max} and *Y* calculated by nonlinear regression in each six various combinations of the first simulation case

Model	μ_{max} (day^{-1})	Y(gr^{-1})
MVKH1-MVKH1s-P1	5.6	70.96
MVKH2-MVKH2s-P1	3.46	45.16
MVKH3-MVKH3s-P1	0.488	5.8
MVKH4-MVKH4s-P1	0.9	12.9
MVKH5-MVKH5s-P1	0.162	2.09
MVKH6-MVKH6s-P1	0.17	2.25

In order to solve the microbial growth kinetics and substrate consumption models (mentioned in section 2), which are ordinary differential equations, the forth order Runge–Kutta method (using *ode45* function in MATLAB software environment (version of 2013a)) has been employed with variable time step size. The mentioned solving method is called "linear method". The initial value of biomass concentration is 4×10^8 cells gr^{-1}, and the initial value of substrate concentration is 500 μgr gr^{-1}.

3. Results and discussion

As mentioned before, in the first case of simulation, each growth kinetic models (*MVKH1-MVKH6*) was considered with the corresponding substrate consumption model (*MVKH1s-MVKH6s*) and Haldan's specific growth rate constant model (identified as *P1*). For the best fitting with experimental data, some of the adjustable parameters in above mentioned models were computed through nonlinear regression in MATLAB environment (version 2013a) using '*nlinfit*' function and presented in Tables 4 and 5.

The values of maximum specific growth rate constant (μ_{max}) and yield factor (Y) (calculated by nonlinear regression) in six different combinations of the first case simulations were presented in Table 5.

The results achieved from the simulations of the first case, and the experimental data extracted from Beolchini *et al.* (2010) work are illustrated in Figures 1 and 2 for growth of prokaryotic communities on aliphatic and polycyclic aromatic hydrocarbon contaminants. As can be seen in Figure 1 (a and b), the variation of biomass in the bioremediation process shows an increasing-decreasing trend. This is due to the fact that the microorganisms will initially grow, and the amount of biomass will increase by passing of the time and substrate consumption. After passing specified time, the microorganisms will inter the death phase and the amount of biomass will take a decreasing trend. The variations trend of substrate (contaminant) consumption is shown in Figure 2 (a and b). It is clearly evident that, the substrate is consumed during the time and its value is decreased. According to the obtained results in Figures 1 and 2, accessibility of microorganisms to substrate has been reduced by decreasing the amount of substrate from a certain value, which is a vital factor for stopping the cells growth.

As can be seen from Figures 1 and 2, *MVKH2* growth kinetic model along with its related substrate consumption model (*MVKH2s*) has the best agreement with experimental data. The sum of squared error (SSE) of each proposed models for the biomass growth and substrate consumption along with the total SSE of each six existing combinations are given in Table 6. Clearly, the combination of *MVKH2–MVKH2s* has the least total error in the first case of simulations.

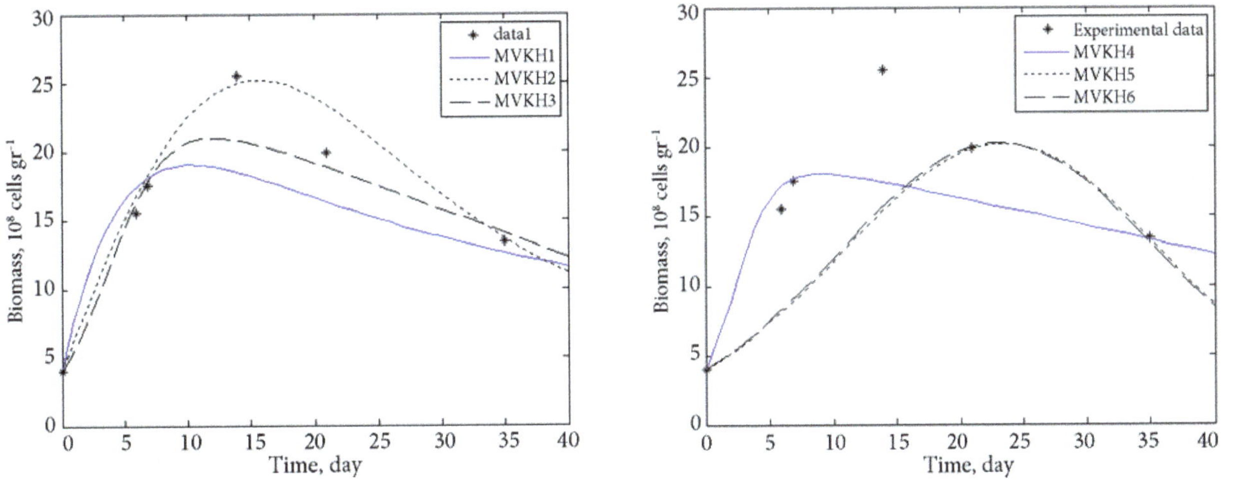

a) b)

Fig. 1. a) – predictions of the biomass variations versus time using *MVKH1* to *MVKH3* growth kinetic models accompanied with Beolchini *et al.* (2010) experimental data for growth of prokaryotic communities on aliphatic and polycyclic aromatic hydrocarbon contaminants in the first case of simulation; b) – predictions of the biomass variations versus time using *MVKH4* to *MVKH6* growth kinetic models accompanied with Beolchini *et al.* (2010) experimental data for growth of prokaryotic communities on aliphatic and polycyclic aromatic hydrocarbon contaminants in the first case of simulation

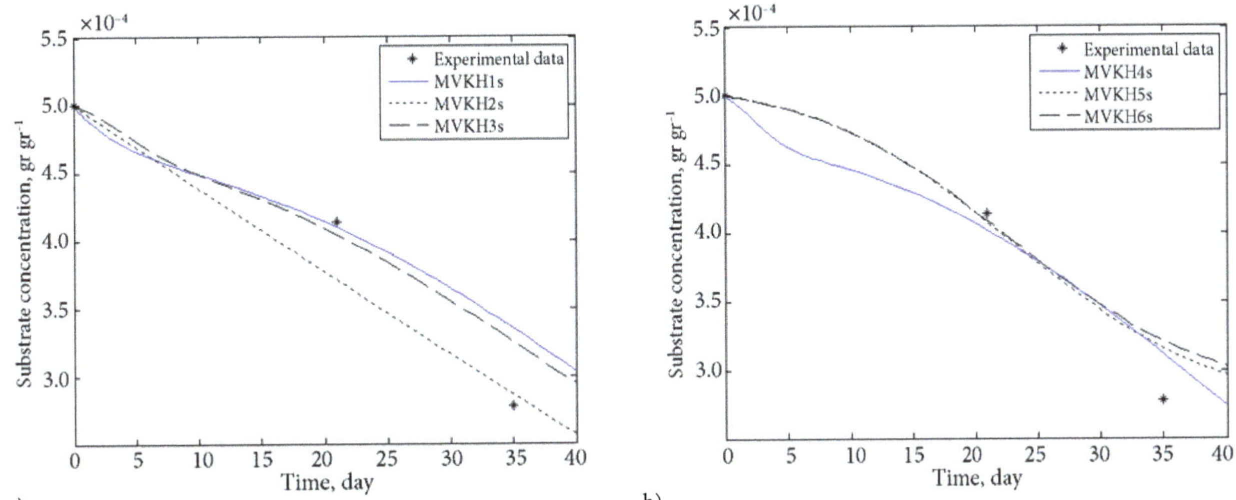

a) b)

Fig. 2. a) – predictions of the contaminant variations trend versus time using *MVKH1s* to *MVKH3s* substrate consumption models accompanied with Beolchini *et al.* (2010) experimental data for growth of prokaryotic communities on aliphatic and polycyclic aromatic hydrocarbon contaminants in the first case of simulation; b) – predictions of the contaminant variations trend versus time using *MVKH4s* to *MVKH6s* substrate consumption models accompanied with Beolchini *et al.* (2010) experimental data for growth of prokaryotic communities on aliphatic and polycyclic aromatic hydrocarbon contaminants in the first case of simulation

Table 6. The SSE of microbial growth kinetic and contaminant consumption models, as well as total SSE obtained from simulations of the first case

Model	SSE of growth kinetic model	SSE of contaminant consumption model	Total SSE
MVKH1-MVKH1s-P1	64.22	1.12	65.34
MVKH2-MVKH2s-P1	15.77	0.6	16.37
MVKH3-MVKH3s-P1	22.02	0.84	22.86
MVKH4-MVKH4s-P1	87.81	0.46	88.27
MVKH5-MVKH5s-P1	326.63	2.02	328.65
MVKH6-MVKH6s-P1	209.45	0.59	210.04

In the second case of the simulation, each growth kinetics (*MVKH1-MVKH6*) and corresponding substrate consumption model (*MVKH1s-MVKH6s*) was considered with Aiba's specific growth rate constant model (identified as *P2*). Like the fist case, for the best fitting with experimental data, some of the adjustable parameters in proposed models were obtained through the nonlinear regression in MATLAB environment and presented in Tables 7 and 8. Similar to the previous stage, the values of μ_{max} and Y (calculated by nonlinear regression) in six different combinations of the second case simulations were given in Table 8.

Table 7. The parameters value calculated by nonlinear regression in the second simulation case

Parameter	Value	Unit
K_s	43.9	μM
K_i	9137.7	μM
γ	0.04	–
k_d	-0.006	day^{-2}

Table 8. The values of μ_{max} and Y calculated by nonlinear regression in each six various combinations of the second simulation case

Model	μ_{max} (day^{-1})	$Y(gr^{-1})$
MVKH1-MVKH1s-P2	0.48	3.87
MVKH2-MVKH2s-P2	3.24	20.96
MVKH3-MVKH3s-P2	0.51	4.83
MVKH4-MVKH4s-P2	0.705	5.8
MVKH5-MVKH5s-P2	0.165	1.45
MVKH6-MVKH6s-P2	0.172	1.45

The predictions of the second case modeling and the experimental data of Beolchini *et al.* (2010) work were indicated in Figures 3 and 4 for growth of prokaryotic cells on aliphatic and polycyclic aromatic hydrocarbon contaminants. As can be shown, the trend of boimass variations and contaminant degradation is almost similar to the first case of simulation.

As depicted in Figures 3 and 4, *MVKH2* growth kinetic model with corresponding substrate consumption model (*MVKH2s*) has the best agreement with experimental data similar to the first case of simulations. The SSE of each proposed models for biomass growth and substrate consumption along with the total SSE of each six available combinations are given in Table 9. Again, the combination of *MVKH2-MVKH2s* has the least total error in the second simulation case.

According to the presented results in Tables 6 and 9, the total SSE value in combination of *MVKH2-MVKH2s-P1* is 16.37, while this value in combination of *MVKH2-MVKH2s-P2* is 14.45. Consequently, by comparing the achieved results from the first and second case of simulations, and the calculated error values it can be specified that the Aiba model has better performance in predicting specific growth rate constant of the microorganisms.

It is worthy to mention that the presented models in this paper are general, and they can be applied to different contaminants and bioremediation processes, but the focus of this study is on aromatic hydrocarbons degradation, because this group is one of the most important environmental contaminants.

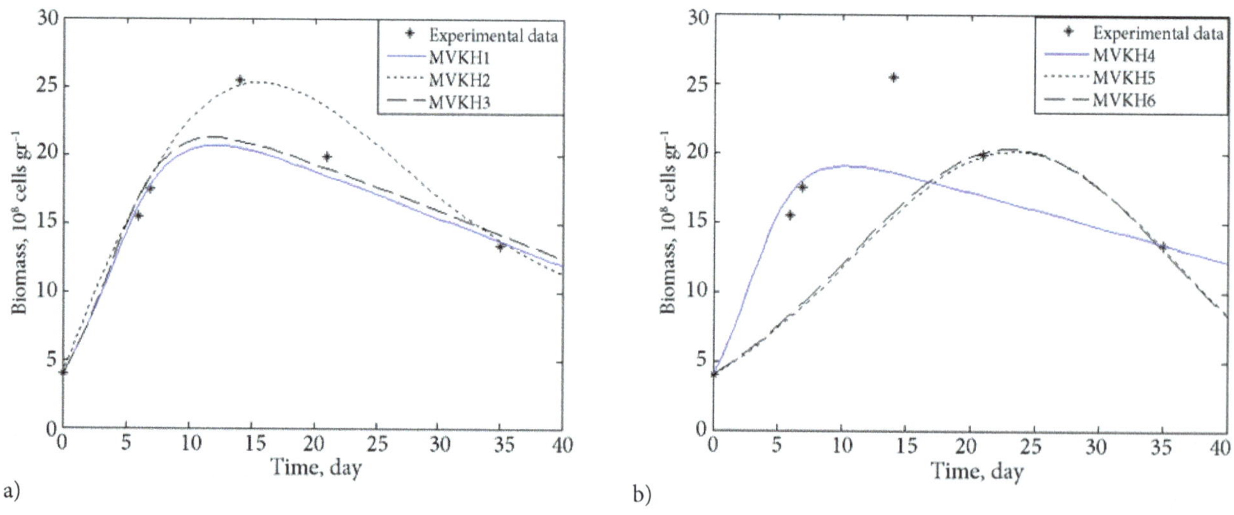

Fig. 3. a) – predictions of the biomass variations versus time using *MVKH1* to *MVKH3* growth kinetic models accompanied with Beolchini *et al.* (2010) experimental data for growth of prokaryotic communities on aliphatic and polycyclic aromatic hydrocarbon contaminants in the second case of simulation; b) – predictions of the biomass variations versus time using *MVKH4* to *MVKH6* growth kinetic models accompanied with Beolchini *et al.* (2010) experimental data for growth of prokaryotic communities on aliphatic and polycyclic aromatic hydrocarbon contaminants in the second case of simulation

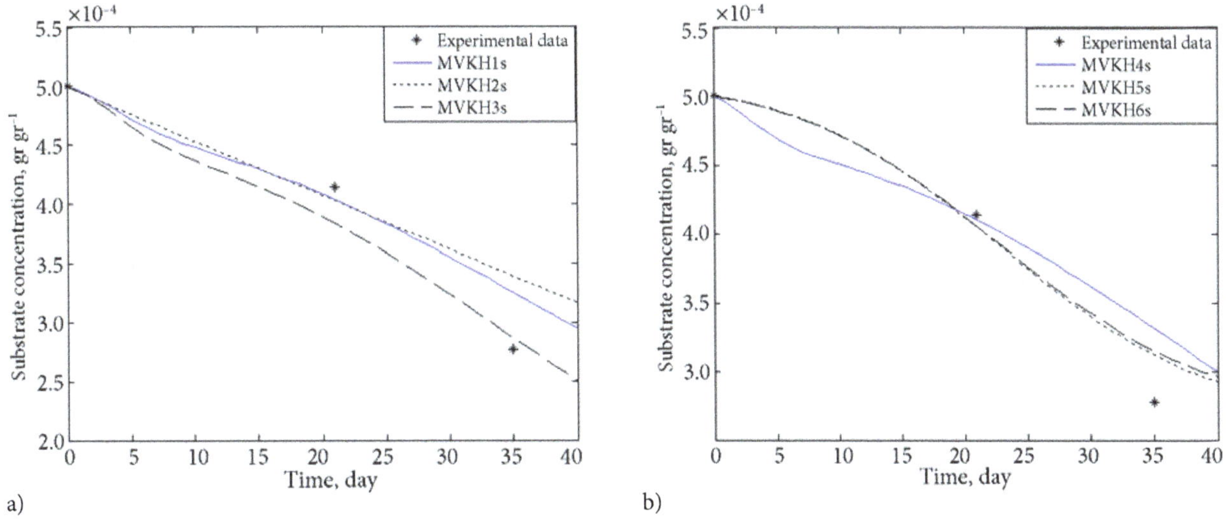

Fig. 4. a) – predictions of the contaminant variations trend versus time using *MVKH1s* to *MVKH3s* substrate consumption models accompanied with Beolchini *et al.* (2010) experimental data for growth of prokaryotic communities on aliphatic and polycyclic aromatic hydrocarbon contaminants in the second case of simulation; b) – predictions of the contaminant variations trend versus time using *MVKH4s* to *MVKH6s* substrate consumption models accompanied with Beolchini *et al.* (2010) experimental data for growth of prokaryotic communities on aliphatic and polycyclic aromatic hydrocarbon contaminants in the second case of simulation

Table 9. The SSE of microbial growth kinetic and contaminant consumption models, as well as total SSE obtained from simulations of the second case

Model	SSE of growth kinetic model	SSE of contaminant consumption model	Total SSE
MVKH1-MVKH1s-P2	27.28	0.98	28.26
MVKH2-MVKH2s-P2	13.18	1.27	14.45
MVKH3-MVKH3s-P2	23.41	6.36	29.77
MVKH4-MVKH4s-P2	59.34	0.89	60.23
MVKH5-MVKH5s-P2	18.94	0.44	19.38
MVKH6-MVKH6s-P2	222.9	0.014	222.914

Conclusion

Bioremediation is an effective, simple, and economic method for cleaning the contaminated soils and waters with petroleum compounds, industrial solvents, pesticides, and heavy metals by microorganisms. The experimental implementation of the bioremediation for evaluating different aspects of the process including biomass growth trend and contaminant degradation is extremely time-consuming and expensive. Therefore, modeling and simulation of the process could play a significant role in achieving the desired objectives. Among the presented models in the current study, *MVKH2*, *MVKH2s* and *P2* (Aiba) were introduced as the best models for predicting biomass growth, contaminant degradation and the specific growth rate constant, respectively. The achieved predictions from combination of mentioned models (*MVKH2–MVKH2s–P2*) had an apropriate agreement and the least errors to the experimental data, which indicated their efficient performance. Results obtained from this investigation could be applied to design, optimize and scale-up the bioremediation units.

Refrences

Abbassi, B. E.; Shquirat, W. D. 2008. Kinetics of indigenous isolated bacteria used for ex-situ bioremediation of petroleum contaminated soil, *Water, Air, & Soil Pollution* 192: 221–226. http://dx.doi.org/10.1007/s11270-008-9649-4

Beolchini, F.; Rocchetti, L.; Regoli, F.; Dell Anno, A. 2010. Bioremediation of marine sediments contaminated by hydrocarbons: Experimental analysis and kinetic modeling, *Journal of Hazardous Materials* 182: 403–407. http://dx.doi.org/10.1016/j.jhazmat.2010.06.047

Clarke, K.; Pugsley, T.; Hill, G. A. 2009. Gas-solid packed and fluidized bed models for bioremediation of volatile organic compounds in air, *Biochemical Engineering Journal* 46: 34–43. http://dx.doi.org/10.1016/j.bej.2009.04.009

Cristina Souza, E.; Christina Vessoni-Penna, T.; Pinheiro de Souza Oliveira, R. 2014. Biosurfactant- enhanced hydrocarbon bioremediation: an overview, *International Biodeterioration and Biodegradation* 89: 88–94. http://dx.doi.org/10.1016/j.ibiod.2014.01.007

Das Saha, P. 2013. Mathematical modeling of the reduction of safranin onto chemically modified rice husks in stirred tank reactor using response surface methodology and artificial neural network, *Bioremediation Journal* 17: 52–60. http://dx.doi.org/10.1080/10889868.2012.731445

Fallgren, P. H.; Jin, S.; Zhang, R.; Stahl, P. D. 2010. Empirical models estimating carbon dioxide accumulation in two petroleum hydrocarbon–contaminated soils, *Bioremediation Journal* 14: 98–108. http://dx.doi.org/10.1080/10889861003767084

Geng, X.; Boufadel, M. C.; Wrenn, B. 2013. Mathematical modeling of the biodegradation of residual hydrocarbon in a variably-saturated sand column, *Biodegradation* 24: 153–163. http://dx.doi.org/10.1007/s10532-012-9566-5

Guo, H.; Luo, S.; Chen, L.; Xiao, X.; Xi, Q.; Wei, W.; Zeng, G.; Liu, C.; Wan, Y.; Chen, J.; He, Y. 2010. Bioremediation of heavy metals by growing hyperaccumulaor endophytic bacterium Bacillus sp. L14, *Bioresource Technology* 101: 8599–8605. http://dx.doi.org/10.1016/j.biortech.2010.06.085

Jankaite, A.; Vasarevicius, S. 2005. Remediation technologies for soils contaminated with heavy metals, *Journal of Environmental Engineering and Landscape Management* 13: 109–113.

Jeyasingh, J.; Somasundaram, V.; Philip, L.; Murty Bhallamudi, S. 2011. Pilot scale studies on the remediation of chromium contaminated aquifer using bio-barrier and reactive zone technologies, *Chemical Engineering Journal* 167: 206–214. http://dx.doi.org/10.1016/j.cej.2010.12.024

Khataee, A. R.; Dehghan, G.; Zarei, M.; Ebadi, E.; Pourhassan, M. 2011. Neural network modeling of biotreatment of triphenylmethane dye solution by a green macroalgae, *Chemical Engineering Research and Design* 89: 172–178. http://dx.doi.org/10.1016/j.cherd.2010.05.009

Kindred, J. S.; Celia, M. A. 1989. Contaminant transport and biodegradation, 2. Conceptual model and test simulations, *Water Resources Research* 25: 1149–1160. http://dx.doi.org/10.1029/WR025i006p01149

Koreiviene, J.; Valciukas, R.; Karosiene, J.; Baltrenas, P. 2014. Testing of *Chlorella/Scenedesmus* microalgae consortia for remediation of wastewater, CO2 mitigation and algae biomass feasibility for lipid production, *Journal of Environmental Engineering and Landscape Management* 22: 105–114. http://dx.doi.org/10.3846/16486897.2013.911182

Li, X.; Shuang, Sh.; Li, X.; Kong, L.; Xu, L.; Tai, P.; Lin, X.; Jia, Ch.; Gong, Z. 2013. The effect of concentrations and properties of phenanthrene, pyrene, and benzo(a)pyrene on desorption in contaminated soil aged for 1year, *Journal of Soils and Sediments* 13: 375–382. http://dx.doi.org/10.1007/s11368-012-0629-3

Louati, H.; Ben Said, O.; Got, P.; Soltani, A.; Mahmoudi, E.; Cravo-Laureau, C.; Duran, R.; Aissa, P.; Pringault, O. 2013. Microbial community responses to bioremediation treatments for the mitigation of low-dose anthracene in marine coastal sediments of Bizerte lagoon (Tunisia), *Environmental Science and Pollution Research* 20: 300–310. http://dx.doi.org/10.1007/s11356-012-0860-x

Mohajeri, L.; Aziz, H. A.; Isa, M. H.; Zahed, M. A.; Mohajeri, S. 2013. Effect of remediation strategy on crude oil biodegradation kinetics and half life times in shoreline sediment samples, *International Journal of Marine Science and Engineering* 3: 99–104.

Moscoso, F.; Deive, F. J.; Longo, M. A.; Sanromán, M. A. 2012. Technoeconomic assessment of phenanthrene degradation by Pseudomonas stutzeri CECT 930 in a batch bioreactor, *Bioresource Technology* 104: 81–89. http://dx.doi.org/10.1016/j.biortech.2011.10.053

Moussa, M. S.; Hooijmans, C. M.; Lubberding, H. J.; Gijzen, H. J.; Van Loosdrecht, M. C. M. 2005. Modeling nitrification, heterotrophic growth and predation in activated sludge, *Water Research* 39: 5080–5098. http://dx.doi.org/10.1016/j.watres.2005.09.038

Nakamura, Y.; Sawada, T. 2000. Biodegradation of phenol in the presence of heavy metals, *Journal of Chemical Technology and Biotechnology* 75: 137–142. http://dx.doi.org/10.1002/(SICI)1097-4660(200002)75:2<137::AID-JCTB194>3.0.CO;2-0

Nikakhtari, H.; Kumar, P.; Nemati, M.; Hill, G. A. 2008. Mass transfer and bioremediation of aromatics from NAPL in a baffled roller bioreactor, *Process Safety and Environmental Protection* 86: 252–258. http://dx.doi.org/10.1016/j.psep.2008.04.007

Sheoran, A. S.; Sheoran, V.; Choudhary, R. P. 2010. Bioremediation of acid-rock drainage by sulphate-reducing prokaryotes, *Minerals Engineering* 23: 1073–1100. http://dx.doi.org/10.1016/j.mineng.2010.07.001

Tarighian, A.; Hill, G.; Headley, J.; Pedras, S. 2003. Enhancement of 4-chlorophenol biodegradation using glucose, *Clean Technologies and Environmental Policy* 5: 61–65.

Vaezihir, A.; Zare, M.; Raeisi, E.; Molson, J.; Barker, J. 2012. Field-scale modeling of benzene, toluene, ethylbenzene, and xylenes (BTEX) released from multiple source zones, *Bioremediation Journal* 16: 156–176. http://dx.doi.org/10.1080/10889868.2012.687415

Van Ginneken, L.; Meers, E.; Guisson, R.; Ruttens, A.; Elst, K.; Tack, F. M. G.; Vangronsveld, J.; Diels, L.; Dejonghe, W. 2007. Phytoremediation for heavy metal contaminated soils combined with bioenergy production, *Journal of Environmental Engineering and Landscape Management* 15: 227–236.

Vasiliadou, I. A.; Molina, R.; Martínez, F.; Melero, J. A. 2013. Biological removal of pharmaceutical and personal care products by a mixed microbial culture: Sorption, desorption and biodegradation, *Biochemical Engineering Journal* 81: 108–119. http://dx.doi.org/10.1016/j.bej.2013.10.010

Wang, Y.; Riess, R.; Nemati, M.; Hill, G.; Headley, J. 2008. Scale-up impacts on mass transfer and bioremediation of suspended naphthalene particles in bead mill bioreactors, *Bioresource Technology* 99: 8143–8150. http://dx.doi.org/10.1016/j.biortech.2008.03.070

Behrooz MAHMOODZADEH VAZIRI. Assistant professor in Chemical Engineering, PhD (2013), over 10 scientific papers, and one patent. Research interests: modeling and simulation of chemical, petrochemical and biological processes; computer-aided design; artificial neural networks.

Azadeh KHOSHDEL. Master of Science in Chemical Engineering (2014). Research interests: modeling and simulation of processes; environment engineering; biotechnology.

PERMISSIONS

All chapters in this book were first published in JEELM, by Vilnius Gediminas Technical University; hereby published with permission under the Creative Commons Attribution License or equivalent. Every chapter published in this book has been scrutinized by our experts. Their significance has been extensively debated. The topics covered herein carry significant findings which will fuel the growth of the discipline. They may even be implemented as practical applications or may be referred to as a beginning point for another development.

The contributors of this book come from diverse backgrounds, making this book a truly international effort. This book will bring forth new frontiers with its revolutionizing research information and detailed analysis of the nascent developments around the world.

We would like to thank all the contributing authors for lending their expertise to make the book truly unique. They have played a crucial role in the development of this book. Without their invaluable contributions this book wouldn't have been possible. They have made vital efforts to compile up to date information on the varied aspects of this subject to make this book a valuable addition to the collection of many professionals and students.

This book was conceptualized with the vision of imparting up-to-date information and advanced data in this field. To ensure the same, a matchless editorial board was set up. Every individual on the board went through rigorous rounds of assessment to prove their worth. After which they invested a large part of their time researching and compiling the most relevant data for our readers.

The editorial board has been involved in producing this book since its inception. They have spent rigorous hours researching and exploring the diverse topics which have resulted in the successful publishing of this book. They have passed on their knowledge of decades through this book. To expedite this challenging task, the publisher supported the team at every step. A small team of assistant editors was also appointed to further simplify the editing procedure and attain best results for the readers.

Apart from the editorial board, the designing team has also invested a significant amount of their time in understanding the subject and creating the most relevant covers. They scrutinized every image to scout for the most suitable representation of the subject and create an appropriate cover for the book.

The publishing team has been an ardent support to the editorial, designing and production team. Their endless efforts to recruit the best for this project, has resulted in the accomplishment of this book. They are a veteran in the field of academics and their pool of knowledge is as vast as their experience in printing. Their expertise and guidance has proved useful at every step. Their uncompromising quality standards have made this book an exceptional effort. Their encouragement from time to time has been an inspiration for everyone.

The publisher and the editorial board hope that this book will prove to be a valuable piece of knowledge for researchers, students, practitioners and scholars across the globe.

LIST OF CONTRIBUTORS

Anne-Katrin Große
Department of Maritime Systems, Faculty of Interdisciplinary Research, Universität Rostock, Justus-von Liebig Weg 6, 18059 Rostock, Germany
Chair of Geotechnics and Coastal Engineering, Universität Rostock, Justus-von-Liebig Weg 6, 18059 Rostock, Germany

Stefan Cantré and Fokke Saathoff
Chair of Geotechnics and Coastal Engineering, Universität Rostock, Justus-von-Liebig Weg 6, 18059 Rostock, Germany

Rafał Strachel, Jadwiga Wyszkowska and Małgorzata Baćmaga
Department of Microbiology, Faculty of Environmental Management & Agriculture, University of Warmia and Mazury in Olsztyn, Plac Lodzki 3, 10-727 Olsztyn, Poland

Ruben Sakrabani
Cranfield Soil and Agrifood Institute, School of Water, Energy and Environment, Cranfield University, MK43 0AL United Kingdom

Jürgen Kern
Leibniz Institute for Agricultural Engineering and Bioeconomy, Max-Eyth Allee 100, 14469 Potsdam, Germany

Utra Mankasingh
Department of Life and Environmental Sciences, University of Iceland, Sturlugata 7, Reykjavik 107, Iceland

Costanza Zavalloni
California State University Stanislaus, Agricultural Studies Program, One University Circle, Turlock, CA 95382, USA

Giulia Zanchettin
University of Udine, Ecosystems Ecology, Via delle Scienze 206, Udine, 33100, Italy

Ana Catarina Bastos
Department of Biology and Centre for Environmental and Marine Studies (CESAM), University of Aveiro, 3810-193 Aveiro, Portugal

Priit Tammeorg
Department of Agricultural Sciences, Plant Production Sciences), FIN- 00014 University of Helsinki, Finland

Simon Jeffery
Crop and Environment Sciences Department, Harper Adams University, Newport, Shropshire, TF10 8NB, United Kingdom

Bruno Glaser
Institute of Agronomy and Nutritional Sciences, Martin Luther University Halle-Wittenberg, von-Seckendorff-Platz 3, 06120 Halle/Saale, Germany

Frank G. A. Verheijen
Department of Environment and Planning and Centre for Environmental and Marine Studies (CESAM), University of Aveiro, 3810-193 Aveiro, Portugal

Chih-Yu Liu, Cheng-Yu Ku, Jing-En Xiao and Shih-Meng Hsu
Department of Harbor and River Engineering, National Taiwan Ocean University, Keelung, Taiwan

Chi-Chao Huang
Central Geological Survey, Ministry of Economic Affairs, Taiwan

Isaac I. Akinwumi
Department of Civil Engineering, College of Engineering, Covenant University, Ota, Nigeria

Colin A. Booth
Construction and Property Research Centre, Faculty of Environment and Technology, University of the West of England, BS16 1QY Bristol, UK

Ahmed A. El Baroudy, Talaat R. El-Beshbeshy and Abd Elhalim
Soils and Water Department, Faculty of Agriculture, Tanta University, Tanta, Egypt

Mohamed S. Shokr
Soils and Water Department, Faculty of Agriculture, Tanta University, Tanta, Egypt
The University of Wolverhampton, Wolverhampton WV1 1LY, UK

Michael A. Fullen
The University of Wolverhampton, Wolverhampton WV1 1LY, UK

Ramadan R. Ali
Soils and Water Use Department, National Research Centre, Giza, Egypt

Antonio J. T. Guerra and Maria C. O. Jorge
Department of Geography, Federal University of Rio de Janeiro, Brazil

Muhammad Abbas Aziz
Fauji Fertilizer Company, Dera Ghazi Khan, Pakistan

Hamaad Raza Ahmad and Muhammad Sabir
University of Agriculture Faisalabad, 38040 Faisalabad, Pakistan

Dennis L. Corwin
USDA Agricultural Research Service, 20250 Washington, United States

Khalid Rehman Hakeem
Department of Biological Sciences, Faculty of Science, King Abdulaziz University, Jeddah-21589, Saudi Arabia

Münir Öztürk
Botany Department & Centre for Environmental Studies, Ege University, Izmir, 35100 Bornova, Turkey

Sharif Moniruzzaman Shirazi
Department of Civil Engineering, World University of Bangladesh, Dhaka, Bangladesh

Md Ibrahim Adham, Faridah Othman and Zubaidah Ismail
Department of Civil Engineering, Faculty of Engineering, University of Malaya, 50603 Kuala Lumpur, Malaysia

Noorul Hasan Zardari
Centre for Environmental Sustainability and Water Security (IPASA), Department of Hydraulics and hydrology, Faculty of Civil Engineering, Universiti Teknologi Malaysia, Skudai 81310, Johor Bahru, Malaysia

Thomas D. Bucheli and Isabel Hilber
Environmental Analytics, Natural Resources and Agriculture, Agroscope ISS, Reckenholzstrasse 191, 8046 Zurich, Switzerland

Ana Catarina Bastos and Susana Loureiro
Department of Biology & Centre for Environmental and Marine Studies (CESAM), University of Aveiro, 3810-193 Aveiro, Portugal

Gerhard Soja and Aleksandra Marsz
Environmental Resources and Technologies, AIT Austrian Institute of Technology GmbH, 3430 Tulln, Austria

Gerard Cornelissen
Department of Environmental Engineering, Norwegian Geotechnical Institute NGI, Ullevål Stadion, 0806, Oslo, Norway
Department of Environmental Sciences (IMV), Norwegian University of Life Sciences (NMBU), 1432 Ås, Norway
Department of Applied Environmental Sciences (ITM), Stockholm University, 10691 Sweden

Rimantė Zinkutė, Valentinas Baltr Ūnas, Ričardas Taraškevičius, Bronislavas Karmaza, Rimutė Stak Ėnienė, Vaida Šeirienė and Dalia Kisielienė
Nature Research Centre Institute of Geology and Geography, T. Ševčenkos g. 13, 03223 Vilnius, Lithuania

Valentina Iurchenko
Kharkiv National Automobile and Highway University, 25 Petrovskoho str., Kharkiv, Ukraine

Larysa Mykhailov
Kharkiv National Automobile and Highway University, 25 Petrovskoho str., Kharkiv, Ukraine
Brandenburg University of Technology Cottbus-Senftenberg, Central Analytical Laborator, Konrad-Wachsmann-Allee 6, 03046 Cottbus, Germany

Thomas Fischer
Brandenburg University of Technology Cottbus-Senftenberg, Central Analytical Laborator, Konrad-Wachsmann-Allee 6, 03046 Cottbus, Germany

Omer Frenkel and Yigal Elad
Department of Plant Pathology, The Volcani Center - Agricultural Research Organization, Rishon Lezion, Israel

Amit K. Jaiswal
Department of Plant Pathology, The Volcani Center - Agricultural Research Organization, Rishon Lezion, Israel
Department of Plant Pathology and Microbiology, Faculty of Agriculture, Food and Environment, The Hebrew University of Jerusalem, Rehovot, Israel

Claudia Kammann
Hochschule Geisenheim University, Geisenheim, Germany

Beni Lew and Ellen R. Graber
Soil Chemistry, Plant Nutrition and Microbiology, The Volcani Center - Agricultural Research Organization, Rishon Lezion, Israel

Galia Shulga, Sanita Vitolina, Brigita Neiberte and Anrijs Verovkins
Laboratory of Lignin Chemistry, Latvian State Institute of Wood Chemistry, Dzerbenes iela 27, LV-1006 Riga, Latvia

Talrits Betkers
Faculty of Building and Civil Engineering, Riga Technical University, Azenes iela 16/20, LV-1048 Riga, Latvia

Olga Anne, Dalia Ambrazaitiene and Audrone Zukauskaite
Faculty of Natural Sciences and Mathematics, Klaipėda University, H. Manto g. 84, LT-5808 Klaipėda, Lithuania

Audrone Zukauskaite
Faculty of Marine Engineering, Klaipėda University, Bijūnų g. 17, LT-91225 Klaipėda, Lithuania

Enrico Mancinellia and Giorgio Passerinia
Department of Industrial Engineering and Mathematical Sciences, Università Politecnica delle March e, via Brecce Bianche 12, 60131 Ancona, Italy

Edita Baltrėnaitė, Pranas Baltrėnas and Eglė Marčiulaitienė
Vilnius Gediminas Technical University, Saulėtekio al. 11, LT-10223 Vilnius, Lithuania

Marika Viisimaa and Anna Goi
Department of Chemical Engineering, Tallinn University of Technology, Ehitajate tee 5, 19086 Tallinn, Estonia

Corneliu Tanase
Faculty of Biology, Al. I. Cuza University of Iasi, 20 A Bd. Carol I, 700505 Iasi, Romania

Irina Volf
Environmental Engineering and Management Department, Gheorghe Asachi Technical University of Iasi, 71 Bd. Prof. D. Mangeron, 700050 Iasi, Romania

Valentin I. Popa
Institute of Research, Development and Innovation in Technical and Natural Sciences, Aurel Vlaicu University of Arad, 77 Bd Revoluţiei, 310130 Arad, Romania

Asta Kazlauskaitė-Jadzevičė
Traku Voke Branch of the Lithuanian Research Centre for Agriculture and Forestry, Žalioji a. 2, Trakų Vokė, LT-02232 Vilnius, Lithuania

Jonas Volungevi čius
Faculty of Environmental Sciences, Vilnius University, M. K. Čiurlionio g. 21/27, LT-03101 Vilnius, Lithuania

Virginija Gregorauskienė
Lithuania Geological Survey, S. Konarskio g. 35, LT-03123 Vilnius, Lithuania

Saulius Marcinkonis
Vilniaus kolegija/University of Applied Sciences, Saltoniškių 58, LT-08105 Vilnius, Lithuania

Azadeh Khoshdel and Behrooz Mahmoodzadeh Vaziri
Department of Chemical Engineering, Quchan Branch, Islamic Azad University, Quchan, Iran

Index

www.ingramcontent.com/pod-product-compliance
Lightning Source LLC
LaVergne TN
LVHW060608210926
848485LV00005B/1387
* 9 7 8 1 6 4 7 4 0 0 8 2 8 *